PRIMARY BATTERIES

PRIMARY BATTERIES
Recent Advances

Robert W. Graham

NOYES DATA CORPORATION
Park Ridge, New Jersey, U.S.A.
1978

Published in the United States of America by
Noyes Data Corporation
Noyes Building, Park Ridge, New Jersey 07656

FOREWORD

The detailed, descriptive information in this book is based on U.S. patents, issued since the Autumn of 1975, that deal with primary batteries and their commercial technology.

This book serves a double purpose in that it supplies detailed technical information and can be used as a guide to the U.S. patent literature in this field. By indicating all the information that is significant, and eliminating legal jargon and juristic phraseology, this book presents an advanced, technically oriented review of primary batteries, their manufacture and use.

The U.S. patent literature is the largest and most comprehensive collection of technical information in the world. There is more practical, commercial, timely process information assembled here than is available from any other source. The technical information obtained from a patent is extremely reliable and comprehensive; sufficient information must be included to avoid rejection for "insufficient disclosure." These patents include practically all of those issued on the subject in the United States during the period under review; there has been no bias in the selection of patents for inclusion.

The patent literature covers a substantial amount of information not available in the journal literature. The patent literature is a prime source of basic commercially useful information. This information is overlooked by those who rely primarily on the periodical journal literature. It is realized that there is a lag between a patent application on a new process development and the granting of a patent, but it is felt that this may roughly parallel or even anticipate the lag in putting that development into commercial practice.

Many of these patents are being utilized commercially. Whether used or not, they offer opportunities for technological transfer. Also, a major purpose of this book is to describe the number of technical possibilities available, which may open up profitable areas of research and development. The information contained in this book will allow you to establish a sound background before launching into research in this field.

Advanced composition and production methods developed by Noyes Data are employed to bring these durably bound books to you in a minimum of time. Special techniques are used to close the gap between "manuscript" and "completed book." Industrial technology is progressing so rapidly that time-honored, conventional typesetting, binding and shipping methods are no longer suitable. We have by-passed the delays in the conventional book publishing cycle and provide the user with an effective and convenient means of reviewing up-to-date information in depth.

The Table of Contents is organized in such a way as to serve as a subject index. Other indexes by company, inventor and patent number help in providing easy access to the information contained in this book.

Some of the illustrations in this book may be less clear than could be desired; however, they are reproduced from the best material available to us.

v

15 Reasons Why the U.S. Patent Office Literature is Important to You

1. The U.S. patent literature is the largest and most comprehensive collection of technical information in the world. There is more practical commercial process information assembled here than is available from any other source.

2. The technical information obtained from the patent literature is extremely comprehensive; sufficient information must be included to avoid rejection for "insufficient disclosure."

3. The patent literature is a prime source of basic commercially utilizable information. This information is overlooked by those who rely primarily on the periodical journal literature.

4. An important feature of the patent literature is that it can serve to avoid duplication of research and development.

5. Patents, unlike periodical literature, are bound by definition to contain new information, data and ideas.

6. It can serve as a source of new ideas in a different but related field, and may be outside the patent protection offered the original invention.

7. Since claims are narrowly defined, much valuable information is included that may be outside the legal protection afforded by the claims.

8. Patents discuss the difficulties associated with previous research, development or production techniques, and offer a specific method of overcoming problems. This gives clues to current process information that has not been published in periodicals or books.

9. Can aid in process design by providing a selection of alternate techniques. A powerful research and engineering tool.

10. Obtain licenses—many U.S. chemical patents have not been developed commercially.

11. Patents provide an excellent starting point for the next investigator.

12. Frequently, innovations derived from research are first disclosed in the patent literature, prior to coverage in the periodical literature.

13. Patents offer a most valuable method of keeping abreast of latest technologies, serving an individual's own "current awareness" program.

14. Copies of U.S. patents are easily obtained from the U.S. Patent Office at 50¢ a copy.

15. It is a creative source of ideas for those with imagination.

CONTENTS AND SUBJECT INDEX

INTRODUCTION

The continuing development of portable electrically powered devices of compact design such as tape recorders and playback machines, radio transmitters and receivers, shavers, watches, and motion picture or still cameras creates a continuing demand for the development of reliable, compact batteries for their operation.

The power needs of such devices are varied. Thus, a watch requires a battery which will perform uniformly for at least a year at low drain; recorders and radios require batteries which will operate intermittently for perhaps a half hour to several hours at substantially higher drains followed by longer periods of nonuse. A motion picture camera in which a battery may operate exposure control means as well as drive a motor usually requires the battery to operate in a repetitive series of relatively short periods of time in a given day, but may not be used for weeks or months. A still camera in which a battery may be used to fire a flash bulb and in some cases to control exposure means and advance the film after each exposure requires the battery to deliver a series of pulses of rather high current, frequently in rapid succession.

Common primary dry cells are composed essentially of a consumable metal anode, a cathode depolarizer usually manganese dioxide and an electrolyte. The familiar Leclanche primary dry cell conventionally used as the power source in flashlights and other portable electric devices, comprises a zinc anode, a cathode depolarizer mix cake containing manganese dioxide and a conductive material such as carbon black or graphite, and an electrolyte consisting of an aqueous solution of zinc chloride and ammonium chloride (i.e., sal ammoniac). Various inhibitors such as mercuric chloride, chromates, etc., may also be used in relatively small amounts within the electrolyte.

Another type of primary dry cell which has attracted considerable attention is the magnesium dry cell. This dry cell system is very similar to the conventional Leclanche dry cell. Basically, the magnesium dry cell comprises a magnesium anode, a cathode depolarizer mix cake containing manganese dioxide and a conductive material and an electrolyte. The electrolyte consists essentially of an aqueous solution containing a magnesium salt such as magnesium chloride, magnesium perchlorate or magnesium bromide.

In most recent years, considerable research effort has been directed to higher energy density cells such as the silver-zinc, air-zinc and other alkaline type batteries. The use of alkali metal anodes has lead to the development and refinement of nonaqueous and solid electrolyte batteries based on lithium and sodium and various cathode active materials.

Special environmental and performance requirements have led to improvements in batteries which are activated by seawater or thermal energy as well as batteries which provide long life in the reserved state and can be activated as required.

In the medical implantation field, pacemakers are powered by miniature batteries which must be extremely reliable and compatible with body tissue. Film packs, for instant photograph development, require extremely delicate flat type batteries and considerable specialized progress has been achieved in this area.

While the battery industry has been quite successful in providing batteries to satisfy many diverse demands, the vast majority of the batteries commonly produced are cylindrical. They may range in height from the familiar button cells to as much as one-half inch to one inch or even more, and in diameter from roughly about one-half inch to one inch or more. Although they are excellent sources of electricity, their shape has limited to some extent the size and shape of the devices for which they are intended. As design concepts change there is also an increasing emphasis placed on thin, flat shapes. Devices of thin, flat shape cannot be made to accommodate the familiar cylindrical battery without devoting more space to the battery than is desired. Accordingly, there is an increasing research effort devoted to the development of thin, flat batteries.

This book describes 215 processes relating to all phases of primary battery technology as presented in the recent patent literature of the United States. This truly worldwide research and development effort covers the design fabrication and performance characteristics of hundreds of battery systems and includes almost 100 detailed process illustrations.

This book is largely devoted to primary batteries, but it is recognized that virtually all chemical-electrochemical reactions, in principle, can be reversed and many cells now considered primary may well be developed as secondary cells in the future.

In this context, the reader is referred to another very timely Noyes Data Corporation publication:

Secondary Batteries—Recent Advances, by R.W. Graham, 1978.

ZINC DRY CELLS

ELECTRODES

Acid Leaching to Provide Chloride Free Zinc Anode

According to a process described by *E.J. Curelop and N. Marincic; U.S. Patent 3,926,672; December 16, 1975; assigned to P.R. Mallory & Co. Inc.* an anode structure for an electrochemical cell, is formed by mixing specific portions of the zinc powder, mercuric oxide, and a chloride free filler selected from the group consisting of sodium acetate, sodium borate and sodium oxalate, and by leaching out the chloride free filler with an aqueous solution of the corresponding chloride free acid selected from the group consisting of acetic acid, boric acid and oxalic acid.

Example 1: The following reactive mixture was utilized to produce the anode of the process. Five thousand grams zinc powder, sifted through 60 mesh onto and through 100 mesh; 2,100 g anhydrous sodium acetate sifted through 60 mesh; 780 g HgO fine dust, battery grade; 30 ml kerosene which is inert and utilized to hold these types of particles together. The blending was carried out for 10 to 12 minutes at speed of 15 to 30 rpm. About 1.68 to 1.70 g of the above mixture was pressed into a green anode 77 to 80 mils thick. The pressure required was about 12,000 psi.

The leaching of the anodes was carried out with acetic acid diluted in the ratio of 1 part by weight acid to 3 parts by weight water. The amount of acid should be in a controlled relation to the weight and number of anodes, at least during the first 15 minutes of leaching. About 2 ml of the above solution is the maximum allowed for the above anodes, in other words approximately 1 ml of the above solution per each gram of the green anode of any size and shape. The excess of the above leaching solution can be added for faster leaching, but only after the initial 15 minute period.

The acetate leaching process is generally faster than the chloride leaching process. It is completed in 5 to 6 hours at room temperature when the above size

3

anodes are processed. The leached anodes were washed in distilled water, until the effluent showed a pH value of 6 or higher. The anodes were then washed with alcohol and dried in air. They had a dry weight of 1.14±0.05 g, a porosity of 70% and contained 12% mercury.

Example 2: Anodes of various porosities can be produced using the same procedure except for the mix composition. The following is another variation of the above mix, used for the production of large anodes; 5,000 g of zinc powder, 1,300 g of sodium acetate, 780 g HgO, and 30 ml kerosene.

48.5 g of the above mix were pressed into a rectangular anode 3.2 inches by 1.6 inches with a force of 30 tons (5.86 tons per square inch). The resulting anode produced by the chloride free acetic acid leaching process weighed 36.5 g and contained 12% mercury; it was 4.2 mm thick and 64% porous. The same type of anode was pressed with a copper screen in the middle as a current collector for the rechargeable cell application. The copper screen was amalgamated during the leaching process and provided a good contact to the porous anode body over the entire anode cross section.

Example 3: The following mixture of 5,000 g zinc powder, 2,100 g sodium oxalate, 780 g HgO, and 30 ml kerosene was treated with correspondingly diluted oxalic acid in the ratio of 1 ml/g of green anode as in Example 1 and produced a zinc anode of about the same weight, porosity, and mercury content as in Example 1.

Example 4: The following mixture of 5,000 g zinc powder, 1,300 g sodium oxalate, 780 g HgO, and 30 ml kerosene was treated with the same quantity of oxalic acid as in Example 3 and produced a zinc anode of about the same weight, porosity, and mercury content as in Example 2.

Zinc particles in anodes produced by the above examples measured between 25 and 40 microns. A zinc anode prepared according to each of the above examples was tested as follows. The anode, wherever necessary, was reshaped into the form of a disc pellet having a diameter of 0.5 inch and a thickness of 0.1 inch. The pellet was then inserted as an anode into an alkaline cell of flat cylindrical construction in contact with an absorbent spacer impregnated with an alkaline electrolyte, such as 35 to 40% KOH, 3.5 to 6.5% ZnO, and the balance water. In contact with the opposite surface of the spacer was a suitable depolarizer such as the metal oxide HgO, MnO_2, or Ag_2O containing from 5 to 25% graphite. This sealed cell was subjected to 180°F for 24 hours with the result that about 0.05 to 0.07 cm^3 of gas were produced.

When zinc pellets of about the same size, weight, porosity and mercury content were produced by the use of ammonium chloride, the gassing rate was determined to be 0.20 cm^3 to 0.25 cm^3 for a 24 hour period at 180°F.

Thus the gassing rates for the chloride free leaching are very satisfactory relative to the gassing rate for the chloride process and indicate the superiority of the process over conventional anode structures.

Coil Anode

According to a process described by *N. Marincic, R. Merz and R.H. Kelsey; U.S.*

Patent 4,007,054; February 8, 1977; assigned to P.R. Mallory & Co., Inc. a zinc wire coil is wound as a helix on a conducting support such as a copper or brass wire or rod, and the unit amalgamated in a mercuric salt solution to form a rigid anode element. In a cell the anode maintains substantially uniform spacing between anode and cathode to maintain substantially uniform current flow distribution.

Using a multistrand cable, or multiple layers of zinc wires for the helix permits predetermined percentage ratio of porosity to volume of anode.

A feature of the anode structure is the fact that the current collector, as a support, and the anode wire conductor are joined into an integral part before the mercury is applied to the anode for amalgamation. In fact the basic important feature of the process is that the zinc wire, with its inherent strength as a cohesive wire element, is used, and can be used as an active material. A further feature, of course, is the making of the anode with the current collector serving as an integral part of the anode structure and providing structural support and strength to the anode, whereby the anode may be handled and positioned in assembly during manufacture, with greater assurance that the anode will remain in its predesignated position, and thereby maintain an optimum physical disposition relative to the cathode, to establish a substantial uniformity in current distribution between the cathode and the anode.

Zinc Chloride in Cathode Depolarizer Mix

L.F. Urry; U.S. Patent 3,996,068; December 7, 1976; assigned to Union Carbide Corporation describes a primary dry cell system comprising a zinc anode, a cathode depolarizer mix cake containing a mixture of manganese dioxide and an electrolyte absorptive conductive material and an inner electrolyte consisting essentially of an aqueous solution containing a metallic salt of a halogen-containing acid especially, though not exclusively, a zinc salt such as zinc chloride, the inner electrolyte constituting from about 60 to 71% by volume of the total cathode depolarizer mix cake.

The process is predicated on the discovery, which is an outgrowth of earlier experimental work with the magnesium dry cell, that a primary dry cell utilizing a zinc anode and an aqueous metallic halide or perchlorate salt electrolyte can be made having a high service capacity if the cathode mix formulation is devised using a high solution volume, i.e., percent by volume of inner electrolyte. In the case of a primary dry cell using an aqueous zinc chloride electrolyte, for example, it has been found that during the electrochemical process which takes place on discharge, water is consumed or tied up in the form of a zinc hydroxide reaction product. This reaction product in turn will react with the zinc chloride electrolyte to form a hard, dense material having the formula $ZnCl_2 \cdot 4Zn(OH)_2$.

Because the cell does dry out on discharge, it has been found that the solution volume of the cathode depolarizer mix cake must be high enough to satisfy the above reaction and to insure electrolyte paths from the anode to cathode throughout the useful life of the cell.

Flexible Plates

A process described by *L.M. Gillman and D.W. Walker; U.S. Patent 3,918,989;*

November 11, 1975; assigned to The Gates Rubber Company pertains to flexible electrodes which are capable of being wound in a jellyroll configuration to be used in alkaline cells.

The process involves a two component additive consisting of a water-soluble resin and a compatible plasticizer to be incorporated in an electrode paste formulation. The additive is mixed intimately with the electrochemically active material required for the electrode plate, and applied to a flexible electrically conductive substrate material to which the paste mixture becomes securely bonded. The resulting electrode plate is preferably dried under controlled humidity conditions, thus imparting required flexibility to the electrode plate.

The electrode plate has particular utility in alkaline galvanic cells in which the electrode plate takes on a bent configuration although it is also useful in parallel stacked plate arrangements. The electrode plate may be spirally wound on a suitable mandrel to produce the so-called jellyroll plate and separator configuration.

Examples of suitable binder materials include cellulose derivatives having various degrees of substitution such as cellulose esters, exemplified by cellulose acetate; mixed cellulose esters exemplified by cellulose acetate propionate; carboxymethylcellulose and its salts, preferably its alkali metal salts; cellulose ethers exemplified by lower alkyl ethers, including methyl and ethyl, and carbocyclic, including benzyl, ethers; other nonionic cellulose and anionic cellulose compounds exemplified by hydroxypropyl methylcellulose; starch and its derivatives exemplified by dextrin, starch acetates, starch hydroxyethyl ethers, and ionic starches where carboxyl, sulfonate or sulfate groups are introduced into the molecule.

Compatible copolymers or mixtures of the abovementioned binders may also be employed, e.g., starch or sodium alginate and polyvinyl alcohol. Particularly preferred materials as binders include polyvinyl alcohol, polyacrylamides and cellulose materials such as hydroxypropyl methylcellulose, methylcellulose and the like. Partially hydrolyzed grades may be employed. These compounds have particularly good adhesion in potassium hydroxide, exhibiting minimum dustiness, and impart good flexibility to the finished electrode plate.

As an illustration of a preferred electrode paste mixture, the following formulation is utilized for a zinc-containing flexible electrode plate, adapted to be spirally wound on a suitable mandrel. It consists of preferably from 25 to 99% and more preferably 35 to 85% by weight of relatively pure zinc powder, and 15% by weight or less of mercuric oxide, from 10 to 60% by weight zinc oxide, and an aqueous solution of binder and plasticizer. The metallic zinc which accounts for the majority of the paste mixture may be provided in any desired form, although relatively pure zinc in the form of a powder is preferred.

Alternatively zinc may be provided by cathodically reducing zinc oxide to sponge zinc, which gives a very porous mass. The mercuric compound is present to reduce corrosion coupling with zinc by raising the hydrogen overvoltage. This compound may be a reducible compound of mercury such as mercuric oxide or a functionally equivalent compound. It is also desirable to use an excess of a reducible zinc-active material, e.g., zinc oxide, with respect to the amount of oxidizable material present, to minimize hydrogen evolution from the zinc plate and possible cell rupture during charge and overcharge.

Preferably from 0.1 to 5% by weight of the dry, finished paste formulation is composed of the water-soluble resin binder defined above. In general, enough binder solution should be employed to blend the components into a smooth paste and impart adherence of the active material to the substrate, although drier or wetter consistencies may be employed.

The paste is preferably formulated by preparing an aqueous solution (e.g., from 0.1 to 35% by weight binder). To this solution is added the plasticizer, in an amount preferably from at least about 1% by weight and more preferably in the range from 5 to 50% by weight based on the binder utilized. This binder/plasticizer solution is then added to the dry metal powder mix and the paste is spread evenly on a cleaned substrate. The paste may be applied to the substrate in any convenient manner.

It has been found that flexibility of the resultant pasted plate is vastly improved by the controlled drying of the plate in an atmosphere in which the humidity is preferably maintained in the range of from 30 to 70%, more preferably from 40 to 60% at temperatures in the range of 50° to 90°F. After drying, the electrode plate according to the process readily bends and is spirally wound on a ¼ inch mandrel. No dusting is evident; a number of evenly spaced very shallow cracks extending parallel to the axis of the bend are sometimes present, but present no drawbacks in the operation of the cell.

A number of plates in the following examples are prepared in the manner described above. These are tested for the properties of flexibility, adhesion in KOH, and dustiness when binders, plasticizers, or mixtures thereof are incorporated into the plates. The percents are based on the finished dry paste formulation unless otherwise noted.

Example 1: Plate A is prepared solely with a binder consisting of a 50% by weight water solution of 20,000 average molecular weight polyethylene glycol (when used in lower concentrations this polymer exhibits plasticizer properties, while at high concentrations, it behaves more like a binder). This plate is dried under the controlled conditions mentioned previously. While the dry plate is dusty (i.e., particles do not rub off the plate onto the fingers under moderate pressure) and when immersed in a 35% solution of KOH, the paste exhibits good adhesion to the substrate; the plate thus prepared exhibits poor flexibility (i.e., the plate does not bend easily around a ¼ inch mandrel). The cracks induced in the plate are deep and the paste is so hard that when the plate is bent it tends to sever the expanded metal substrate.

Example 2: Plate B is prepared solely with a plasticizer consisting of a 2% by weight aqueous solution of 20,000 average molecular weight polyethylene glycol. This plate is dried under the controlled conditions mentioned previously. The dry plate is dusty and exhibits only fair flexibility, and it demonstrates poor adhesion in concentrated KOH.

Example 3: Plate C is prepared with a binder consisting of 2.6% by weight methylcellulose. The plate is dried under the controlled conditions mentioned previously. The dry plate is not dusty, exhibits excellent adhesion in KOH, but has only fair flexibility, i.e., when the plate is wound around a ¼ inch mandrel, resultant cracks are induced which are uneven and cause portions of the active material to slough off.

Example 4: Plate D is prepared with a binder consisting of 3% by weight methyl-cellulose and 33% by weight (based on the binder) of polyethylene glycol of 20,000 average molecular weight as a plasticizer according to the process. This plate is dried at 50% relative humidity and 71°F for about 20 hours. When examined, this plate lacks dustiness and demonstrates excellent flexibility and adhesion in KOH.

Example 5: Plate E is prepared with a binder consisting of 2.5% by weight methylcellulose and 40% by weight (based on the binder) of low molecular weight polyvinyl acetate as a plasticizer. This plate is dried at 45% relative humidity and 72°F for about 18 hours. When examined, this plate lacks dustiness and demonstrates excellent adhesion in KOH, and very good flexibility.

SEPARATORS

Electrolyte Gelling Agent Powder and Fibers

According to a process described by *M. Harada, M. Takeda, M. Ichida, S. Nozaki and K. Miyahara; U.S. Patent 3,905,834; September 16, 1975; assigned to Matsushita Electric Industrial Co., Ltd., Japan* a dry cell comprises a positive electrode mix containing manganese dioxide as a depolarizer, a negative zinc electrode and a separator layer consisting of powders of electrolyte gelling agent and fibers. The separator layer is in a paper form wherein the powders and the fibers are integrally mixed together. The separator layer is interposed between the positive electrode mix and the negative zinc electrode. The dry cell has less internal resistance and is excellent in discharge capacity, preservable life and electrolyte leakage resistibility, as compared with dry cells using a conventional separator layer.

Example: 100 parts by weight of natural starch and 5 to 30 parts by weight of kraft pulp are mixed and dispersed in water, and 0.1 to 0.005 part by weight of a 1% aqueous solution of sodium methacrylate is added as an adsorbent medium to the aqueous dispersion and further 0.01 to 0.5 part by weight of a nonionic surface active agent is added thereto on the basis of one part by weight of water of the aqueous dispersion. A paper is made from the aqueous dispersion by means of a combined cylinder and Fourdrinier type paper machine. After suction and draining, the paper is compressed under a pressure of 10 to 30 kg/cm^2 and then dried at a temperature at which the powders of the electrolyte gelling agent undergo no gelation, that is, below 70°C, whereby a separator paper (a) is prepared.

For comparison, a starch dispersed in an electrolyte liquid containing ammonium chloride, zinc chloride and water as principal ingredients is coated onto one side of a sheet of kraft paper, and the thus coated kraft paper is dried, whereby the conventional separator paper (b) is prepared.

In Figure 1.1a, changes in water retainability of the separator papers **a** and **b** when AA type dry cells are set up from these separator papers respectively, and stored at 45°C are shown.

In Figure 1.1b, changes in the amounts of water absorbed when 10 g of the separator papers **a** and **b** absorb the electrolyte liquid are shown.

In Figure 1.1c, changes with time in the degree of swelling of the separator papers a and b when the separator papers are immersed in the electrolyte liquid are shown.

FIGURE 1.1: DRY CELL PERFORMANCE DATA

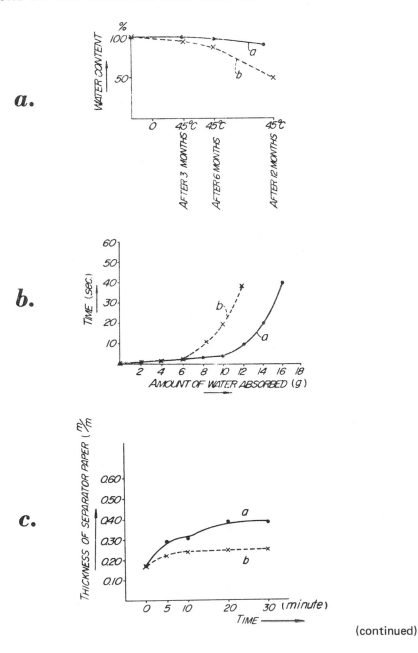

(continued)

FIGURE 1.1: (continued)

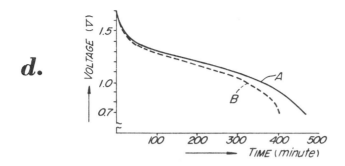

d.

(a) Changes with time in the water retainability of the separator paper when the dry cells are stored.
(b) Amount of water absorbed by the separator paper.
(c) Changes in the degree of swelling of the separator paper.
(d) Intermittent discharge capacity of the dry cell.

Source: U.S. Patent 3,905,834

When the AA cell-type dry cells A and B are set up using the separator papers **a** and **b** respectively, the discharge properties and the rate of electrolyte leakage of these dry cells are determined and the results are described below.

The discharge duration time down to 0.85 V in the case of intermittent discharge using a 10 Ω load (30 min/day) was as follows: dry cell A—immediately after preparation, 430 minutes; after 6 months' storage at 45°C, 400 minutes; dry cell B—immediately after preparation, 380 minutes; after 6 months' storage at 45°C, 330 minutes.

The number of voltage-defect dry cells among a batch of 100 dry cells that were stored at 45°C was as follows: dry cell A—after 6 months' storage at 45°C, 0; after 12 months' storage at 45°C, 1; dry cell B—after 6 months' storage at 45°C, 2; after 12 months' storage at 45°C, 5.

The number of leaked dry cells in the case where 50 dry cells are discharged for 24 hours at 20°C, with an external load of 10 Ω and then left for standing at 30°C for the period indicated was as follows: dry cell A—after standing 30 days, 0; after standing 60 days, 8; dry cell B—after standing 30 days, 10; after standing 60 days, 32.

The discharge curves of the dry cells A and B that had been subjected to intermittent discharge (30 minutes per day with 10 Ω load) are shown in Figure 1.1d.

It is evident from the foregoing results that the separator paper has a high rate of water absorption and is excellent in water retainability and degree of swelling, and that the dry cell using the separator paper has less internal resistance, a greater discharge capacity, less deterioration in performance even after storage and good electrolyte leakage resistibility.

Viscose Rayon and Polyvinyl Alcohol Fiber

Y. Uetani, Y. Taniguchi, T. Ashikaga and K. Mizutani; U.S. Patent 3,915,750; October 28, 1975; assigned to Hitachi Maxell, Ltd. and Kuraray Co., Ltd., Japan describe a separator for a battery such as an alkaline manganese cell or a Leclanche cell, which comprises a nonwoven fabric consisting of 30 to 60% by weight of viscose rayon fiber, 10 to 65% by weight of slightly water-soluble polyvinyl alcohol fiber and 5 to 30% by weight of water-soluble polyvinyl alcohol fiber. The viscose rayon fiber and the slightly water-soluble polyvinyl alcohol fiber are bound with each other by water-soluble polyvinyl alcohol fiber. The separator has large mechanical strength and high chemical resistance and retains a large amount of the electrolyte therein, and also improves the electrical discharge capacity of the battery.

Example 1: 42.5 parts by weight of viscose rayon fiber having a fiber size of 1.5 denier and a staple length of 10 mm, 42.5 parts weight of slightly water-soluble polyvinyl alcohol fiber having a fiber size of 1 denier and a staple length of 3 mm, and 15 parts by weight of water-soluble polyvinyl alcohol fiber consisting of polyvinyl alcohol fiber having a fiber size of 1 denier and a staple length of 3 mm which dissolves in water at a temperature of 70°C were weighed and were dispersed into water. The dispersion was made into paper by a paper machine. Thus obtained wet paper was passed at a rate of 40 m/min along a curved surface of a roll which has a large diameter and is heated to a temperature of 100° to 120°C in order to dry the paper. The dried paper has one smooth surface and one napped coarse surface and is used as a separator.

In this example, the viscose rayon fiber and slightly water-soluble polyvinyl alcohol fiber were obtained in the following manner. The viscose rayon fiber was produced by spinning of viscose consisting of linter or dissolved pulp, and the slightly water-soluble polyvinyl alcohol fiber made from polyvinyl alcohol having a mean polymerization degree of 1700 and a degree of saponification of 99.0%, after treating the polyvinyl alcohol fiber with heat in air at a temperature of 200° to 240°C for a few minutes.

Example 2: Forty parts by weight of viscose rayon fiber consisting of Polynosic fiber having a fiber size of 2.0 denier and a staple length of 51 mm, 50 parts by weight of slightly water-soluble polyvinyl alcohol fiber having a fiber size of 1.5 denier and a staple length of 38 mm, and 10 parts by weight of water-soluble polyvinyl alcohol fiber having a fiber size of 1.5 denier and a staple length of 38 mm which dissolves in water at a temperature of 80°C were mingled by a hopper mixer and then were made into a web having constant thickness by twining the fibers with each other by means of a carding machine. The obtained web was sprayed with steam which was heated to a temperature of 80° to 120°C and then was dried by passing through heat rolls. Thus a separator both sides of which were smooth was obtained.

The separators having a size of 120 mm x 74 mm and thickness of 0.35 mm and containing viscose rayon fiber in an amount as shown in Table 1 were tested with respect to their swelling percentage facing the direction of their thickness, their shrinking percentage facing the direction of their length, the rate of decrease of their weight and the amount of the electrolyte absorbed therein at the time of their manufacture and then after having been immersed in an aqueous solution of potassium hydroxide (42% by weight) for 24 hours at 45°C. The results obtained are shown in Table 1.

TABLE 1

Content of viscose rayon fiber (weight percent)	Swelling percentage (volume %)	Shrinking percentage (volume %)	Rate of decrease of weight(%)	Amount of electrolyte(g)
10	110	0.1	0.01	7.7
30	120	0.3	0.1	8.7
45	140	1.0	2.1	9.4
60	170	2.7	3.9	9.6
70	230	7.6	6.1	10.3
90	340	10.5	8.4	12.4

Table 1 shows that it is preferred to use viscose rayon fiber in an amount of 30 to 60% by weight. Otherwise, if the content of viscose rayon fiber is more than 60% by weight, it becomes a case of increasing the swelling percentage, the shrinking percentage and the rate of decrease of weight. On the other hand, if the content of viscose rayon fiber is less than 30% by weight, it causes the decrease of the amount of the electrolyte absorbed in the separator. As a result, it is impossible to retain a sufficient amount of the electrolyte necessary for the electrical discharge reaction in the separator.

Although it is necessary to have a certain degree of swelling in order to adhere the separator with an electrolyte, if the swelling percentage is extremely large, it becomes a case of decreasing the amount of active material to be filled in the hollow portion of the cup shaped separator used for the battery. Further, more, it is also required to lessen the shrinking percentage and the rate of decrease of the weight in order to increase the stability of the separator.

Table 2 shows the soaking time and tensile strength when wet of the separators containing water-soluble polyvinyl alcohol fiber in the amounts as shown. The soaking time is the time required for the separator to absorb completely 0.05 ml of the electrolyte consisting of 42% by weight of an aqueous solution of potassium hydroxide.

TABLE 2

Content of soluble polyvinyl alcohol fiber (weight %)	Soaking time (seconds)	Tensile strength under wet state (g./15mm×150mm)
3	10	47
5	29	99

(continued)

TABLE 2: (continued)

Content of soluble polyvinyl alcohol fiber (weight %)	Soaking time (seconds)	Tensile strength under wet state (g./15mm×150mm)
15	75	250
30	358	305
45	1800	361

Crosslinked Polyacrylamide

A process described by *C. Davis, Jr., U.S. Patent 3,905,851; September 16, 1975; assigned to Union Carbide Corporation* relates to the use of polyacrylamide as a separator material for dry cell batteries.

The process specifically involves a method for crosslinking polyacrylamide by adding to an aqueous solution of polyacrylamide a chromium-containing compound wherein a major portion of the chromium is present at a valence other than +3 and a compound which will react with the chromium-containing compound to yield chromium ions at a valence of +3. Exemplary of such compounds are potassium dichromate and potassium thiocyanate, respectively.

Example: A series of D size test cells were prepared employing separators in accordance with the process. These cells were fabricated from 10 gauge zinc cans which served as the cell anode and 52 g of a cathode mix formulation consisting of powdered electrolytic manganese dioxide, powdered acetylene black and a 32% by weight aqueous zinc chloride solution in an approximate weight ratio of 6:1:3.

The method of construction used was the so-called "spin paste" method. The zinc cans were individually rotated and 4 ml of separator-forming mixture were injected from a syringe into each can. Continued rotation for a few seconds caused the separator-forming mixture to uniformly coat the bottom and side of the can with excess mixture being thrown out of the can. The coated cans were then placed in an oven at 70°C for 10 minutes to dry out the liner, α-cellulose paper liners were inserted, the cathode mix was injected into the lined cans and a carbon rod cathode collector inserted in the center of the mix. Two milliliters of 32% by weight aqueous zinc chloride solution were added to each cell and the cell was sealed.

The separator-forming mixture was formed from 100 ml of a 3% by weight solution of zinc chloride in water to which was added 0.40 g of potassium dichromate, 0.40 g of potassium thiocyanate, polyacrylamide resin and minor amounts of commonly employed corrosion inhibitors. Additionally, some of the mixtures contained talc as a bulking agent. The polyacrylamide resin used in each test cell was Cyanamer P 250 Polyacrylamide which has a molecular weight in the range of 5 to 6 million. Table 1 sets forth the composition of each of the tested separators.

TABLE 1: SEPARATOR-FORMING MIXTURES

| | Each consisted of 4 ml of a mixture of 100 ml of 3% aqueous $ZnCl_2$, O.40 g. $K_2Cr_2O_7$, 0.40 g. KSCN, and: | |
Cell Designation	Polyacrylamide (% by weight)	Talc (% by weight)
A	3	None
B	3	20
C	3	23
D	3	28
E	5	None
F	5	23
G	5	17
H	7	None
I	7	23
J	7	17

Discharge data for fresh cells using the various separator-forming mixtures of Table 1 continuously discharged at 25°C is shown in Table 2.

TABLE 2: 2.25 OHM CONTINUOUS DISCHARGE OF FRESH CELLS

Cell Designation	Open Circuit Voltage (volts)	Flash Current (amps)	Closed Circuit Voltage (volts)	Service, Minutes to 1.1V.	1.0V.	0.9V.	0.65V.
A	1.76	11.3	1.61	170	256	351	565
B	1.75	9.4	1.60	183	295	415	595
C	1.77	9.0	1.61	170	278	385	570
D	1.76	9.3	1.61	179	285	404	618
E	1.76	10.4	1.62	200	280	387	592
F	1.75	9.2	1.59	195	320	422	608
G	1.75	8.9	1.58	163	258	357	558
H	1.76	10.2	1.61	183	274	373	570
I	1.76	8.8	1.59	165	261	350	535
J	1.63	9.4	1.50	192	305	420	650

From the above it can be seen that good performance is obtained from cells formed from each of the various separator-forming mixtures both with and without the addition of talc. Even cell J, a relatively low voltage cell, displayed good performance for a sustained period of time.

ELECTROLYTES

Starch-Thickened Electrolyte

L.L. Belyshev, A.V. Chuvpilo, V.V. Trizno, V.A. Naumenko, L.F. Penkova, V.I. Gorokhov, E.G. Apirina and S.A. Gantman; U.S. Patent 4,038,466; July 26,

1977 describe a method of producing thickened electrolyte by mixing, at room temperature, aqueous solutions of calcium chloride, zinc chloride, and ammonium chloride with a tanning agent, such as chromium sulfate and starch as a thickening agent, and aging the resulting mixture till the formation of thickened electrolyte with a viscosity sufficient for the adhesion of an electrolyte layer to the negative electrode of a cell is obtained. The mixture is aged for a time period necessary for the formation of thickened electrolyte with a compressive strength of 0.05 to 0.85 kg/cm^2, the thickened electrolyte being pressed, after aging, through at least one draw-hole at a rate of at least 0.05 m/sec, the resulting viscosity being sufficient to apply an electrolyte layer to the negative electrode of a primary cell.

It is expedient to age the mixture for at least 24 hours. It is also expedient to carry out the mixing within the temperature range from 21° to 23°C.

Example 1: A solution of chromium sulfate is added to an aqueous solution of calcium chloride, zinc chloride and ammonium chloride in an amount of 20 ml per 1 liter of the solution, and starch, with a moisture content of 6%, is used as a thickening agent added in amounts of 250 g per 1 liter of the solution, the resulting mixture being stirred at a temperature of 14°C until a viscosity of 25 poises is reached.

The resulting electrolyte solution is kept at a temperature of 18°C for 48 hours. When the resulting thickened electrolyte is compressed, its ultimate strength is equal to 0.12 kg/cm^2 and adhesion to 0.0005 kg/cm^2.

After pressing through a draw-hole at a rate of 0.05 m/sec with the ratio of the flow rate per second to the volume of one draw-hole channel equal to 50, the electrolyte is ready for application to the inner surface of the negative electrode for those primary cells in which the positive electrode is formed under a reduced pressure on a layer of the thickened electrolyte.

Example 2: A solution of chromium sulfate is added to an aqueous solution of calcium chloride (27%), zinc chloride (5%), and ammonium chloride (7.5%) in amounts of 20 ml per 1 liter of the solution. Starch with a moisture content of 18% is used as the thickening agent in amounts of 200 g per 1 liter of the solution. The mixture is stirred at 20°C until a viscosity of 150 poises is reached.

The resulting electrolyte solution is kept at 12°C for 60 hours which results in the formation of thickened electrolyte with a compressive strength of 0.3 kg/cm^2 and adhesion of 0.03 kg/cm^2.

After pressing through a draw-hole at a rate of 0.4 m/sec and ratio between the flow rate per second and the volume of a draw-hole channel equal to 400, the electrolyte thus obtained is suitable for direct application to the inner surface of the negative electrode of primary cells in which the positive electrode is formed under a reduced pressure on an applied thickened electrolyte layer.

For cells where the positive electrode is formed under a higher pressure, the negative electrodes with an applied layer of thickened electrolyte are first kept for 1.5 to 2 hours at room temperature, thus enhancing the hardness of the applied electrolyte layer.

Reticulated Starch Gelling Agent

P. Croissant and R. Coudrin; U.S. Patent 3,969,147; July 13, 1976; Société des Accumulateurs Fixes et de Traction, France describe a gelled electrolyte for electrochemical generators, of the type comprising an alkaline solution to which is added a gelling agent. The agent comprises modified or reticulated starches.

Such modified or reticulated starches, which are starches processed in known manner with certain reagents, such as epichlorohydrin, have a far greater resistance than the native starches to the effect of alkaline solutions, irrespective of the concentration of such solutions. Examples of such starches are described in U.S. Patent 2,748,183.

With the modified or reticulated starches, it is possible to use potassium hydroxide solutions having a normality less than 11 N without occurrence of liquefaction of the gels.

Gelled electrolyte of the process is useful, for example, in electrochemical generators of the alkaline type, depolarized by oxygen. The positive electrode is supplied suitably with oxygen. The negative electrode comprises gelled electrolyte containing powdered zinc as negative active material. Gelled electrolyte separates the positive and negative electrodes.

Magnesium Perchlorate

J. Augustynski, F. Dalard, J.-Y. Machat and J.-C. Sohm; U.S. Patent 3,902,921; September 2, 1975; assigned to Anvar, France describe an electric cell of the Leclanché type, i.e., comprising a solid depolarizing agent such as manganese dioxide or mercury oxide and includes the combination of a zinc anode and an electrolyte having a base of magnesium perchlorate. The depolarizing agent is conveniently pretreated by heat in the presence of the electrolyte.

Furthermore, the separator paper inserted between the "bobbin", i.e., the cathode and depolarizer unit, and the cylindrical can-shaped zinc anode is merely impregnated with the electrolyte without any organic gel liable to generate hydrolyzed products which in turn evolve CO_2 when oxidized by the depolarizer. and without any mercury salt to amalgamate the zinc anode, such amalgamation being unnecessary with magnesium perchlorate. Owing to the absence of any gas evolution, the cell may be sealed perfectly gas-tight.

Zinc Perchlorate

J.-Y. Machat and J.-C. Sohm; U.S. Patent 3,926,677; December 16, 1975; assigned to Anvar, Agence Nationale de Valoirisation de la Recherche, France describe a primary electric cell in which mercuric oxide is used as a depolarizer in combination with a substantially pure zinc anode and an uncorrodible cathode. The electrolyte is formed of an aqueous solution of zinc perchlorate $Zn(ClO_4)_2$.

Tests have demonstrated that aqueous solutions of zinc perchlorate are practically noncorrosive for a zinc anode, provided, however, that the pH of the solution is properly selected. Any possible gas evolution can in this way be made quite negligible. This noncorrosive character of zinc perchlorate may be enhanced by adding to the electrolyte the proper amount of zinc oxide (ZnO) or zinc hydroxide $Zn(OH)_2$ required to adjust the pH of the solution to an optimum level.

A principal advantage of the cell is that its voltage remains constant within one millivolt, even when the cell has been at rest for a long period of time.

CORROSION INHIBITORS

Polyethylene Glycol Monolauryl Ether

G. Jung and J. Bauer; U.S. Patent 3,928,074; December 23, 1975; assigned to Varta Batterie AG, Germany describe a primary cell and battery having a polyethylene glycol ether as a corrosion inhibitor.

The polyethylene glycol monoalkyl ethers which are useful may be designated by the general formula

$$HO-CH_2-CH_2-(O-CH_2-CH_2-)_x O-C_y H_{2y+1}$$

where the values for x and y are not critical, but for satisfactory results, x can be between 6 and 100 and y representing the alkyl residue, can be between 1 and 20. For best results, there are selected soluble ethers having a fatty acid ether residue where y is between 8 to 22, usually between 8 and 18. It is desirable that x represent a value between 10 and 30. Typical ethers are the following:

> polyethylene glycol monolauryl ether, x = 23
> polyethylene glycol monocetyl ether, x = 20
> polyethylene glycol monooctyl ether, x = 35
> polyethylene glycol monostearyl ether, x = 15
> polyethylene glycol monoeicosanoyl ether, x = 10

Preferred at this time is polyethylene glycol monolauryl ether or PEL, (wherein x = 23 and y = 12).

The weight of the inhibitor, which is added, can range from 0.001 to 2% preferably from 0.01 to 0.2% by weight of the depolarizer mass. Greater or smaller amounts tend to be wasteful or inadequate for best results. Typical test results, obtained with the primary cell containing the mixture according to the process, are illustrated below. The most effective results are found with types of electrolytes from weakly acid to weakly basic characteristics.

Example 1: Negative Magnesium Electrode — A primary cell is made up using the alloy $Mg(Al)_2Zn$ (2% aluminum, 1% zinc) for the electrode. It is tested as follows.

Ungreased tins of the above alloy (1.2 g; 1 mm thickness) are positioned into tightly sealable weighing bottles, 40 ml of the respective electrolyte are added and finally the bottles are stored for one month at ambient temperature. The tins are dried and weighed again after etching in a CrO_3 bath.

The bath composition used removed only the adherent layers and did not attack the material itself.

The following observations were made.

		- - - - Loss of Weight - - - -	
Electrolyte	Addition	mg	%
26% $Mg(ClO_4)_2$	–	1,200	100
26% $Mg(ClO_4)_2$	0.5% PEL	200-250	16-21

The useful effect of the PEL is readily apparent. Similarly satisfactory results are obtained using 0.2% of PEL. Also when a polyethylene glycol monostearyl ether (x = 15) is used, a comparable satisfactory effect is observed.

Example 2: System Zinc/NH_4Cl-$ZnCl_2$-Electrolyte/Pyrolusite — The effect of the inhibitors, used according to the process has been tested in Leclanche cells of the above system. The results obtained with cells of size IEC R20 are described by way of illustration only, as follows. The batteries were manufactured with the following materials:

> The negative electrode: Zn-alloy (of customary cells),
> The positive electrode: depolarizer mix, e.g., of
> > 50% pyrolusite
> > 13% conducting material (carbon black and graphite)
> > 14% ammonium chloride
> > 1% zinc oxide
> > 5% zinc chloride
> > 17% water
> > + inhibitor (solved in the electrolyte; if necessary, dis-
> > solved before in a little amount of alcohol)
> Weight of the positive electrode: about 50 g,
> Positive abductor: carbon rod, and
> Separator: lining paper coated with electrolyte.

Test Results

		. . . . Inhibitor		
	Storage*	$HgCl_2$	PEL	PEL
Concentration in percent by weight of the mix	–	0.04	0.04	0.08
Cell-voltage, V	0	1.74	1.72	1.72
	3	1.69	1.69	1.68
Current in short circuit, A	0	25	24	23
	3	19	19	18
Discharge by lamp, min**	0	932	948	948
	3	880	890	900
Discharge by transistor, hr***	0	88	85	84
	3	83	82	79
Discharge by motor, min†	0	285	238	255
	3	237	235	234

*0 = fresh condition and 3 = 3 months storage at 45°C.
**Daily four minute discharge in hourly intervals of 1 to 8 hours at a load of 4 ohms until a discharge cut-off voltage of 0.9 V was reached.
***Daily four hour discharge at a load of 20 ohms until a discharge cut-off voltage of 0.9 V was reached.
†Daily one hour discharge, five times per week, at a load of 500 mA until a discharge cut-off voltage of 1.0 V was reached.

The beneficial effect of PEL is apparent under various conditions, even upon storage for several months or elevated temperatures. When polyethylene glycol monoctyl ether is substituted in the above tests for PEL comparable results are obtained. Corrosion resistance is likewise increased when instead of PEL there is used polyethylene glycol monocetyl ether in an amount of 0.2% of the depolarizer mass.

Ascorbic Acid

According to a process described by *P. Cerfon; U.S. Patent 3,970,476; July 20, 1976; assigned to Compagnie Industrielle des Piles Electriques Cipel, France* in an electrochemical cell having a zinc negative electrode, a positive electrode, and a saline electrolyte, a corrosion inhibitor is added to the saline electrolyte to prevent corrosion of the zinc electrode by the electrolyte. The inhibitor is ascorbic acid, which is added to the saline electrolyte preferably in proportions comprising between 700 and 900 mg/l. The mixture of the electrolyte and inhibitor is generally in paste form. The resultant inhibitor-containing electrolyte in practice does not corrode the zinc negative electrode of such cells during storage. Thus, prolonged storage of such cells, even in adverse tropical environments is possible.

Oleic Acid Diethanolamide

J. Bauer and A. Winkler; U.S. Patent 3,963,520; June 15, 1976; assigned to Varta Batterie AG, Germany describe a primary cell with a corrosion inhibitor in the form of a saturated or unsaturated monocarboxylic acid with at least two ethanolamide radicals.

Monocarboxylic acids containing more than 12 carbon atoms in the molecule, particularly those present in the form of diethanolamides are preferably used. They may, however, also be present in the form of polyethanolamides.

Suitable saturated monocarboxylic acids include, for example, palmitic acid and stearic acid; suitable unsaturated monocarboxylic acids include, for example, linoleic acid and oleic acid.

Tests have shown that good results may be obtained by using oleic acid diethanolamide $C_{22}H_{43}O_3N$. Compounds of the above groups are suitable for use as inhibitors provided that they are soluble in or miscible with particularly water, the electrolyte or an organic solvent so that the introduction of the material into the element or cell may be facilitated.

Pulverulent or granular zinc conventionally used in the production of alkaline primary cells is usually amalgamated with from 8 to 15% of mercury based on the total quantity. It has been found that the mercury content may be substantially reduced to less than 5% provided that one of the above substances is added in an amount of 0.005% to 1%, based on the weight of the zinc electrode. Amounts exceeding 1% result in an undesirable decrease in active zinc material. At concentrations of less than 0.005%, the activity suffers. Concentrations of from about 0.01 to 0.3% are preferred.

The inhibitor is initially dissolved or dispersed in a solvent (for example, water) and a predetermined quantity of the resulting solution or dispersion is then applied to the dried zinc powder amalgamated with from 3 to 5% of mercury.

The solvent is then allowed to evaporate. The rate of evaporation is advantageously increased by drawing a vacuum or by introducing the treated material into a stream of dry air.

Potassium Cyanide and Mercury

According to a process described by *D.V. Louzos; U.S. Patent 3,905,833; September 16, 1975; assigned to Union Carbide Corporation* the inhibition of the corrosion of a zinc anode in an alkaline galvanic cell is accomplished by the presence of an inhibiting amount of at least one cyanide compound and mercury in the cell. The cyanide compound provides a source of soluble cyanide ions in the electrolyte. This inhibitor does not interfere with the normal operation of a zinc anode in a primary alkaline galvanic cell and yet, in combination with mercury, is effective in inhibiting nonproductive corrosion of the zinc, reducing gassing associated therewith and suppressing the formation of zinc crystals during storage.

It is believed that the inherent higher throwing power and more uniform, finer grained deposits obtained from cyanide-containing zincate plating baths result in the cyanide additive serving as a corrosion inhibitor in galvanic cells since it provides an efficient means for attaining a more equipotential zinc surface. Cathodic sites can be formed on the zinc anode surface, e.g., by the deposition of metals less electronegative than zinc arising from heavy metal impurities in the cell. Such a cathodic site in electronic contact with the remainder of the zinc anode results in a corrosion couple in which zinc is anodically dissolved and replated on the cathodic site. This reaction may result in excessive zinc crystal formation (and accompanying hydrogen evolution at the cathodic sites), which if not suppressed may extend through the separator of the cell, thus creating an internal electronic short.

It is believed that the cyanide functions in the cell to form particularly stable metal ion complexes which do not give the same ionic reactions as the heavy metals they contain. Thus, when cyanide compounds are in the presence of mercury, the strong complexing ability of the cyanide ion is believed to result in formation of soluble $Hg(CN)_4^=$ complex which is believed to act as a reservoir for supplying mercury for uniform amalgamation of the zinc anode. This will result in an equipotential surface on the zinc anode, thereby substantially eliminating the formation of zinc crystals. Furthermore, it can be reasoned that the strong complexing ability of the cyanide ion may prevent heavy metal impurities originating in the cell cathode and elsewhere from forming cathodic gassing sites on the zinc anode.

For example, metal ion impurities such as Fe^{++}, Fe^{+++}, Cu^+ and Cu^{++} are expected to be essentially tied up in the stable complexes, $Fe(CN)_6^=$, $Fe(CN_6^=$, $Cu(CN)_4^=$ and $Cu(CN)_2$, respectively having instability constants 10^{-35}, 10^{-42}, 10^{-32} and 10^{-24}, respectively.

Suitable cyanide compounds for use in this process include potassium cyanide [KCN], sodium cyanide [NaCN], zinc cyanide [$Zn(CN)_2$], lithium cyanide [LiCN], magnesium cyanide [$Mg(CN)_2$], calcium cyanide [$Ca(CN)_2$], barium cyanide [$Ba(CN)_2$], strontium cyanide [$Sr(CN)_2$] and aluminum cyanide [$Al(CN)_3$].

Example: Thirty-two G-size alkaline-MnO_2 cells were fabricated with 2.5% by weight potassium cyanide added to an alkaline zincate electrolyte containing 35% by weight potassium hydroxide, 3% by weight zinc oxide, and 8% by weight mercury, such weight percents being based on the weight of the electrolyte, except for mercury which was based on the weight of the anode. The zinc anodes for the cells were composed of zinc powder suspended in a carboxy-methylcellulose-KOH gel.

One hundred twenty similar type cells were produced, but without the addition of the potassium cyanide. The two types of G-size alkaline-MnO_2 cells were stored for 5 years at a temperature of 72°F (22.2°C). During the 5-year storage period, the average bottom bulges of the cyanide-containing cells and the test control cells were measured. The table below shows the comparison in the degree of bottom bulge for the cells during different time periods.

Time in Storage (yr)	Bottom Bulge in Cyanide-Containing Cells (inch)	Bottom Bulge in Test Control Cells (inch)
½	0.022	0.019
2⅓	0.022	0.044
5	0.022	0.056

As shown in the table, the bottom bulge of the cyanide-containing cells remained rather constant while in storage between the time period of ½ year and 5 years. Contrary to this, the bottom bulge of the test control cells increased almost linearly with time during this period. For the full time period of 5 years, the average bottom bulge of the cyanide-containing cells measured 0.022" (0.056 cm) while the average bulge of the test control cells measured 0.056" (0.142 cm). Since bottom bulge is presumably caused by internal gas pressure resulting from corrosion, the bulge data indicated that the cyanide-containing cells had far less internal corrosion and thus good shelf life characteristics.

SEALING AND LEAK PREVENTION

Resilient Edge Ring

D. Naylor and F.J. Harris; U.S. Patent 3,904,438; September 9, 1975; assigned to Mallory Batteries Limited, England describe an electrochemical cell having a metal housing, a metal top and a resilient sealing ring compressed between the edge region of the top and the internal surface of the housing. The edge ring embraces the edge region of the top, and has a peripheral annular projection on its radially outer surface which projection engages the internal surface of the housing whereby the compressive sealing forces in the ring are localized in an annular region of the ring.

The cell shown in Figure 1.2b has a cup-shaped metal housing 9 containing a cathode depolarizer 8, a permeable barrier 7, an absorbent 6 containing an alkaline electrolyte, anode material 5, and a metal top closure disc 1. This disc is dished so that a portion stands above the top of the housing 9 in order to form an anode contact for the cell. The metal top is made of, or has on its internal surface a layer of, a material compatible with the anode material and electrolyte. The edge of the metal disc 1 lies within the upper region of the housing 9, and a thermoplastic sealing ring or collar 2 is injection molded about the edge of the top disc. The top disc and sealing ring accordingly form an integral top closure member, as shown in Figure 1.2a, which facilitates handling during assembly of the cell.

Initially, the side wall of the housing **9** is rectilinear. The constituents of the cell are placed in the housing and the composite top closure member **1,2** is fitted into the housing so that the sealing ring is trapped between the edge region of the disc **1** and the internal surface of the upper region of the wall of the housing. The edge **10** of the housing is then crimped inwardly so as to press down on the peripheral region of the top outer surface of the sealing ring.

FIGURE 1.2: CLOSURE FOR ALKALINE CELL

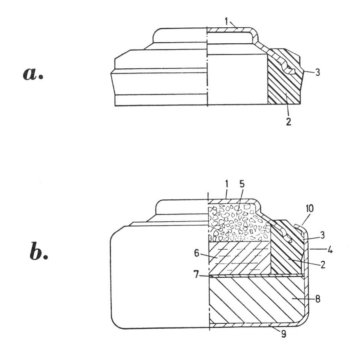

a.

b.

(a)   Side view, half in section, of a composite top closure member
        for an alkaline cell.
(b)   Similar view of a completed cell.

Source: U.S. Patent 3,904,438

As can be seen in Figure 1.2a, an annular projection **3** is provided on the external peripheral side surface of the sealing ring. The outermost diameter of this projection is greater than the internal diameter of the housing **9**. Other regions of the surface are of less diameter than the internal surface of the housing. Consequently, when the composite top closure member is inserted into the housing, the seal is subjected to radial compression localized in the radially outermost region of the projection. This region is aligned with the edge of the metal disc **1**, which resists radial compression of the sealing ring, with the result that the annular region of the sealing ring between the edge of the disc and the internal surface of the housing is under concentrated radial compression and in particular there is high radial pressure at the interface of the sealing ring and the internal surface of the housing, over a relatively small area **4** extending round the periphery of the sealing ring. This compression is independent of the pressure exerted on the sealing ring by the crimped edge of the housing.

To facilitate insertion of the top closure member into the housing, the projection **3** preferably has a frustoconical external surface tapering downwardly as seen in Figure 1.2a. Alternatively, the projection may have a convex arcuate axial cross section.

The housing and top disc should be sufficiently rigid to ensure that the radial compression is maintained in the sealing ring, after sealing. Also, it is desirable that the operation of crimping the edge of the housing should not unduly reduce the radial pressure between the housing and the sealing ring in the area **4**; it will be understood that the crimping tends to bow the housing onwards in the region below the crimp. Consequently, unless the housing is sufficiently rigid to resist this effect, it may be necessary to support the housing on its outer surface, e.g., by means of a shroud or collar, during the crimping of upper edge of the housing, to ensure that the internal diameter of the housing does not increase as a result of the crimping.

An advantage of the sealing arrangement described, in addition to the excellent sealing itself, is that the principle sealing pressure is radial and is localized in a particular region of the sealing ring. This means that the sealing pressure can be as great as desired, whereas in conventional sealing arrangements the sealing pressure is at least partly in the axial direction, and can impose excessive pressure on the constituents of the cell, in particular the barrier **7**, and may cause fracture of the latter. In the arrangement described, the pressure on the cell constituents and in particular on the barrier depends on the force exerted during the crimping operation and can be controlled by control of the crimping force; consequently a relatively low pressure can be applied to the barrier, without detriment to the sealing of the cell.

Cardboard Sleeve and Resilient Plastic

A process described by *H. Füllenbach, F. Christof and A. Fränzl; U.S. Patent 3,953,240; April 27, 1976; assigned to Varta Batterie AG, Germany* relates to a sealing element for an electric primary cell having a steel casing and more particularly to a sealing element of resilient material which is disposed in the primary cell between the flanged upper end of a cup electrode, a cardboard sleeve, and the outer edge of an upper cover disc.

The sealing element may be a self-contained ring having a shape conforming to the upper edge of the primary cell. This sealing element ring may be placed on the edge of a cup electrode which is rolled over at the top or on the edge of a prism-shaped electrode which is already inserted into the steel casing. Thereafter, the upper cover disc is put in place and subsequently, the cell is resiliently closed by rolling the edge of the steel casing over the cover disc which has an outer edge which is slotted accordingly.

According to a further example of the process, the sealing element is not put in place as a self-contained ring on the flanged-over edge of the cup, but rather, the sealing element material is sprayed, squeezed, or otherwise dispensed or applied directly into a wedge or gap which exists between the edge of the cup and the outer casing, and the sealing element material is subsequently cured. The closing operation proper is initiated by positioning the cover disc in place on the cell.

Figure 1.3a shows a partial cross section through an electric primary cell before the latter has been closed. The sealing element **1** according to the process, is in a form conforming to the upper closing edge of the cell, and may be, for example, a circular ring or a rectangular gasket made of an elastic material. The sealing element is inserted between a rolled-over edge of a cup electrode **2** and its outer enclosure which consists of a cardboard sleeve **4** and a steel casing **5**. By way of example the cup electrode may be made of zinc. A separator or paper lining **3** separates the cup electrode from a depolarizer substance **8** which surrounds a current take-off element **9**, the latter usually consisting of a carbon rod. A bottom disc designated **7** is also shown in the drawings.

FIGURE 1.3: PRIMARY CELL-SEALING ELEMENT

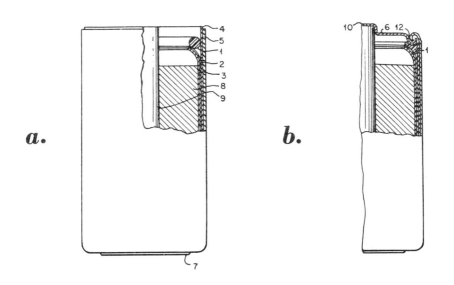

(a)   Elevational view, partly broken away and in cross section, of an
       electric primary cell before the cell has been closed.
(b)   View similar to Figure 1.3a showing the electric primary cell after
       the latter has been closed.

Source: U.S. Patent 3,953,240

Figure 1.3b shows a partial cross section of the primary cell after closing. To effect closing, an upper cover disc **6** having a terminal cap **10** is placed on the current take-off element **9** and subsequently, the enclosures consisting of the cardboard sleeve **4** and the steel casing **5** are folded into a groove **12** in the cover disc **6**. Due to the resilient pressure of the rolled-over steel casing **5**, the sufficiently elastic sealing element **1** lying on the edge of the cup electrode **2** is deformed in such a manner that through contact over a circular area, an excellent seal with respect to the escape of gas or electrolyte is obtained.

If the internal pressure within the cell becomes excessively high due to unintentionally occurring gas development within the cell, this internal pressure counteracts the spring or resilient tension of the rolled-over flange of the steel casing and lifts the cover disc slightly off the sealing surface on the cup electrode. The gas under pressure can then escape to the outside between the cup electrode and the sealing element. Subsequently after the internal pressure is thus relieved, the system closes again through the elastic springback of the rolled-over flange of the steel casing.

Starch Sealing Agent

J. Miyoshi, Y. Kajikawa, A. Ota, J. Asaoka and Y. Kino; U.S. Patent 4,001,044; January 4, 1977; assigned to Matsushita Electrical Industrial Co., Ltd., Japan describe a dry cell enclosing an electrolyte mainly consisting of zinc chloride, wherein a filler or sealing agent is used to seal the carbon electrode extended out of the plastic top seal or closure and the portion of the plastic tube or jacket covering the top surface of the plastic top seal in order to improve both the shelf life and leak-proofness.

In order to investigate the effects of various starches upon electrolytes, extensive studies and experiments were made, and the results are shown in Figure 1.4a. Along each swelling characteristic line, the original volume of the starch is increased twofold at 20°C. It is seen that the higher the content of zinc, the more the swelling of the starches is increased.

FIGURE 1.4: DRY CELL

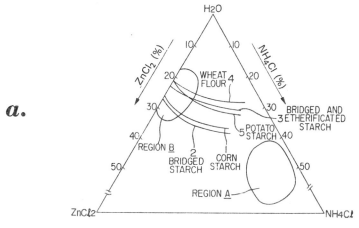

FIGURE 1.4: (continued)

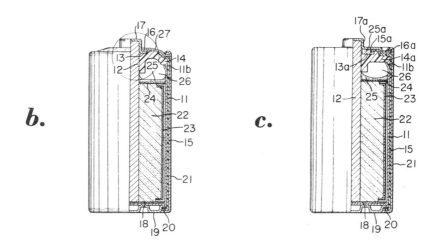

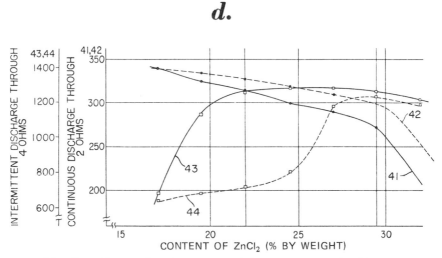

(a)   Phase diagram of an electrolyte consisting of water, zinc chloride
       and ammonium chloride.
(b)   Partially sectional view of a prior art dry cell.
(c)   Partially sectional view of a dry cell in accordance with the process.
(d)   Diagram illustrating the relationship between the minutes to a final
       voltage and the concentration of zinc chloride.

Source: U.S. Patent 4,001,044

It has been confirmed that some starches are swelled about 30 times as much as the original volume. In Figure 1.4a, the swelling characteristic curve **1** is that of corn starch; **2**, that of bridged starch consisting of corn starch, 0.25 mol percent of which is bridged with epichlorohydrin; **3**, that of bridged and etherificated corn starch, 0.05 mol percent being bridged with epichlorohydrin and the starch being etherificated with ethylene oxide at the degree of substitution of 0.15; **4**, that of wheat flour; and **5**, that of potato starch. The degrees of bridging and etherification may be suitably selected, but the volumetric expansion of the starches used in the dry cells is limited. The higher the zinc ion concentration, the greater the volumetric expansion of the starches becomes, and finally the molecular bond becomes so weak that the starch particles are collapsed.

Next, referring to Figure 1.4b, a prior art dry cell will be described. Reference numeral **11** denotes a zinc can with a top opening **11b**; **12**, a carbon electrode; **14**, a plastic top seal provided with a center hole through which is extended the carbon electrode **12**; **13**, an annular groove; **15**, a plastic tube or jacket; **17**, a cathode cap attached to the plastic top seal **14** with an insulating ring **16** fitted into the annular groove **13** thereof; **18**, an insulating paper; **19**, a bottom reinforcement; **20**, a sealing ring; **21**, an outer jacket; **22**, a depolarizing mix; **23**, a separator; **24**, a wax layer supporting washer; **25**, a wax layer; **26**, an air or expansion chamber; and **27**, a metal sealed top.

In the prior art dry cell of the type shown in Figure 1.4b, oxygen penetrating into the dry cell through the joint between the carbon electrode **12** and the plastic top seal **14**, that is the portion encircled by a circle in Figure 1.4b, contributes about 80% of the degradation in the shelf life of the dry cell.

Figure 1.4c shows one preferred example of a dry cell in accordance with the process, and in Figure 1.4c the component parts which are similar to those shown in Figure 1.4b, but are improved in order to improve the shelf life are designated by the same reference numerals plus a suffix a.

The top opening **11b** of the zinc can **11** is closed with a plastic top seal **14a** provided with a center hole through which is extended the carbon electrode **12** so as to be centered and a circular recess **13a** formed in the top surface of the plastic top seal **14a** coaxial with its center hole. The plastic tube or jacket **15** which covers the top surface of the plastic top seal **14a** extends over the boundary or ridge of and into the circular recess **13a** so that the edge portion **15a** may be embedded into the circular recess with a suitable filler of a high viscosity such as pitch or rubber. A metal top seal **17a** formed integral with the cathode cap is fitted over the plastic top seal **14a**, and extending along the periphery of the metal top seal **17a** is placed an insulating ring **16a**.

The outer jacket **21** is fitted over the zinc can **11** and has its upper edge portion folded inwards 180° toward and pressed against the top of the insulating ring **16a**, whereby the zinc can may be sealed. Since the edge portion of the outer jacket is pressed against the insulating ring, the undersurface of the metal top seal **17a** is firmly pressed against the upper surface of the filler **25a**. Therefore, the joint between the plastic top seal **14a** and the plastic jacket **15a** as well as the joint between the plastic top seal and the carbon electrode **12** are securely sealed with the filler **25a** so that the leakage of the electrolyte through these joints may be completely prevented. Thus, the leak-proofness or liquid-tightness of the dry cell and other dry cell characteristics may be improved. The dry cell operation characteristics are in general dependent upon the composition of the electrolyte in the depolarizing mix, the quantity thereof and the quantity of the starch (the degree of swelling). Some examples are shown in the table below, and their test results are shown in Figure 1.4d.

	. . . . Sample Numbers													
	1	2	3	4	5	6	7	8	9	10	11	12	13	14
Depolarizing mix														
MnO₂, g	. . . . 500.													
Acetylene black, g	. . . . 100													
Zinc oxide, g	. . . . 5													
Electrolyte, ml	. . . . 326.1													
ZnCl₂, wt %	17.0		19.5		22.0		24.5		27.0		29.5		32	
NH₄Cl, wt %	4		4		4		4		4		4		4	
H₂O, wt %	79		76.5		74		71.5		69.0		66.5		64	
H₂O/MnO₂ weight ratio	0.598		0.590		0.582		0.574		0.565		0.556		0.546	
Starches	C	CE	C	CE	C	CE	C	CE	C	CE	C	CE	C	CE
Volumetric expansion ratio of starch	1	1	1	1.5	1	3.0	1.5	5.6	2.1	8.9	5.0	10.4	8.0	13.8

In the table, C designates a bridged starch, and CE, a bridged and etherificated starch. In Figure 1.4d, curves **41** and **42** indicate continuous discharge through a load of 2 ohms for the dry cells containing CE and a gelatinous C-containing paste, respectively; **43** the discharge curve when the dry cell with the gelatinous paste containing CE was intermittently discharged through a load of 4 ohms at a rate of 4 minutes per hour and 8 hours per day; **44**, the discharge curve when the dry cell with the gelatinous paste containing C was intermittently discharged at the above rates. When the quantity of the electrolyte is too much, the paste becomes too soft while when the quantity is too little, the paste collapses when it is formed. Therefore, the quantity is so selected as to attain the best formability.

The separator **23** which is in contact with the depolarizing mix was prepared in the following manner. A double-ply kraft paper sheet consisting of a first ply or layer of 50 μ in thickness made of the pulp of the degree of beating of 300 to 600 cc (measured by the Canadian freeness instrument) and a second ply or layer of 50 μ in thickness made of the pulp of the degree of beating 800 to 850 cc was prepared by the conventional paper sheet making method. Upon the surface of the layer of the low degree of beating was applied a layer of C or CE shown in Figure 1.4a at a rate of 36 to 44 g/m².

The dry cell enclosures were prepared in accordance with R-20, IEC Standards, and the top enclosure of the type shown in Figure 1.4b was employed. Referring to Figure 1.4d, it is seen that the lower the concentration of the electrolyte, the better the continuous discharge characteristic becomes and that the smaller the volumetric expansion ratio of starch, the better the continuous discharge characteristic is. The intermittent discharge characteristic is closely related with the volumetric expansion ratio of starch, and is not satisfactory unless the ratio is in excess of 3.0.

In both the continuous and intermittent discharge tests with the loads of 2 and 4 ohms, the terminal voltage was 0.9 V. Similar characteristics were found when load current was drawn continuously through the load of 4 ohms, and when load current was drawn intermittently through the load of 2 ohms or 4 ohms more than an hour a day. The intermittent discharge characteristic of the dry cell which was discharged through a 10 ohm load was similar to the intermittent discharge characteristic of the dry cell which was discharged intermittently through a 4 ohm load at a rate of 4 min/hr and 8 hr/day.

The satisfactory discharge characteristics may be obtained when the concentration of zinc chloride is 22.5 to 29.5% by weight and when the volumetric expansion ratio is between 3.0 and 10.4 at 20°C. It was found that the concentration of ammonium chloride less than 2% by weight results in the greater local corrosion of zinc and that when the concentration exceeds 5.5% by weight, the intermittent discharge is adversely affected.

Plastic Seal

A process described by *R.B. Affeldt; U.S. Patent 3,967,977; July 6, 1976; assigned to Union Carbide Corporation* relates to a seal closure for the open end of a cylindrical container used in dry cells. The closure has a centrally disposed tubular neck adapted to be snugly slid over and adhesively secured to the current collector (carbon rod) of the cell and a peripherally disposed tubular skirt or apron adapted to be snugly slid on and adhesively secured to the upper rim of the container thereby providing a seal for the container.

Referring to Figures 1.5a and 1.5b, there is shown a galvanic dry cell. The dry cell includes a cylindrical container 2 which is made of an electrochemically consumable metal such as zinc and which serves as the anode of the cell. Included in container 2 are an insulator disc disposed at the bottom of the container, a separator 4 lining the vertical wall of the container, a cathode electrode comprising a depolarizer mix 6 disposed within but separated from the container by the insulator disc and the separator and a current collector 8 centrally embedded in the depolarizer mix. The insulator disc could be made of plastic or any other suitable insulating material. The separator may comprise a thin electrolyte paste layer or may be a thin film separator containing electrolyte, e.g., a thin bibulous paper coated with an electrolyte gel paste.

The cathode depolarizer mix is an electrochemically active cell component and may contain, for example, manganese dioxide, a conductive material such as carbon black or graphite and an electrolyte. The depolarizer mix is usually molded around a central carbon cathode collector rod 8 before being inserted into the cell container or the depolarizer mix could be the first inserted into the container whereupon the current collector could then be forced into the mix. A conventional center washer could be employed to maintain the current collector in an upright axial position.

The top of the cathode depolarizer mix is disposed a fixed distance below the open end of the cell container to provide the usual airspace 10 to accommodate any liquid spew that may be formed during cell storage and/or cell discharge.

FIGURE 1.5: CLOSURE FOR GALVANIC CELLS

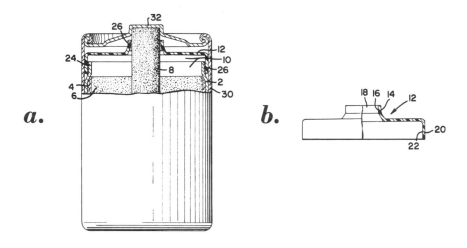

a.　　　　　　　　　　　*b.*

(a) Elevational view, partly in cross section of a primary galvanic dry
     cell seal closure.
(b) Vertical view, partly in cross section, of the seal closure used in
     the dry cell shown in Figure 1.5a prior to cell assembly.

Source: U.S. Patent 3,967,977

An electrically insulating closure **12** is provided for sealing the open cell of the
cell container **2**. As best shown in Figure 1.5b, closure **12** is formed as a disc-
like member having an upward projected tubular neck **14** whose internal wall
16 defines an opening **18**. At the periphery of the closure is a downward depending tubular skirt or apron **20** having a vertical internal wall **22**. As shown
in Figure 1.5a, the diameter of the neck of the closure is made such as to provide a snug sliding fit over current collector **8** and the diameter of the depending skirt is made such as to provide a snug sliding fit with the upper rim portion
24 of the container.

Preferably the rim portion of the container should be crimped or necked in to
allow the thickness of the skirt so as to provide an overall uniform raw cell
diameter as shown in Figure 1.5a. A layer of suitable adhesive **26** is disposed
between wall **16** of neck **14** and the outer surface of current collector **8** and
between wall **22** of skirt **20** and the outer surface of crimped rim portion **24**
of container **2**. The adhesive may be applied to inner walls **16** and **22** of closure **12** and/or to the corresponding surface of the current collector and the
crimped rim of the portion of the container, respectively.

It should be noted that the adhesive joint for the skirt can preferably be made
with the outside surface of the container which is clean, or can easily be made
so, as compared with the inside surface of the container which could be con-
taminated with the ingredients of the depolarizer mix during its insertion into
the container.

The length of the skirt **20** and the neck **14** must be long enough to insure a reliable seal area. For example, the length of the skirt for D size cells could vary between about $\frac{3}{16}$" and $\frac{5}{8}$" and preferably be about $\frac{3}{8}$" while the length of the skirt for C size cells could vary between about $\frac{3}{16}$" and $\frac{5}{8}$" and preferably be about $\frac{3}{10}$". The length of the closure neck could vary depending on the diameter of the current collector rod. For example, the length of the closure neck could vary between about $\frac{1}{16}$" and about $\frac{1}{4}$" for current collector having a diameter of about $\frac{5}{16}$".

However, when vacuum drawing techniques are employed to fabricate the closure of this process, the length of the neck must be drawn too deeply since it may result in material thinning at the neck radius which could decrease the strength of the neck to a degree that moisture retention within the cell would be jeopardized. Preferably, the neck radius could be chamfered so as to provide space for receiving a reservoir of sealant so as to further insure a good seal.

The material thickness of the closure can vary although a thickness between 0.005" and 0.06" has been found acceptable with a thickness about 0.02" being preferable. Closure material thickness less than about 0.005" would be insufficient to provide a good seal while a material thickness above about 0.06" would only result in decreasing the airspace in the cell while increasing the cost of the cell.

As shown in Figure 1.5a, with the closure **12** in place an effective fluid-tight seal around the open end of the cell container **2** is formed which effectively seals the raw cell against the escape of electrolyte and/or moisture by evaporation and seals it against the ingress of air and oxygen from the atmosphere. As stated above, the sealed raw cell could vent through the porous current collector such as a carbon rod.

The cell proper (i.e., raw cell) with the closure locked in place at the open end of the cell container is finished by encasing it within an outer cell assembly including an outer tubular noncorrodible jacket **30** suitably made of a fibrous material such as kraft paper. The upper end of the tubular jacket extends beyond the closure and is locked in engagement with the outer peripheral edge of a one piece metallic top cover plate **32** in the conventional manner.

The outer cell assembly includes means for venting any gas released from inside the cell container to the outer atmosphere. Such means may be provided, for example, by making the locked engagement between the top cover plate and the jacket permeable to gas.

Thus the process provides a seal closure for the open end of a cylindrical container used in sealed galvanic dry cells which requires the very minimum number of parts and which is, therefore, relatively inexpensive to manufacture. The closure can be easily and accurately assembled during manufacture of the dry cells and gives highly reliable and reproducible results in containing and confining gases and cell exudate during cell storage and discharge. Moreover, the closure when used in a dry cell employing an outer cell assembly including a tubular jacket, is not exposed to external pressure or impact and is, therefore, less prone to accidental damage.

Plastic Bag

C.W. Pun and C.C. Poon; U.S. Patent 4,027,078; May 31, 1977 describe a dry cell comprising a negative electrode formed by a zinc tube which is open at both ends and is so dimensioned that it is substantially totally consumed when the cell voltage has fallen to a predetermined minimum value, e.g., in the range 0.70 to 0.85 V, a sealed bag of flexible plastics material enclosing the zinc tube and surrounded by a cylindrical jacket, a carbon positive electrode projecting through the bag in sealing relation thereto, and at least one tongue extending from the zinc tube through the bag in sealing relation thereto into contact with a negative terminal of the cell.

In order to conserve zinc, the height of the zinc tube will generally be substantially the same as that of the depolarizing dolly within the cell. The jacket may be of any suitable material, e.g., metal, paper, or cardboard; the base of the cell will ordinarily comprise a metal negative terminal which may be attached to the cylindrical jacket. The carbon electrode and the tongue extending from the zinc tube must form a seal with the bag effective to prevent passage of liquid during storage and use of the cell.

According to a further aspect of the process, the above-defined dry cell may be made by a process comprising the steps of sealing the zinc tube containing a dolly and electrolyte in a bag of flexible plastics material, surrounding the bag by the cylindrical jacket and inserting a carbon electrode into the dolly through the bag.

The bag of flexible plastics material is preferably of polyolefin material such as polyethylene or polypropylene or of vinyl material such as PVC. It may incorporate an aperture to admit the tongue projecting from the zinc tube or the tongue may pierce an aperture when the tube, dolly and electrolyte assembly is placed in the plastics bag. The bag may be preformed and the tube, dolly and electrolyte assembly placed therein or the bag may be formed around the assembly, e.g., by sealing both ends of a length of plastics tubing.

Encapsulating Resin

F.L. Ciliberti, Jr.; U.S. Patent 3,986,894; October 19, 1976; assigned to P.R. Mallory & Co., Inc. describes a construction which minimizes damage to alkaline electrolyte batteries due to electrolyte leakage caused by extreme variations in temperature. The batteries comprise a multiplicity of individually sealed cells, arranged in electrically connected stacks, each stack being disposed in a moisture-proof plastic container and/or cup filled with an encapsulating resin in the entire space within the cup, with the encapsulated assembly being housed in a metal outer container.

The cells within each stack and each stack itself are thus isolated, thereby minimizing any damage that might result from intercell leakage, from interstack leakage, and from leakage that would corrosively affect associated equipment in an assembled apparatus, due to extreme temperature variations, especially under high humidity conditions.

Dish Structure

D.L. Sanchez; U.S. Patent 3,951,690; April 20, 1976; assigned to Pilas Secas Juipter SA, Spain describes a method and device for providing leak-proof sealing of dry electrochemical cells allowing excellent tightness to be obtained. Onto the bottom of a metallic cup-shaped negative electrode of the cell is fitted a dish having a central protrusion and apertures positioned in a peripheral zone outwardly of the protrusion. A hot thermoplastic material is then employed for blocking or sealing the apertures in the dish, thus creating a mass of the plastic material on both sides of the apertures. At the same time the mass is bonded to the thermoplastic casing, inside which the cup of the cell is enclosed.

The material used for blocking the apertures may be injected from the outside, placed in advance in the dish prior to fitting the latter to the cup, or be provided by a part of the casing itself, which part extends for a sufficient length beyond the bottom of the cup. The cell sealed in this manner does not need to be encircled by a metallic ring. The cell bottom contains an electric contact formed by the protrusion of the dish. The remainder of the dish, covered with a plastic material, is uniformly joined with the casing.

Separator Above Depolarizer Cake Level

T. Kamai; U.S. Patent 3,973,995; August 10, 1976 describes a dry cell of the kind having a protective outer casing, a zinc anode in the form of a sleeve or canister within the casing, a cake of a depolarizer and an electrolyte within the anode, and a separator between the anode and the cake. The dry cell is characterized in that the upper surface of the cake is above the upper extremity of the anode, and the separator extends above the upper surface of the cake and is in close cooperation with the protective outer casing.

With this arrangement, because the separator extends above the anode and the cake, the attraction of the positive ions, such as zinc ions and ammonium ions, to the upper face of the cake, and therefore the depositing of electrolyte on the upper face of the cake, is prevented and the deterioration of the separator by the zinc chloride, which is produced with the decomposition of the zinc, is obviated.

Further, because the upper extremity of the anode is lower than the upper surface of the cake, the discharge reaction from the cake in its upper part takes place somewhat later than in the lower part. Thus, the semitransparent layer of zinc diammonium chloride is not produced in the upper part of the cake and therefore the electrolyte which rises from the lower part is impeded by the closely contacting separator and outer casing. The electrolyte flows in the layer of the cake and is drawn into it so that escape of the electrolyte is prevented.

In order to prevent further the leakage of the electrolyte it is advantageous to prepare the electrolyte solution from zinc chloride and ammonium chloride such that the zinc chloride is the main substance. With this arrangement no zinc chloride occurs with the discharge, no fine crystals such as zinc diammonium chloride form and thus there is little tendency for the electrolyte solution to be pressed upwards due to osmotic pressure.

Advantageously the amount of the anode zinc which contributes to the discharge is such that a useable voltage can be obtained until the depolarizer, which is preferably manganese dioxide, loses its oxidizing ability. However, if zinc still remains after the manganese dioxide has lost its oxidizing effect, it has the effect of producing hydrogen gas on the electrolyte side and thereby pressing electrolyte liquid into the outer casing. Preferably, therefore, the minimum amount of zinc in order to produce the current is provided. This reduces considerably the manufacturing cost of the cell.

The escape of any electrolyte via the surface of the cake would lead to the result that the substance producing the electricity is outside the reaction area and thus the output of the battery would be considerably reduced.

With the cell according to the process, however, the electrolyte remains in the material producing electricity and, therefore, the cell has a higher discharge capacity. With the cell according to the process, the zinc preferably disappears in the last stage of the discharge process and in this case the zinc cannot maintain the shape of the cell. Therefore, the cell is preferably provided with a metal covering as an outer container.

The process is applicable to dry cells having a paste construction whereby the separator contains starch and meal in the sticky state as well as to dry cells of the so-called paper lined construction in which gummed paper spread with meal or other adhesive material is used.

Reactive Acid Layer

T. Tsychida, K. Shinoda, T. Yasuda and T. Takeshima; U.S. Patent 3,909,295; September 30, 1975; assigned to Fuji Electrochemical Co., Ltd., Japan describe an alkaline cell comprising an anode active mass, a cathode active mass, an anode terminal member electrically connected to the anode active mass, a cathode terminal member electrically connected to the cathode active mass, and an insulator disposed between the terminal members. An acid layer is provided on an inner surface of the anode terminal member where an alkaline electrolyte creeps by electrocapillary action. The acid layer is reactive to the alkaline electrolyte to produce a salt.

In Figure 1.6, a cathode metal casing **1** of the cell is coated with nickel and contains therein an annular shaped cathode active mass **2** which consists of a mixture of manganese dioxide MnO_2 with phosphorus, graphite and binders such as carboxymethylcellulose, polyvinyl alcohol, or polyisobutylene. The cathode active mass is separated from an anode active mass **5** by a separator **3** made of nonwoven fabric of polypropylene. At one end of the separator, which is an upper end in the drawing, adhered is an insulating disc **4** of nylon, polypropylene, on polyethylene. The anode active mass inside of the separator and the insulating disc **4** are made of a mixture of zinc oxide, water, carboxymethylcellulose, potassium hydroxide and amalgamated zinc powder containing 3 to 4% of mercury.

A current collector **6** made of brass and the like extends into the anode active mass along the axis and is spot-welded to an anode terminal plate **8** at the center portion. The cathode casing **1** and the anode terminal plate **8** are electrically separated by two sealing insulators **9,10**. The first sealing insulator **9** is made of nylon, polypropylene or the like and disposed inside of the anode terminal plate **8** and extends radially inwardly from the marginal portions between the anode terminal plate **8** and the cathode casing **1** to the circumferential portion of the current collector **6**. The second sealing insulator **10** is made of rubber or the like having elasticity higher than the first sealing insulator and is constricted between free ends of the cathode casing and the anode terminal plate.

The first sealing insulator **9** has an annular wall portion at the center portion thereof through which the current collector penetrates and which is constricted against the current collector by an annular cap member **7**. The cathode metal casing has at least a gas venting aperture **11** at an end portion to which the first sealing insulator **9** abuts. The venting aperture allows the gas, developed in the cell to a predetermined high pressure, to escape through to the outside of the cell.

Provided between the first sealing insulator **9** and the anode terminal plate adjacent the current collector is a thin acid layer **12**, the acid of which reacts with the alkaline electrolyte to form a salt. Accordingly, the alkaline electrolyte creeping, due to the electrocapillary action, through an interspace between the current collector and the annular wall portion of the first sealing insulator **9** is changed into salt as it comes to the inner surface of the anode terminal plate where the acid layer is provided. Thus, the leakage of the alkaline electrolyte outside of the cell is completely interrupted.

The acid layer is preferably provided in combination with a known sealing structure of cells in which the creeping way of the alkaline electrolyte is elongated. The provision of the acid layer shows remarkable advantages in small-sized cells in which a long creeping way cannot be formed such as mercury cell, silver oxide cell, and alkaline manganese cell.

FIGURE 1.6: ALKALINE CELL

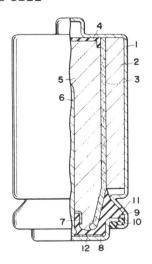

Source: U.S. Patent 3,909,295

The acid employed includes organic and inorganic acids which actively neutralize the alkaline electrolyte and produce salts. The organic acids are acidic organic compounds. Though the acidic organic compounds include those soluble or insoluble in water, they react easily with the alkaline electrolyte such as sodium hydroxide or potassium hydroxide and produce soluble salts.

The organic acid is preferably mixed with a bonding agent or viscous material and attached or coated to the inner surface of the anode terminal plate 8 to form a thin layer 12. At this time, in order not to corrode the metal surface of the anode terminal plate 8 by the organic acid, the layer of the organic acid is made as dry as possible.

As the inorganic acid to be used to react with the alkaline electrolyte to produce salts, a boric acid is preferable for handling. The alkaline electrolyte leakage was examined with respect to 180-AA size alkaline-manganese cells and no leakage was found on the surfaces of the cells, whereas the same type of cells using a bonding agent or vaseline alone experienced the leakage with respect to more than 50% of all the cells examined.

Examples 1 through 11: In the following experiments, all of the cells were shelved in an atmosphere at a temperature of 55°C and humidity of 95% for 1 year. As is apparent from the experiments, Examples 1 through 8 using the organic acids and Example 9 using inorganic acid in the acid layer showed no leakage at all after one year shelving under such severe conditions. It was found that the alkaline electrolyte in the cells of Examples 1 to 9 was completely prevented from creeping at the inner surface of the anode terminal plate where the acid layer is provided.

On the other, hand, it was found that even in the cells where no leakage appeared, in Examples 10 and 11, the alkaline electrolyte crept up to the inner circumferential portion of the anode terminal plate and that it was a matter of time before the alkaline electrolyte crept out of the cells.

Example	Composition of Acid Layer	Cells with Leakage/ Cells Examined
1	Oxalate + vaseline (1:1)	0/20
2	Phthalic acid + vaseline (1:1)	0/20
3	Citric acid + vaseline (1:1)	0/20
4	Tartaric acid + vaseline (1:1)	0/20
5	Stearic acid + vaseline (1:1)	0/20
6	Salicylic acid + vaseline (1:1)	0/20
7	Maleic acid + vaseline (1:1)	0/20
8	Succinic acid + vaseline (1:1)	0/20
9	Boric acid + vaseline (1:1)	0/20
10	Vaseline (conventional type)	10/20
11	Epoxy resin (conventional type)	12/20

Multicell Construction

K.V. Anderson; U.S. Patent 3,954,505; May 4, 1976; assigned to ESB Incorporated describes an alkaline primary battery having at least two cells. The cells are of the type where the anode material (zinc) is in the form of a powder suspended in a gel and is located in the central portion of the cell. The anode current collector for the second cell is made as an integral portion of the first cell and takes the form of a metal rod dependent from the base of the first cell.

The cells are built without anode current collectors. In the assembly of a battery from such cells, the anode current collector attached to a first cell is forced through the cover of a second cell and embeds itself in the anode structure of the second cell. The anode collector for the first cell is in the form of a loose metal rod and is similarly forced through the cover and into the anode structure of the cell. The end cell opposite to the first cell is similar to the other cells except that it does not have a current collector attached to its base.

Referring to Figure 1.7, items **10**, **12** and **14** represent the three cylindrical cells of a three-cell battery prior to the assembly thereof. Cells **10** and **14** are shown in full. Cell **12** is shown in diametric cross section. In cell **12**, a container **16** and cover **18** contain the active cell ingredients. These include the cathode mix **20** of tubular form in electrical contact with the can **18**; the anode mix **22** in cylindrical form, and a tubular separator **24** separating the cathode mix from the anode mix.

An electrolyte permeates the pores of the cathode mix, the anode mix and the separator. In the design shown, the cover **18** is a continuous piece of stiff molded plastic such as polyethylene, polypropylene, methyl methacrylate, nylon, etc. It is clamped and sealed between the bead **26** and the flange **28** both formed at the upper part of the container **16**. Other cover designs and clamping means known in the art may also serve. The container **16** is made of metal such as steel, plated steel, stainless steel, nickel, etc. In the design shown, it is of drawn construction so that there are no seams.

The cathode mix **20** or positive active material comprises a pressed tubular pellet. In the several electrode systems for which the process is suitable, the cathode active material may be manganese dioxide, mercuric oxide, nickel hydrate, or mono or divalent silver oxide. The cathode materials may be mixed in with conductive material such as carbon powder, graphite, or metal powders as is well known in the galvanic battery art.

The anode mix **22** or negative active material comprises metallic zinc powder or amalgamated zinc powder suspended in a gel such as carboxymethylcellulose. This form of negative is well known in the art. The gel has the consistency of a heavy oil. It is added to the cell as a high viscosity liquid. The material is thixotropic and becomes a gel upon standing. The separator **24** may be a paper, a synthetic fiber felt or other similar article as known in the art. The electrolyte is an aqueous solution of sodium hydroxide, potassium hydroxide or mixed sodium and potassium hydroxides. The strength is from 30 to 40% hydroxide. These electrolytes are all well known in the battery art.

A collar **30** dependent from the cell cover **18** makes an interference fit with the top of the separator **24** and seals the anode compartment from the cathode compartment. Cell **12** as shown is complete except that it has no current collector for the anode. This part comprises the nail-like rod **32** dependent from the bottom of the container **20a** of the first cell directed away from the bottom of the cell and electrically and mechanically attached.

FIGURE 1.7: MULTICELL ALKALINE PRIMARY BATTERY

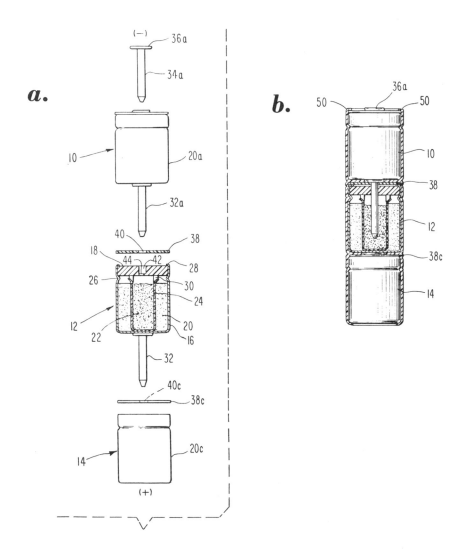

(a)   Elevation and partial cross-section view of the several
       parts required to assemble a 3-cell battery.
(b)   Fully assembled battery.

Source: U.S. Patent 3,954,505

The rod may be tin-plated steel, brass, bronze, etc. It should be a good conductor of electricity and it should be structurally strong. The fastening means might be a solder. However, for best conductivity and maximum strength, it is preferred to weld the current collector **32a** to the bottom of the container **20a**. The container **20** of cell **12** likewise has a current collector **32** attached to the bottom thereof to serve as the current collector of cell **14**.

Cells **10** and **14** are end or terminal cells. Because cell **14** is a terminal cell, it does not have a rod dependent from the bottom thereof. The positive terminal of the battery is the can **20c** itself. However, if a fastener-type contact should be required, it could be welded to the bottom of the can **20c**. Cell **10** requires a current collector. This is provided by a separate piece **34a** having the same size and shape as collectors **32** and **32a**. The collector **34a** shown has a head **36a** to serve as the negative terminal of the battery. If a fastener-type terminal (negative) is required for the battery, it can form a part of or be attached to the upper end of the collector **34a**.

In a multicell battery assembly, it is important that the metallic containers of adjacent cells do not touch as this would directly short out a cell. As one means of insulating the cell pile, discs **38** and **38c** made of a dielectric material are supplied. These discs are about the same diameter as the several cells and have a centrally located hole **40** and **40c**.

To assemble a battery from the parts listed above, collector **34a** is pressed through the cover of cell **10** until the head **36a** rests on the cover. The disc **38** is placed upon collector **32a**. **32a** is pressed through the cover **18** and into the anode gel **22** until disc **38** rests firmly on the flange **28** and the head of collector **32** rests on the disc **38**. The current collector **32a** thus passes through the cover of cell **12** and is embedded in and is in electrical contact with the anode mix **22** of cell **12**. It may be desirable to form the cover **18** with a depression **42** so that the collector **32a** may be properly centered and directed in the assembly.

It is to be noted that whether the cover is flat or includes the depressions **42** with a web **44** at the bottom thereof it is continuous and seals the cell against loss of moisture to the atmosphere. Because the seal is present even when cells are not assembled as batteries, individual cells may be stored for considerable periods of time. This in turn allows the cells to be built in large quantities and assembled into batteries at later times in accordance with any particular order.

In the same manner, cell **14** is assembled onto the collector **32** of cell **12**. Figure 1.7b shows the assembly as described. To further strengthen the battery and insulate it, the cylindrical sides of the battery may be covered with an insulating casing such as a paper or plastic wrapping, a pasteboard tube or a plastic tube shrunk over the assembly as shown in cut-away form by **50**.

A very critical part of this assembly and one that is vital to its success is the seal between the collectors **34a**, **32a** and **32** and the covers of cells **10**, **12** (item **18**) and **14**. There are a number of coatings known in the art such as bitumen, heavy grease, etc., that can be placed on the upper portions of the collectors to improve the seal. Cement materials such as certain epoxy compounds can be applied. The fit between the collector and depression **42** can be interference fit, and the collector can be slightly tapered so as to slightly compress the cover material against the metal collector. U.S. Patent 3,713,896 describes a number of sealing means suitable for the assembly.

VENTING AND PRESSURE CONTROL

Cover Deformation

M.G. Rosansky and I. Michalko; U.S. Patent 3,909,303; September 30, 1975; assigned to Power Conversion, Inc. describe a battery capable of safely venting its contents without exploding which comprises a casing defining a cavity in which the electrochemical contents of the battery is confined. A rigid partition is disposed across an opening in the casing, and a relatively bendable cover is disposed with its under surface overlying the partition. A means is provided for attaching the cover to the partition at at least one location. Preferably, this means takes the form of a spot weld.

A passageway, which may be an aperture in the partition, connects the under surface of the cover with the cavity whereby any internal pressure generated by the contents is applied against the cover which will deform and release the contents without becoming detached from the partition when a predetermined internal pressure limit is reached.

A battery construction capable of safely venting its contents without exploding is shown in Figure 1.8a. It includes a cylindrical steel battery casing **10** which is closed at its bottom end **12**. The casing **10** is necked about its perimeter to form a ridge **13** along the interior of the casing slightly spaced from its opened top end **14**. A rigid annual metallic thrust ring **18** is inserted from the open end **14** of the casing **10** and seated against this ridge **13**.

Disposed across the open end **14** of the casing **10** is a rigid metal disc-shaped partition **20**. A disc-shaped cover **22** of the same size is positioned with its under surface **23** overlying the rigid partition **20**. The perimeter of the partition **20** and cover **22** is surrounded and engaged by an annual elastomeric insulator or sealing member **24** which is pressed upwardly by the thrust ring **18** and downwardly by a crimped edge **25** formed by the inwardly bent perimeter of the open end **14** of the casing **10**.

This construction normally seals a cavity **26** within the casing **10** wherein the electrochemical system of the battery is confined. (In Figure 1.8a, only fragments of the vertical side walls are shown, but it will be understood that the casing **10** forms an elongated cylinder.) A cap **27** welded atop the center of the cover **22** forms one terminal of the battery.

A bottom view of the partition **20** is shown in Figure 1.8b. The partition includes a centrally located aperture **28** which is aligned with a seam **30** in the cover member **22** which has been closed by cold welding. The cover member **22** is securely attached to the partition **20** by two diametrically opposed spot welds **32**. While the partition **20** is rigid, the cover **22** is softer and more ductile, and thus relatively bendable. Any internal pressure within the cavity **26** is applied against the under surface **23** of the cover **22** because that surface **23** is in communication with the cavity **26** through the passageway formed by the aperture **28**.

If this internal pressure exceeds a predetermined limit, the bendable cover **22** will be permanently deformed so that it pulls away from the crimped upper edge **25** of the casing **10** near the points **A** on its periphery most distant from the welds **32**. Alternatively, the cover may permanently deform and tear near the weld. Thus, the deformation will allow the contents to escape and release the pressure while the cover **22** remains safely and securely attached to the partition **20**, thus eliminating the danger from flying projectiles which would otherwise be associated with the release of the internal pressure.

FIGURE 1.8: BATTERY CONSTRUCTION

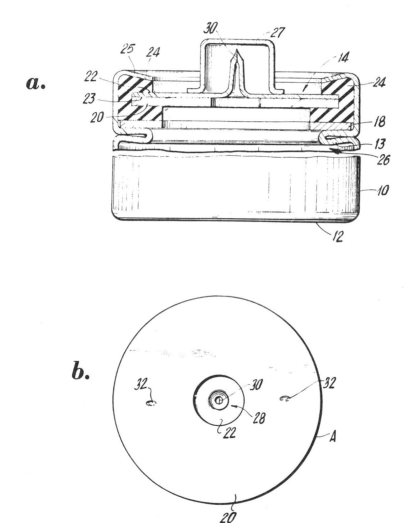

(a)   Cross-sectional view of battery
(b)   Bottom view of a portion of the battery of Figure 1.8a

Source: U.S. Patent 3,909,303

The specific configuration depicted in Figure 1.8b is shown here only by way of example. It is advantageous, because the pressure limit by which the battery vents can be selected by properly positioning the welds **32**. If the battery is intended to vent at a relatively high pressure, the welds **32** are positioned relatively close to the center of the battery, whereas, if the battery is intended to vent at a relatively low pressure, the welds **32** will be moved farther apart toward the periphery of the partition **20**. There are, however, many other desirable configurations. By way of example, this electrochemical system may include two electrodes, one of which is lithium, and sulfur dioxide as a depolarizer. The process may be practiced, however, using any sealed electrochemical system presenting a danger of explosion.

Resealable Vent Closure

A process described by *H. Heinz, Jr.; U.S. Patent 4,020,241; April 26, 1977; assigned to Union Carbide Corporation* relates to sealed galvanic dry cells, and more particularly to a low-pressure resealable vent for releasing excessive gas pressure from inside the dry cell. The improvement comprises a gas-impermeable, resiliently compressible elastomeric sponge gasket compressed between the upper wall of the container and the cover.

The gasket should be sufficiently resilient such that gas buildup within the cell in the range of about 5 to 75 psi will provide a sufficient force to temporarily deflect or further compress the gasket at the gasket-container interface and/or gasket-cover interface thus enabling the gas to vent. When used in a galvanic dry cell employing a conventional-type cathode collector rod, the gasket may be constructed with a suitable opening or aperture at its center for sliding over and in some cases maintaining contact with the cathode collector rod so as to effectively provide another venting path between the cathode collector rod and the gasket in addition to the venting paths between the gasket-container and gasket-cover interfaces.

One or more of these venting paths or selected areas of these venting paths may be permanently sealed so as to effectively direct the venting of gas from within the cell along preselected venting paths or areas.

Elastic Gas Valve

A process described by *G.S. Bell and J. Bauer; U.S. Patent 3,923,548; Dec. 2, 1975; assigned to Varta Batterie AG, Germany* relates to a gas valve for galvanic cells, especially primary cells, and more particularly to such a gas valve that functions through the use of elastic materials and that is provided with a carbon rod which serves as a current conductor.

In the process, there is provided at the bordering surfaces of the sealing disc and the current conductor of a galvanic cell, a weak region formed of elastic material which yields and thereby opens the valve when subjected to a given pressure. The sealing disc or at least a portion thereof at a radially inner region must therefore be formed of a material having elastic properties and which maintains these properties together with a stability of form or shape over a wide temperature range substantially from −40° to +80°C. A suitable material for this purpose has been found to be a polyolefin, for example, having a rigidity and hardness of a polyolefin of medium to high density and having a tensile strength and elongation or stretchability close to that of polypropylene.

In the examples shown in Figure 1.9, the sealing disc, whose inner opening fits stiffly on the current conductor, is always pressed onto the latter so that the upper end of the latter extends freely therefrom. Significantly, the current conductor, which is preferably a carbon rod, has a surface which is as smooth as possible, so that a virtually gas-tight seal is attained. The sealing disc or ring must tightly close the negative electrode vessel which is illustrated in the figures as a round cup. To provide this sealing action, any and all conventional and successful methods employed in practice are usable.

FIGURE 1.9: ELASTIC GAS VALVE

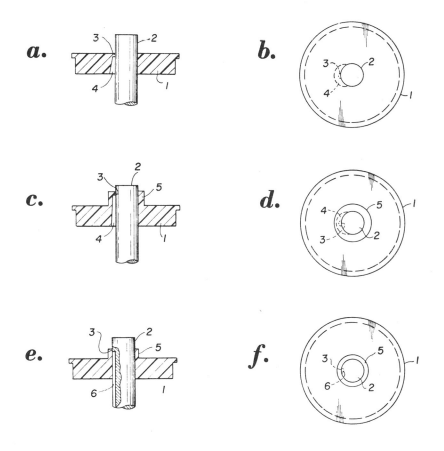

(a)(c)(e)   Longitudinal sectional views
(b)(d)(f)   Respective top plan views of gas valves

Source: U.S. Patent 3,923,548

The gas valve of Figures 1.9a and 1.9b is formed of a sealing disc **1** with a thinned-out or weakened portion **3** having a wall thickness usually of from 0.05 to 1.0 mm, preferably 0.1 to 0.6 mm, and a length of engagement with the current conductor **2** of from 0.1 to 10 mm, preferably 0.5 to 2.0 mm, depending upon the material of the sealing ring or disc **1**. A tapering or conical groove **4** is formed at the inner peripheral surface of the sealing ring **1** and permits the weakened gas valve portion **3** to be raised under the action of gas pressure within the groove **4**.

Through the sealing ring **1** having an extension in the form of a collar **5** according to Figures 1.9c and 1.9d, the weakened valve portion **3** is more accurately adjustable for smaller excess pressures because the collar **5** is somewhat yieldable elastically outwardly. For this reason, the thickness of the tonguelike weakened portion **3** can be greater than in the example of Figures 1.9a and 1.9b, and can be as thick as 2 mm but preferably about 0.5 mm.

In the gas valve of Figures 1.9e and 1.9f, the path of the escaping gas does not go through the sealing ring **1** per se to the weakened or thinned valve portion **3**, but rather through a longitudinal groove **6** formed on the surface of the current conductor **2**. The thin-walled collar **5** formed, for example, by diecasting on the sealing disc **1**, and which is dimensioned to correspond to the gas pressure to which it is expected to be subjected, is forced to a side by the pressure of the gas and, after the gas pressure has subsided, returns to its initial position by its inherent elasticity.

Through the spacing between the end of the gas feed groove **6** formed in the current conductor-carbon rod **2** and the sealing valve portion **3** located thereabove, a pressure adjustment is also possible, in addition in fact also with respect to the thickness of the collar.

As illustrated, the cover and gas valve, i.e., sealing ring **1** and valve portion **3**, are fabricated as a single piece with the gas valve an integral part of the cover. In practice, by means of this gas valve construction and by suitable selection of the wall thickness and the surface of the weakened portion **3** which acts as an elastic tongue, adjustment to any desired gas pressure can be effected. The valve of this process is suitable especially for mass production because of the simplicity of manufacture.

Longitudinal Groove

L.F. Urry; U.S. Patent 3,940,287; February 24, 1976; assigned to Union Carbide Corporation describes a magnesium dry cell which comprises a metal anode cup having a cathode mix therein, separated from the side wall of the anode cup by a porous, ionically permeable separator, and having an open end which is gastightly sealed by a closure including a vent.

A gas-venting passageway is provided around the cathode mix which comprises a longitudinal groove in the side wall of the anode cup communicating with the void space defined between the cathode mix and the seal closure. Gas that is generated near the bottom of the cell is vented through the cathode mix via the passageway into the void space from whence the gas escapes through the vent. In the preferred case, the longitudinal groove in the side wall of the anode cup, forming the gas-venting passageway, also communicates with a radial groove in the bottom wall of the anode cup.

Passageway Through Cathode Mix

V.S. Alberto and D.G. Clash; U.S. Patent 3,932,196; January 13, 1976; assigned to Union Carbide Corporation describe a primary dry cell comprising a metal anode cup having a cathode mix and having an open end which is gas-tightly sealed by a closure including a vent. A gas-venting passageway extends through the cathode mix into communication with the void space defined between the cathode mix and seal closure.

Referring to Figures 1.10a and 1.10b, there is shown a primary dry cell comprising a cylindrical anode cup 10 made of magnesium metal or a magnesium alloy, and having an upper open end and a close bottom end. Within the anode cup 10 there is a cathode mix 12 comprising particles of an oxide depolarizer, such as manganese dioxide, finely divided conductive material, such as acetylene black, and an electrolyte. Suitably, the electrolyte may be an aqueous magnesium perchlorate solution, for example.

The cathode mix 12 is separated from the side wall of the anode cup 10 by a porous, ionically permeable separator 14, suitably a porous kraft paper. The cathode mix 12 is also separated from the anode cup bottom by a porous paper or cardboard washer 16. The washer 16 is also permeable to the electrolyte and renders the anode cup bottom anodically active along with the side wall of the anode cup 10. A central carbon electrode element 18 is embedded in the cathode mix 12 and protrudes slightly beyond the upper open end of the anode cup 10.

As best illustrated in Figure 1.10b, four longitudinal, equidistantly-spaced apart holes 20 are pierced through the cathode mix 12 at the approximate midcircle of the annular area between the carbon electrode element 18 and the paper separator 14. These holes 20 are substantially parallel to the carbon electrode element 18 and terminate in the cathode mix 12 a short distance above the paper or cardboard washer 16.

The seal closure for the cell comprises an annular insulating disc 22 which is gas-tightly sealed within the upper open end of the anode cup 10. Disc 22 is suitably molded from a plastic material and is formed on its top surface with an annular, stepped or raised portion 24 surrounding its outer periphery. The disc 20 is fitted tightly around the protruding end portion of the carbon electrode element 18 and its outer edges abut against the interior side wall of the anode cup 10.

The upper side wall of anode cup 10 is turned or bent inwardly by a seal ring 26, suitably made of steel. This ring 26 is compressed or forced inwardly under a high radial pressure against the side wall of the anode cup 10 to form a tight radial seal between the abutting outer edges of the disc 22 and the interior side wall of the anode cup 10. A metal terminal cap 28 is fitted over the top of the carbon electrode element 18 and serves as the positive terminal of the cell.

A resealable vent is incorporated in the seal closure and is preferably constituted by a small vent aperture 30 which is provided in the insulating disc 22. The vent aperture 30 communicates with the void space 32 which is defined between the disc 22 and the exposed surface on the top of the cathode mix 12.

FIGURE 1.10: MAGNESIUM DRY CELL

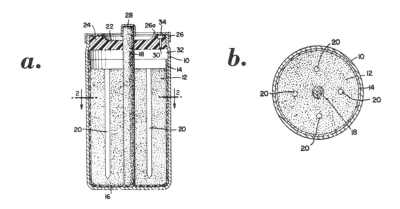

a.   *b.*

(a) Elevational view, in section, of a magnesium dry cell.
(b) Sectional view taken along the line 2–2 in Figure 1.10a.

Source: U.S. Patent 3,932,196

The cathode mix **12** substantially fills the anode cup **10** to just below its upper peripheral edges leaving enough space to accommodate both the seal closure and void space **32**. Overlying the vent aperture **30** is a flat annular seal gasket **34**. This gasket **34** may be made from any suitable elastomeric material, such as Tenite (cellulose acetate or cellulose acetate butyrate), and preferably covers the whole top surface of the annular stepped or raised portion **24** on the disc **22**. The seal ring **26** has one leg element **26a** which extends radially inwardly from the peripheral edge of the anode cup **10** and is mounted in resilient pressure contact against the top of the flat annular seal gasket **34**.

The leg element **26a** constitutes a retaining member which biases the seal gasket **34** into normally sealing relation around the vent aperture **30**. Upon the buildup of a predetermined excessive gas pressure in the void space **32**, the leg element **26a** deflects slightly in a direction away from the gasket **34** and allows gas to escape through the vent aperture **30**. Once the gas pressure has been relieved, the resiliency of the leg element **26a** causes it to again close or seal the aperture **30**. A more detailed discussion of the resealable vent and its method of operation is given in U.S. Patent 3,494,801.

Hydrogen-Absorbing Pellet

A. Kozawa; U.S. Patent 3,939,006; February 17, 1976; assigned to Union Carbide Corporation describes a device which will prevent buildup of unreacted hydrogen gas in sealed cells. More particularly there is provided a hydrogen-gas-absorbing device comprising a discrete-shaped body containing a material which will react chemically with hydrogen gas, a catalyst for the hydrogen gas-consuming reaction and a binder which is compatible with the other materials and will maintain them in a shaped body which is permeable to hydrogen gas. The shaped body is preferably packaged in a material which is permeable to hydrogen gas but impermeable to the cell electrolyte throughout the life of the cell.

Hydrogen-absorbing devices in accordance with the process have three essential components: a substance which will react with hydrogen gas, a catalyst for the hydrogen oxidation reaction, and a binder which will maintain the reactant and catalyst as an integral unit. Preferably, the device will also have an external coating to prevent contamination of the reactant and catalyst by cell components or by-products of the cell reaction.

The material which will react with the hydrogen gas can be any solid compound which will react with or oxidize hydrogen to yield a solid or liquid reaction product. Specifically, suitable materials are those which, at ordinary ambient temperature, will exhibit a negative free energy change (ΔF) associated with the reaction of that material with hydrogen. Preferred reactants include manganese oxides such as manganese dioxide (MnO_2), manganic oxide (Mn_2O_3), manganese hydroxide ($MnOOH$) and hausmannite (Mn_3O_4), cupric oxide, silver oxide, mercuric oxide, manganese phosphate, bismuth trioxide, m-dinitrobenzene and quinone. Of these, manganese dioxide is particularly preferred since it is a relatively inexpensive and readily available material, and its properties and predictability in battery systems are well established.

A catalyst for the hydrogen-consuming reaction is necessary since, while each of the above compounds will react spontaneously with hydrogen, the reaction will not take place at ordinary ambient temperature at a rate sufficient to insure that hydrogen gas will be absorbed from the system before the internal pressure causes rupture of the battery seal.

Suitable noble metals which may be used are palladium, platinum and rhodium. Of these, palladium is preferred since it has a natural tendency itself to absorb and hold hydrogen gas on its surface and would act to retain the gas in the vicinity of the reactant until the reaction between the gas and the reactant can take place.

A preferred binder is an inorganic cement such as Portland cement. This binder is preferred since the cement provides sufficient strength and porosity and yet does not contaminate or poison the metal catalyst. Acetylene black can be added to improve the conductivity of the pellet and to provide the device with the maximum number of hydrogen gas permeable channels to enable the gas to penetrate to the interior of the device and make efficient use of the reactant and catalyst contained therein. The following example illustrates the process.

Example: Into a ball mill were placed, in the dry state, 160 g of electrolytic manganese dioxide powder, 1 g of palladium-coated carbon powder (5% palladium on carbon), 2 g of acetylene black, 80 g of Portland cement, and 5 g of steel wool chopped to $\frac{1}{8}$-inch lengths. These components were mixed thoroughly in the ball mill for about 30 minutes. At the end of this time 16 g of the dry mixture were placed in a beaker to which a few drops of 9 molar potassium hydroxide and, subsequently, 7 ml of distilled water were added to make a spreadable paste. After being well-blended, the paste was filled into half-moon-shaped channels 1½ inch in length and $\frac{3}{16}$ inch in diameter cut in a Lucite block. The paste-filled block was oven-dried at 45° to 65°C for 40 minutes and the pellets were removed from the block and further dried in air overnight at room temperature. The average weight of each pellet was 1.3 g of which 0.85 g was manganese dioxide. The formed pellets were wrapped in a layer of heat-shrinkable polyethylene film about one to two mils thick. The film was then heat-shrunk around the pellet and the edges of the film were heat-sealed using a hot air blower set at 130°C.

These pellets could absorb up to 220 cubic centimeters of hydrogen gas at standard temperature and pressure and, when placed inside the current collector of rechargeable D-size alkaline manganese dioxide cells, substantially improved the high-temperature storage capacity of the cells such that no cell rupture occurred after 14 weeks' storage at 71°C. Fifty percent of control cells containing no hydrogen-absorbing pellet ruptured after similar storage.

Safety-Pressure Switch

According to a process described by *G.R. Tucholski and F.G. Spanur; U.S. Patent 4,025,696; May 24, 1977; and G.R. Tucholski; U.S. Patent 4,028,478; June 7, 1977; both assigned to Union Carbide Corporation* a galvanic cell assembly incorporates an integral safety pressure switch comprising an active member and an insulating member. The active member is an electrically conductive invertible spring member having an inclined deformable first section, a curved second section and a centrally disposed opening. The cell container may be dished inwardly to provide additional room for the switch and to provide added deflection under pressure for operating the switch.

Referring to Figures 1.11a through 1.11c inclusive a typical alkaline galvanic cell **10** is shown comprising an inverted metallic cupped container **12** provided with an outer metal jacket **14** separated by an insulating liner **15** of preferably paper or other fibrous material. Disposed within the container **12** is a tubular anode **16**, a tubular cathode **18**, a separator **20** and an alkaline electrolyte which permeates the anode **16**, cathode **18**, and separator **20** respectively. An anode current pin-type collector **24** extends lengthwise within the cell **10**, parallel to the longitudinal axis of the cell, from a location in contact with the anode **16** to the negative end **25** of the cell **10** where it terminates.

A metallic cover plate **26** having a projected shoulder portion **27** and a raised protuberance **28** is mounted over the bottom end **29** of the container **12** with the raised protuberance **28** centered in substantial alignment with the longitudinal axis of the cell **10**. The raised protuberance represents the positive terminal of cell **10**. The shoulder portion **27** and the raised protuberance **28** leave a void or cavity **31** between the cover plate **26** and the bottom end **29** of the container **12** in which the switch means is located.

The switch means comprises an active member **36** and a passive member **38**. The outer metal jacket **14** is crimped over the cover plate **26** at the positive end of the cell **10** to form a circumferential edge **30** which compresses the cover plate **26** and the switch members **36** and **38** against the metal container **12**. The active switch member **36** is an electrically conductive resilient spring-like member having as more clearly illustrated in Figure 1.11c, an inclined deformable first section **40**, a curved second section **41**, a flat third section **43** and a centrally located opening **42** of any desired shape although a circular geometry is preferred.

FIGURE 1.11: GALVANIC CELL

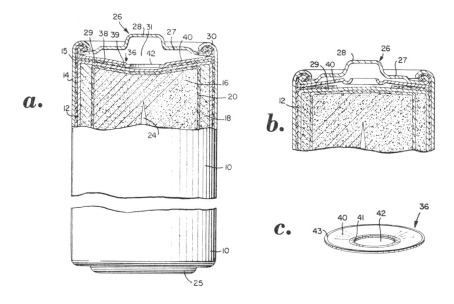

(a)   Elevational view partially in section of an alkaline manganese
      dioxide zinc cell with switch.
(b)   Enlarged fragmentary view of a section of the cell of Figure 1.11a
      showing the switch means in its open configuration.
(c)   Diagrammatic view of the active member of the preferred switch
      means.

Source: U.S. Patent 4,025,696

The passive member **38** is an insulator of preferably a plastic material, although
a fibrous insulator would be acceptable, having a central aperture **39** which is
larger in size than the opening **42** of switch member **36**. The active switch member **36** is positioned in cell **10** with the flat third section **43** slidably contacting
the cover plate **26** and with its opening **42** in substantial alignment with the
raised protuberance **28** of the cover plate **26**. The passive member **38** lies between the active switch member **36** and the bottom end **29** of the metal container **12** with its central aperture **39** arranged concentric to the opening **42** of
switch member **36**.

Since the aperture **39** of the insulator **38** is larger than the opening **42** of the
active member **36** a predetermined portion of the active member **36** will abut
the metallic cupped container **12** to insure electrical continuity between the
metal container **12** and the cover plate **26** during the normal operation of the
cell **10**. The bottom end **29** of the inverted metal container **12** is preferably
dished inwardly to provide additional room for the active switch member **36**
and to increase the bulge or deflection of the bottom end **29** of the container
12 under pressure. The inwardly dished depression in the surface of the bottom
end **29** of the container **12** may be formed during the fabrication of the container **12**.

The active switch member **36** should possess a deflection characteristic which will result in the inclined deformable section **40** being irreversibly displaced from a first stable position to a second stable position upon the application of a predetermined force resulting from a bulge in the bottom end **29** of the container **12**.

The second stable position should preferably represent a substantial geometrical inversion of the first position. It is essential that the deflection be related to the applied force in a manner such that only after reaching the predetermined applied force will the deformable section **40** irreversibly move from the first stable position to the second stable position and preferably with almost no further applied force. The inversion of section **40** to a substantially inside-out geometry as shown in Figure 1.11b will occur when the active switch member **36** is held to within prescribed dimensional limits.

Although the active switch member **36** can be of any configuration having an inclined deformable section **40**, which need not be linear, a frustum geometry as is diagrammatically illustrated in Figure 1.11c is preferred. A spring washer of the conventional Belleville category is typical of such frustum geometry. To exhibit the desired deflection characteristic the metal stock thickness of the switch member **36** should be generally of no more than 0.010 inch.

The deflection sensitivity response to applied force is believed to be increased and made more reproducible by the incorporation of the curved U-shaped second section **41** particularly in combination with the dished bottom end **29** of container **12**. The curved second section **41** is intended to stiffen the active member **36** to provide positive switching action without buckling whereas the dished bottom end **29** provides added movement.

The flat third section **43** serves primarily as an extended lip for readily mounting the spring member **36** in the cell **10** and is not an essential feature of the process. Moreover, both the curved second section **41** and the dished container bottom **29** are required only when space is at a premium and/or when sensitivity of the spring member **36** must be enhanced.

THIN CELLS

Polysulfone Screen

A process described by *K.V. Kordesch; U.S. Patent 3,944,435; March 16, 1976; assigned to Union Carbide Corporation* relates to thin flat cells employing a bonded component assembly and a method for producing such a component assembly. Specifically, the component assembly comprises an anode such as a metal anode, and a cathode, such as an active oxidic material, secured in a spaced-apart relationship by at least one plastic nonconductive network disposed between and adhesively secured at one of its faces to the anode and at its opposite face to the cathode.

The nonconductive plastic material has to be cohesive so that it will strongly cling to itself thereby maintaining its integral structure and must also be stable with respect to the other components of the cell in which it is used. Suitable nonconductive plastic materials for use in this process include such materials as polyethylene, polysulfone, acrylic, polymethacrylic resin, nylon, polyvinyl chloride, polypropylene, polystyrene, polytetrafluoroethylene, or the like. Nylon, polyethylene and polyvinyl chloride are thermoplastic and are relatively hard materials to dissolve and, therefore, should be heated to make them adhesive.

Suitable solvents for use in this process include such materials as ketones, aromatic and aliphatic hydrocarbons, esters and chlorinated hydrocarbons. Preferable plastic and solvent combinations are as follows: (1) polysulfone with dichloromethane, trichloroethylene or chloroform; (2) polymethacrylate with ketones or esters; (3) polystyrene with ketones or aromatic hydrocarbons; (4) polyvinyl chloride with ketones, esters or aromatic hydrocarbons.

In Figure 1.12a there is shown a thin flat cell **2**, typical of a Leclanche-type flat cell, having an expanded zinc anode **4**, electrolyte-impregnated separator **6**, cathode **8** and cathode collector **10**. Disposed between anode **4** and cathode **8** is a plasticizable screen **12** which secures the anode **4** to one of its faces **14** and the cathode **8** to the opposite face **16**. The electrolyte-impregnated separator **6** is positioned between anode **4** and cathode **8** and contacts the inner face of the anode **4** and the inner face of the cathode **8** through the openings in the plasticizable screen **12**. Note that faces **14** and **16** of plasticizable screen **12** are flattened slightly between anode **4** and cathode **8**.

FIGURE 1.12: BONDED COMPONENT FLAT CELL

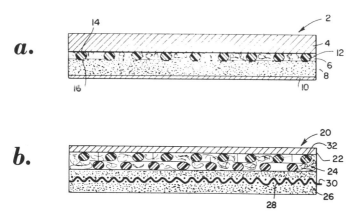

(a) Partial cross section of a cell illustrating a plastic screen disposed between and securing in a spaced-apart parallel relationship the anode and cathode material of the cell.

(b) Enlarged partial cross section of a cell wherein two plastic screens are employed to secure the anode and cathode material in a fixed spaced-apart relationship.

Source: U.S. Patent 3,944,435

Figure 1.12b shows another type of flat cell **20** having an anode **22**, an electrolyte-impregnated separator **24** and cathode **26**. A cathode collector screen **28** is shown disposed in cathode **26** and extending beyond the edge of the flat cell so as to form a tab **30** for providing external electrical contact to the cathode **26**. Disposed between anode **22** and cathode **26** is a two-layer plasticizable screen **32** which secures the anode **22** to cathode **26** in a parallel spaced-apart relationship while allowing electrolyte-impregnated separator **24** to contact the anode **22** and cathode **26** through the openings in the two-layer plasticizable screen **32**. Note that the screens are shown superimposed one on the other with the top screen rotated 45° out of alignment with the bottom screen. The following examples illustrate the process.

Example 1: A Leclanche flat cell was produced similar to the one shown in Figure 1.12b. A polysulfone nonconductive screen, about 2 inches by 2 inches and 0.035 inch thick, was placed between a thin MnO_2 cathode having a similar area as the screen and being about 0.050 inch thick, and a fibrous separator which was soaked with dichloromethane, the separator having a similar area as the screen and being about 0.020 inch thick.

The sandwich assembly was pressed between the plates of a hydraulic press using a force of about 500 lb/in² for 10 seconds which was sufficient to press the solvent-wet, tacky polysulfone screen into the porous MnO_2 cathode and into the fibrous separator. The pressure was released and upon vaporization of the solvent, dichloromethane, a strong permanent bond was obtained between the polysulfone screen and the cathode.

On top of the separator side of the assembled sandwich, a second polysulfone screen was placed substantially superimposed on the first screen. The second plastic screen was wet with dichloromethane and then a zinc screen was superimposed on top of it. The assembly was pressed between the plates of a hydraulic press using a force of about 500 lb/in² for about 10 seconds which was sufficient to press the solvent-wet, tacky polysulfone screen against the zinc screen and the first polysulfone screen. The pressure was released and upon vaporization of the solvent, a complete, strongly bonded cathode-separator-anode assembly was obtained.

The cell was activated by adding the electrolyte KOH through the zinc screen to soak the separator. The fully assembled cell was then abuse-tested by being discharged across a resistive load of 0.5 ohm and then allowed to stand for 48 hours in the completely discharged state after which no noticeable bulge was observed in the cell assembly, the bulge being generally indicative of separation between the anode and cathode of the cell.

Example 2: A similar-type flat cell as in Example 1 was produced except that only one polysulfone screen was used on the anode side of the separator as shown in Figure 1.12a. As in Example 1, the fibrous separator was soaked with dichloromethane before the polysulfone screen was placed between the separator and the zinc screen anode. On the opposite side of the soaked fibrous separator was positioned a porous MnO_2 cathode.

The assembly was pressed between the plates of a hydraulic press using a force of about 500 lb/in² for 20 seconds which was sufficient to partially embed one face of the solvent-wet, tacky polysulfone screen through the separator and into the porous MnO_2 cathode and the opposite face of the screen into the zinc screen. The pressure was removed and upon vaporization of the solvent dichloromethane a strong permanently bonded zinc screen-fibrous separator-MnO$_2$ cathode assembly was obtained. The cell was activated by adding the electrolyte KOH through the zinc screen to soak the separator. The fully assembled cell was then tested as in Example 1 and again no noticeable bulge was observed in the cell.

Colloidal Graphite and Polymeric Binder

A process described by *K.V. Kordesch and A. Kozawa; U.S. Patent 3,945,847; March 23, 1976; assigned to Union Carbide Corporation* provides a dense coherent electrode for use in an electrochemical cell having an aqueous electrolyte. The electrode is composed of particulate manganese dioxide (MnO$_2$) and particulate electrically conductive material both of which are substantially uniformly dispersed in a conductive binder composed of colloidal electrically conductive material having a surface area as defined by the B.E.T. method using N$_2$ absorption at 78°K of less than about 100 m^2/g, preferably between about 10 and 50 m^2/g, and an electrolyte-wettable polymeric binder.

The particulate electrically conductive material is added to increase the electrical conductivity of the electrode since manganese dioxide is a relatively poor conductor, and then when the electrode is assembled in a cell, the conductive material will substantially reduce the internal resistance of the cell. The electrically conductive colloidal material is added to further increase the electrode conductivity of the electrode and to provide particle-to-particle electrical contact between the manganese dioxide particles thereby promoting better utilization of the available manganese dioxide so as to greatly reduce cathode polarization when the electrode is assembled in a cell.

This requirement of surface area for the colloidal particles is essential to insure that the finished electrode will not absorb large quantities of electrolyte when assembled in a cell which would result in excessive swelling and ultimate disintegration of the electrode. In addition, the surface area requirement of the colloidal material will aid in the fabrication of electrodes having a density in excess of about 2½ g per square centimeter which is necessary for producing the relatively hard, well-bonded electrode of this process. Moreover, the colloidal material is necessary to make the polymeric binder conductive since the binder by itself is nonconductive.

The particular polymeric binder for use in this process must be capable of bonding the materials of the electrode together while not completely coating the manganese dioxide particles which would tend to diminish the high activity and capacity of the manganese dioxide particles. Moreover, the polymeric binder must be wettable by the electrolyte the electrode will contact in an assembled cell so that the electrode will absorb sufficient liquid electrolyte to maintain the necessary ionic contact with the active depolarizer material (manganese dioxide).

For those binders that are initially solid, a volatile solvent should be used to dissolve such binders so that an intimate dispersion of the binder can be made with the electrode materials after which the solvent can be removed thereby aiding the development of desirable porosity in the finished electrode so that when the electrode is assembled in a cell, the electrolyte of the cell can be absorbed by the electrode thereby providing sufficient ionic contact between the electrolyte and the active depolarizer material. Since polymeric binders are electrically nonconductors of electricity, the colloidal conductive material is necessary so as to cause the binders to be conductive rather than insulative.

The manganese dioxide electrodes can be produced by a process which comprises: mixing a polymeric binder, a colloidal electrically conductive material, particulate manganese dioxide, and particulate electrically conductive material, to produce a mixture; and forming the mixture into a coherent electrode. To briefly illustrate the primary mode of carrying out this process, 2 grams of an epoxy resin (the diglycidyl diether of bisphenol A) plus amine hardener mixture is dissolved in 14 grams of an alcoholic (isopropyl alcohol) suspension of colloidal graphite to produce a conductive epoxy mixture. The polymeric binder, i.e., epoxy resin, would be nonconductive without the colloidal graphite and could probably insulate some of the manganese dioxide particles if it were mixed directly with the manganese dioxide.

Consequently, the preferred method to employ is to first mix the polymeric binder with the colloidal conductive material so as to render the binder material conductive and thereby minimize the coating of the manganese dioxide particles with an insulating plastic binder which would restrict or limit the high activity and capacity of the manganese dioxide. A mixture containing 90 weight percent particulate manganese dioxide and 10 weight percent particulate electrically conductive material (8% graphite powder plus 2% carbon) is then mixed with the conductive epoxy/colloidal graphite mixture in the ratio of 30 grams of the manganese dioxide mixture to 12 grams of the conductive epoxy/colloidal graphite mixture.

The mixture has a putty-like consistency, which makes it easy to spread on a metal screen current collector. Pressure may be used, if desired, to fill the holes in the screen more uniformly and to insure good contact between the mixture and the collector. The mixture on the screen support is then heated for about one half hour at about 50°C in order to drive off the liquids present in the mixture and to harden the epoxy resin. The coherent electrode is then ready to be incorporated in conventional manner as an electrode in an electrochemical cell.

One of the important aspects of this process is that the coherent manganese dioxide electrodes can be made in the form of thin layers. Thus, flat manganese dioxide electrodes having thicknesses of from about 10 to 50 mils can be fabricated. The importance of this aspect resides in the fact that thin electrodes provide a means for efficient electrochemical utilization of the active material present, and for the fabrication of cells delivering unusually high rates of discharge with relatively efficient utilization of the manganese dioxide. All of the electrodes of the process, thin or otherwise, can also be employed in rechargeable cells having prolonged cycle life.

Polycarbonfluoride Cathode

A process described by *A. Kozawa; U.S. Patent 3,956,018; May 11, 1976; assigned to Union Carbide Corporation* relates to primary electric-current-producing dry cells and more particularly to dry cells utilizing a zinc anode, a cathode composed of a polycarbonfluoride compound and an aqueous alkaline electrolyte. Polycarbonfluoride compounds of the type to which this process refers have the general formula $(CF_x)_n$ wherein x represents the ratio of fluorine atoms to carbon atoms in the compound and n refers to an indefinite number of the recurring (CF_x) groups. The chemical structure of a specific polycarbonfluoride compound may be represented by the formula on the following page. Such polycarbonfluoride compounds may be prepared in accordance with known methods by reacting various forms of carbon material, e.g., graphite, active carbon, carbon black, etc., with fluorine gas at elevated temperatures (e.g., 350° to 400°C).

$$\begin{array}{ccc} F & F & F \\ | & | & | \\ C & C & C \\ | & | & | \\ -C-C-C- \\ | & | & | \\ F & F & F \end{array}$$

Polycarbonfluoride $(CF_x)_n$ is a black, gray or white powder depending upon the x value or the ratio of fluorine atoms to carbon atoms in the compound. When the x value is small or on the order of about 0.2 to 0.7, the color of the material is black. Polycarbonfluorides having an x value of about 0.7 to 0.9 are gray while those materials having an x value of about 1.0 or higher are white.

Broadly, polycarbonfluoride materials having an x value of about 0.3 to 1.1 are useful in the process. Since the more highly fluorinated compounds exhibit higher energy densities, the gray or white polycarbonfluorides having an x value of between about 0.8 and 1.1 are preferred. The most preferred polycarbonfluoride is one having an x value of about 1.0.

Example: Experimental test cells were made using a zinc screen anode, a polycarbonfluoride cathode and aqueous alkaline electrolytes of different compositions. The cells were assembled using AA-size nickel-plated steel cans as cell containers (i.e., cans conventionally used in AA-size alkaline MnO_2 cells) having an inner diameter of 1.27 cm. Cathodes were prepared for the cells using a mixture of 80% by weight $(CF_{1.0})_n$ powders, 10% by weight carbon black, 5% by weight carboxymethylhydroxyethylcellulose and 5% by weight Solka-Floc, a cellulosic material (Brown Company).

This mixture was uniformly spread onto a nickel screen and molded under a pressure of 5,000 lb/cm^2 (about 2,500 kg/cm^2) to form a cathode sheet. The weight of the nickel screen was 0.0386 g/cm^2. The mix content of the cathode sheet was 0.27 g/in^2 (0.042 g/cm^2). The percent of $(CF_{1.0})_n$ in the cathode sheet was 41.6%. Cathode discs approximately 1.27 cm in diameter were punched out from the cathode sheet. Each cathode disc contained 0.0425 g $(CF_{1.0})_n$ and had a theoretical capacity of 36.8 milliampere-hours (mAh). Two discs were placed inside each of the nickel-plated steel cans and packed together at the bottom under a pressure of approximately 2,000 lb/cm^2 (about 1000 kg/cm^2).

The anode for the cells was a zinc screen measuring ⅝ inch long and ⅜ inch wide (1.53 cm long and 0.92 cm wide). The weight of the zinc screen was 1.5 grams. Various aqueous electrolytes were prepared containing conventional salts such as ammonium chloride, zinc chloride, magnesium perchlorate and magnesium bromide as well as aqueous alkaline solutions including the range of 0.2 to 13 molar (0.2 to 13 M) KOH, 19 M NaOH and 6 M LiOH solutions. Different electrolyte solutions were poured into each cell container to approximately three quarters or more of its height. The containers were closed using a rubber stopper which fit tightly inside the open end.

The zinc screen anode covered by a porous paper separator was suspended from the stopper in contact with the electrolyte by a copper wire encased in plastic. The copper wire was spot-welded to the zinc screen and extended outside the

FIGURE 1.13: ZINC-POLYCARBONFLUORIDE CELL

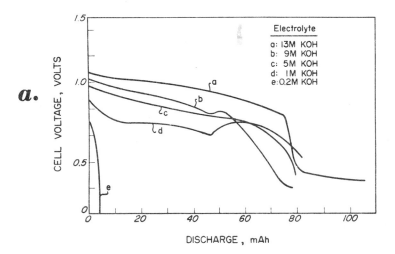

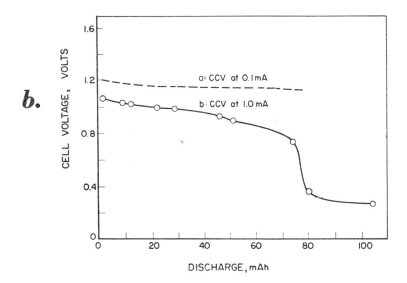

(a)   Curves showing the discharge characteristics of small alkaline
      zinc-$(CF_x)_n$ cells using various concentrations of aqueous
      KOH electrolyte and zinc screen anodes.

(b)   Group of curves showing the discharge characteristics of the
      same type of small alkaline zinc-$(CF_x)_n$ cell at two different
      current drains.

Source: U.S. Patent 3,956,018

cell to form a terminal lead. The cells were tested for open circuit voltage (OCV) and then discharged at low and moderate current drains of 0.1 and 1.0 milliampere, respectively (about 0.08 and 0.8 milliampere/cm²). The closed circuit voltage (CCV) of the cells was recorded throughout the discharge period and the open circuit voltage was recorded after 12 milliampere-hours (mAh) discharge. The table below summarizes the results of the tests.

Cell	Electrolytes	Initial OCV	OCV*	CCV** (1.0 mA)	CCV** (0.1 mA)
1	13 M KOH	1.392	1.317	1.030	1.148
2	9 M KOH	1.370	1.253	0.962	1.049
3	5 M KOH	1.340	1.290	0.900	1.186
4	1 M KOH	1.298	1.225	0.717	0.809
5	0.2 M KOH	1.298	0.124	--	—
6	19 M NaOH	1.083	1.139	1.00	1.024
7	6 M LiOH	1.338	1.041	0.718	0.812
8	2 M MgBr₂ + Mg(OH)₂	0.664	0.511	0	0.076
9	2 M Mg(ClO₄)₂ + Mg(OH)₂	0.995	0.411	0	0.098
10	5 M NH₄Cl + NH₃, pH 8.3	0.737	0.474	0.110	0.281
11	5 M NH₄Cl + 2 M ZnCl₂, pH 4.5	0.856	0.415	0	0.002

*At the 12 mAh discharge point (the initial OCV may not be the true voltage of the system, because O_2 may be adsorbed on the material).

**At the 9 mAh discharge point.

It can be seen from the table above that cells employing the polycarbonfluoride cathode can be successfully discharged in concentrated aqueous alkaline electrolyte. Cells using the $(CF_{1.0})_n$ cathode and a 5 to 13 M KOH electrolyte exhibit open circuit voltages of between 1.2 and 1.3 volts and an average closed circuit voltage of about 1.0 volt. It will be further seen however from the table that cells employing the polycarbonfluoride cathode cannot be successfully discharged in the aqueous salt electrolytes. The closed circuit voltage of the cells with these electrolytes is essentially 0.0 volt.

Curves a through e in Figure 1.13a show the discharge characteristics of these small cells using the $(CF_{1.0})_n$ cathode and 0.2 to 13 M KOH electrolyte when placed on a 1 milliampere drain (0.8 mA/cm²). It will be noted that the discharge curves a through c for each of the cells using the 5 to 13 M KOH electrolyte are relatively flat over most of the discharge period.

Curves a and b in Figure 1.13b show the discharge characteristics of the small cell using the $(CF_{1.0})_n$ cathode and 13 M KOH electrolyte when placed on 0.1 and 1.0 milliampere (mA) current drain. It will be noted from curve b (1.0 mA drain) that the voltage of the cell suddenly drops at about 75 mAh at the end of the discharge period. Since the theoretical capacity of the two disc $(CF_{1.0})_n$ cathode is 73.6 mAh, the sudden drop in cell voltage indicates that nearly 100% utilization of the polycarbonfluoride cathode was attained.

The cathode reaction in a Zn-$(CF_x)_n$ cell produces fluoride ions according to the following reaction: $(CF_x)_n + nxe^- \rightarrow nC + xF^-$. It has been found that the addition of certain compounds such as Al_2O_3 or TiO_2 which will form strong complexes with the fluoride ions in KOH electrolyte, e.g., $[AlF_6]^\equiv$, $[TiF_6]^\equiv$, etc., is beneficial for producing higher cell voltages. Cells were prepared using a zinc screen anode and a $(CF_{1.0})_n$ cathode made in the same manner as described in the foregoing experiment (i.e., with the AA size cell containers). The net $(CF_{1.0})_n$ content in the cathode was 0.104 gram. The electrolyte was a 13 M KOH solution. Each cell was prepared with the KOH electrolyte containing a suspension of about 20% by weight or more of Al_2O_3, SiO_2, TiO_2 or ZnO as an additive based on the weight of electrolyte.

The resulting OCV and CCV at 50% depth of discharge were recorded for each cell. The 50% depth of discharge corresponds to 45 mAh capacity delivered. The results of this experiment are shown in the table below. The closed circuit voltages were higher in the presence of all additives and the open circuit voltages were also higher in the presence of Al_2O_3, TiO_2 and SiO_2.

Cell Number	Additive	CCV at 1.0 mA	CCV at 0.1 mA	OCV
1	None	0.946	1.049	1.203
2	ZnO	0.956	1.062	1.197
3	Al_2O_3	0.978	1.075	1.214
4	TiO_2	0.962	1.077	1.233
5	SiO_2	0.948	1.057	1.212

In the preferred practice of the process, cathodes are made by molding a mixture of the $(CF_x)_n$ powders, particulate electrically conductive material, an electrolyte-wettable polymeric binder and a colloidal electrically conductive material. The polymeric binder and the colloidal electrically conductive material in the range of 0.5 to 20 weight percent and 2 to 10 weight percent respectively, based on the weight of the total cathode mixture, are mixed together and added to the cathode mixture as a suspension containing an organic solvent which readily wets the $(CF_x)_n$ powders.

Solvents such as those based on xylene, isobutanol, isopropanol, etc., are excellent for this purpose. The polymeric binder may be an epoxy, polysulfone or acrylic resin, for example, and is preferably used in amounts of between 0.5 and 5 weight percent of the cathode mixture. Commercial suspensions containing the polymeric binder, colloidal graphite and or organic solvent are available.

In one example of the above, cathodes were prepared using the following mix composition: 2 grams $(CF_{1.0})_n$ powder, 0.1 gram acetylene black and 1 cc of "Dag" dispersion No. 2404 (contains 10% colloidal graphite suspended in mineral spirits). The composition was mixed thoroughly and heated at 90°C in air for two hours to drive off the solvent. The dried powder mix was then molded at a pressure of about 4,000 lb/cm² (about 2,000 kg/cm²) to form the cathode body. Because of the high practical energy densities that are attainable, the alkaline zinc-$(CF_x)_n$ cell system is ideally suited for use in miniature size electric-current-producing dry cells.

Zinc Dust and Swollen Polymeric Binder Particles

A process described by *C.I. Sullivan; U.S. Patent 3,954,506; May 4, 1976; assigned to Polaroid Corporation* relates to thin, flat anodes produced from zinc dust and a small amount of polymeric binder. The process involves utilizing a composition comprising finely divided zinc dust and about 0.5 to 5 weight percent of a polymeric binder, which polymer is in the form of an aqueous latex of swollen polymer particles. This composition is utilized to produce thin anodes by application onto an electrically conductive substrate by a procedure such as silk screening, gravure printing, roll coating, or the like.

The coating is then dried to remove the water and leave a substantially uniform layer of about 1.5 mils thickness with the zinc particles firmly adhered together by polymer particles. The amount of polymer is not sufficient to form a continuous film but rather a sticky discontinuous mass of swollen polymer particles which are characterized by chemists as "jelly fish" is provided which, in effect, when dry, glues the zinc particles together, i.e., "spot" welding the zinc particles to form a stable mass.

A large number of diverse polymers may be utilized as binders in carrying out the process. There may be employed a polymer system whose particles may be swollen by a solvent which is miscible with water, and which polymer system does not cross-link or otherwise react with the zinc dust to a degree which would hinder the electrochemical reactions, or interfere with the battery reactions. Vinyl polymers, acrylates, and elastomers have been suitably employed. The polymer particles may be swollen by known solvents for the polymers provided that the solvent is miscible with water so as not to interfere with the maintenance of the latex.

The substrate upon which the composition is coated is preferably an electrically conductive vinyl film, for example, a vinyl film ("Condulon") which is available in very thin sheets. When a battery is to be formed in a series of stacked cells, a so-called duplex electrode may be produced by coating the zinc-containing composition onto one surface of the substrate and by forming the cathode collector for the adjacent cell on the other side of the substrate. The following examples illustrate the process.

Example 1: A zinc anode composition containing 99 weight percent zinc and 1% polymer binder based on the zinc was prepared by forming an aqueous latex of 1.82 grams of a vinyl acetate-ethylene copolymer (Aircoflex) containing 55 weight percent solids and the remainder isopropanol as a swelling agent, and 20 grams of distilled water. To the resulting mixture, 99 grams of zinc dust (New Jersey Zinc No. 44) were slowly added while stirring. The blended slurry was then coated onto an electrically conductive vinyl film (Condulon) after preferably priming the film with a silane adhesion promoter (Union Carbide A-187). After drying, the product was ready for use as an anode in a cell.

Example 2: Employing the general procedure outlined in Example 1, generally equivalent results were obtained by using as binders 3% of an ethyl acrylate/2-sulfoethyl methacrylate copolymer in the weight ratio of 92:8, and 97% zinc dust (78 μ). The above-indicated anodes were used to prepare Leclanche cells and were found to function satisfactorily in such a system.

Conductive Coated Vented Cathode

T. Kalnoki-Kis; U.S. Patent 3,902,922; September 2, 1975; assigned to Union Carbide Corporation describes a thin flat cell employing a metal anode (zinc), a cathode (manganese dioxide), an electrolyte (ammonium chloride) and a perforated cathode collector (steel). The cathode collector is coated with a continuous layer of a gas-permeable, electrolyte-impermeable polymeric material which allows the venting of undesirable gases formed within the cell while preventing any electrolyte loss from the cell.

Thin flat Leclanche cells were prepared using a positive electrode mix of manganese dioxide, graphite and acetylene black, a negative zinc sheet electrode, and an electrolyte of ammonium chloride and zinc chloride. A cellulosic separator was disposed between the zinc anode and cathode mix of each cell and was saturated with the electrolyte of each cell. A vinyl gas-permeable, electrolyte impermeable conductive paint, comprising a solution of a copolymer of vinyl chloride and vinyl acetate in an organic solvent containing a plasticizer, an epoxy resin stabilizer and a conductive particulate material of acetylene black and graphite, was applied to one surface of a steel cathode collector plate having openings disposed therein.

After the paint dried, the coated cathode collector was assembled with the other cell components and then a hot melt adhesive, obtained commercially as Swift Z-863, was deposited in the peripheral recess formed between the extending anode and cathode plates to produce a thin rectangular flat cell. Each cell measured 1.75 inches wide and 2.75 inches long with the active cathode mix component of each cell measuring 1.48 inches by 2.25 inches.

Each cell was then tested by being successively discharged across a 0.312 ohm load for 0.1 second, a 0.832 ohm load for 1.5 seconds and then across a 6.250 ohm load for 1.0 second. This cycle of discharge was repeated after a three-second rest period until the closed circuit voltage of the cell decreased to 1.08 volts. The data obtained from the cells are shown in the table below.

Conditions	Open Circuit Voltage, volts (average)	Number of Cells*	Number of Cycles (average)
25°C, fresh	1.65	100	35
120°F, 2 weeks	1.6	3	26
140°F, 8 hours	1.6	5	16
32°F, 24 hours	1.65	3	5

*The average impedance of the cells was found to be between 2 and 2.5 ohms at 1-kHz.

It is thus shown by the above examples that using the process, flat cells can be constructed which can function under various temperature conditions to deliver a series of pulses of rather high current in rapid succession.

Deformations in Metal Layer

According to a process described by *J.E. Oltman and R.H. Feldhake; U.S. Patent 3,964,932; June 22, 1976; assigned to ESB Incorporated* a plurality of deformations is provided in a metal layer in a battery to avoid wrinkles in that metal layer. The deformations may be situated in that portion of the metal layer which is part of a peripheral sealing system for the battery and/or in that portion of the metal layer inside the peripheral sealing system. The deformations may be oriented along axes which are (a) substantially concentric with the edges of the peripheral sealing system, (b) substantially perpendicular to the edges of the sealing system, or (c) neither (a) nor (b). The metal layer may have a pocket indented therein, the indented pocket being situated inside the peripheral sealing system and over the electrodes of the battery.

FIGURE 1.14: BATTERY DESIGN—METAL LAYER

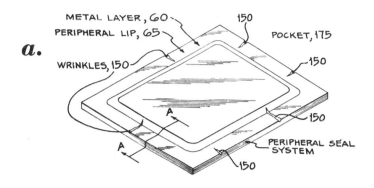

a.

METAL LAYER, 60
PERIPHERAL LIP, 65
WRINKLES, 150
150
POCKET, 175
150
150
150
PERIPHERAL SEAL SYSTEM
A
A

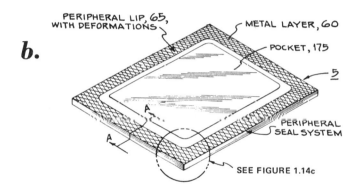

b.

PERIPHERAL LIP, 65, WITH DEFORMATIONS
METAL LAYER, 60
POCKET, 175
5
PERIPHERAL SEAL SYSTEM
SEE FIGURE 1.14c
A
A

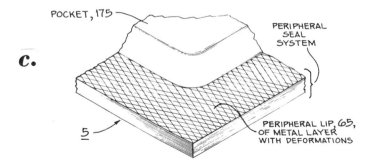

c.

POCKET, 175
PERIPHERAL SEAL SYSTEM
5
PERIPHERAL LIP, 65, OF METAL LAYER WITH DEFORMATIONS

(continued)

FIGURE 1.14: (continued)

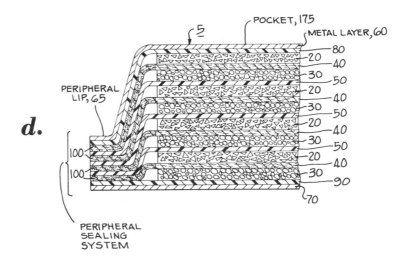

(a) Pictorial view of a battery having a metal layer as its upper
     surface. Wrinkles are shown in the peripheral lip surround-
     ing a substantially rectangular pocket indented in the metal
     layer.
(b) Pictorial view of a battery having a metal layer as its upper
     surface and having a plurality of deformations in the periph-
     eral lip surrounding a substantially rectangular pocket indented
     in the metal layer.
(c) Magnified view of a portion of the battery shown in Figure
     1.14b.
(d) Cross section of the battery shown in Figures 1.14 a or 1.14b
     taken along line **A—A** of either of those figures.

Source: U.S. Patent 3,964,932

Figure 1.14a illustrates a battery **5** having the wrinkles **150** which this process
seeks to prevent. The battery **5** shown in Figure 1.14a has a metal layer **60** as
its upper surface. A peripheral lip **65** extends around the periphery of the battery and forms a component of a peripheral sealing system by being sealed to
at least one other component of the battery. Inside the peripheral lip **65** and
the peripheral sealing system is a pocket **175** which has been indented into the
metal layer **60**. It has been found that the wrinkles **150** tend to provide moisture leakage paths and/or to result in unwanted electrical paths.

Figure 1.14b shows a battery **5** which has been provided with a plurality of deformations to prevent the wrinkles from occurring. These deformations, which
are present in metal layer **60** at the upper surface of the battery, are situated
in the peripheral lip **65** of the metal layer **60**. Inside this peripheral lip **65** is
a pocket **175**, generally rectangular in configuration, which has been indented
in metal layer **60** and which is situated over the electrodes of the battery. The

deformations appearing in Figure 1.14b, which are shown on an enlarged scale in Fiugre 1.14c, are generally diamond-shaped in configuration; this configuration is produced by lines deformed into the metal layer 60 which intersect each other and which are oriented to be neither perpendicular nor parallel to the edges of the indented pocket 175.

Figure 1.14d is a cross section of the battery shown in Figures 1.14a or 1.14b, taken along line A—A of either of those figures. As shown in Figure 1.14d, the battery 5 has four cells each of which comprises a positive electrode 20, a negative electrode 30, and a separator layer 40 which prevents the positive and negative electrodes from contacting each other.

The cells are separated from one another by intercell connector layers 50 made from an electrically conductive plastic. Both the separator layers 40 and the intercell connector layers 50 extend beyond the edges of the electrodes, and those extensions in the separator layers 40 have been impregnated with an electrically nonconductive adhesive 100; for more details of this feature, see U.S. Patent 3,701,690.

At the top of the battery is a laminate of metal layer 60 and a layer of electrically conductive plastic 80, while a similar laminate of a metal layer 70 and conductive plastic 90 is to be found at the bottom of the battery; both of the laminates 60-80 and 70-90 extend beyond the edges of the electrodes, as shown in Figure 1.14d.

Referring to Figures 1.14b through 1.14d, it can be seen that the peripheral lip 65 is sealed to another component of the battery and comprises a member of a peripheral sealing system for the battery. This peripheral sealing system, which provides a moisture barrier around the electrodes to prevent the escape of moisture from the battery, is produced by the seals between the deposits of adhesive 100 and the other components 80, 40, 50 and 90. The deformations made in the metal layer 60 extend to that portion of layer 60 which forms part of this peripheral sealing system.

The deformations which have been described and illustrated may be produced by a variety of processes, including those which are sometimes referred to by names such as embossing, checking, waffling, knurling, drawing, quilting, scoring, scratching, and dimpling.

Seal Testing Method

A.E. Ames, A.G. Kniazzeh and P. Goldberg; U.S. Patent 4,034,598; July 12, 1977; assigned to Polaroid Corporation have found that by subjecting an electrical energy storage device to gaseous oxygen at elevated pressures, the integrity of the container or envelope can be ascertained, thereby providing an accelerated aging evaluation of the device. By subjecting the battery to the oxygen at elevated pressures, the oxygen will seek any pinhole or other disruption in the battery envelope, and rapidly react with the anode material. This reaction, or lack thereof in a well-sealed battery, can be ascertained by measuring the open circuit voltage (OCV) or by weighing the battery.

It is believed that as oxygen enters the battery a change in OCV will occur as a result of an increase in electrolyte pH. This is believed to be due to corrosion of the anode, and, in the case of a zinc anode, the attendant production of zinc oxide, which, when it dissolves in the salt electrolyte causes the pH increase. A decrease in OCV follows according to the known effect of pH on the zinc electrode. Similarly, the oxidation of zinc will effect a weight gain due to oxygen take-up. It is also believed that at relatively short oxygen exposure times, a decrease in OCV is produced by the zinc oxidation rate increasing the zinc polarization voltage. The procedure is particularly suitable for use in planar-type Leclanche batteries, wherein the zinc anode is attacked by the oxygen, forming zinc oxide, which is reflected in measurable OCV change.

OTHER PROCESSES

High Resistance Element Across Terminals

W.P. Conner; U.S. Patent 3,930,887; January 6, 1976; assigned to Hercules Incorporated has found that when a relatively high resistance element is connected across the terminals of a battery, the resulting low current drain inhibits the self discharge of the battery and thus improves the shelf life. The resistance element can be built into the battery or mounted externally.

Referring to Figure 1.15a, there is shown a conventional Leclanche-type dry cell battery. The battery includes a zinc anode **10** in the form of a container lined with a separator **11** and filled with a depolarizer-electrolyte mix **15**, which usually is a paste of manganese dioxide, a solution of zinc and ammonium chlorides and a stiffening agent. The separator **11** prevents shorting of the zinc anode and the manganese dioxide, which is the active cathodic component of the depolarizer-electrolyte mix **15**.

A carbon cathode **12** is positioned in the mix **15** and projects from the insulating cover **14** which seals the container. The projecting end of the cathode **12** serves as the positive terminal of the battery and the bottom of the anode usually serves as the negative terminal, as for most uses the sides of the anode are coated with an insulating material. A high resistance element **13** is mounted outside the container and connects the cathode **12** with the anode **10**.

The resistive value of the element **13** should permit a sufficient current drain on the cell. If the resistance is too high, the current drain may not be sufficient to improve the depolarization of the cell and thus improve the self-discharge rate. On the other hand, if the resistance is too low, too high a current may be drained from the cell and the capacity of the cell may be further reduced. The optimum current drain for improving the shelf life of the batteries is about 20% of the self-discharging current. For a size D dry cell battery at 50°C a resistance load of about 10,000 to 40,000 ohms is desirable.

The self-discharge rate of dry cell batteries is temperature-dependent. At higher temperatures the self-discharge rate is increased and a larger current drain is necessary to improve the capacity after storage. Since the current drawn from a battery is inversely proportional to the resistance load, at elevated temperatures a lower resistance load is required to improve the shelf life. Under actual storage conditions batteries may be exposed to varying temperatures. For this reason, it is particularly preferred to use a resistance element having a high negative temperature coefficient of resistance, i.e., the resistance of the element decreases as the temperature increases. Thus a larger current will automatically be drawn from the battery at higher temperatures and the self-discharge rate of the battery will be improved. Typical of such resistance elements are thermistors prepared from sintered metal oxides. The resistance values of the thermistors can range from 100,000 to 150,000 ohms at 23°C, 25,000 to 50,000 ohms at 50°C and 10,000 to 20,000 ohms at 75°C.

FIGURE 1.15: BATTERY

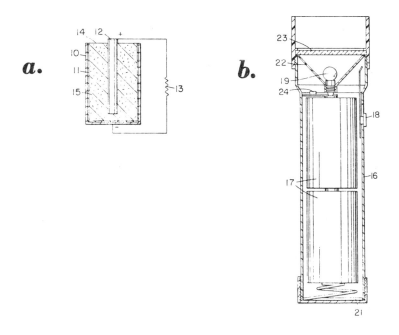

(a) Cross-sectional view of a dry cell battery having a resistance element mounted externally of the battery and connected across the terminals.

(b) Cross-sectional view of a flashlight containing a resistance element positioned to connect across the terminals of the batteries inserted into the flashlight to provide low current drain on the batteries.

Source: U.S. Patent 3,930,887

The resistance element can be externally connected to the battery terminal as illustrated in Figure 1.15a. The external resistance can be made a part of the device in which the battery is to be used, such as in a flashlight, portable radio, battery-operated toy, and the like. Figure 1.15b illustrates a typical flashlight containing a high resistance element **24** positioned to draw a continuous low current drain on the batteries. The flashlight includes a metal casing **16**, two dry cell batteries **17**, an on-off switch **18** which operates to close the circuit and permit a flow of current to a bulb **19**. A spring **21** forces the batteries **17** into contact with the bulb **19**. A reflector **22** surrounds the bulb **19** to concentrate and direct the light from the bulb. A lens **23** further directs the light produced by the bulb. The high resistance element **24** is connected across the terminals of the batteries **17** by grounding the positive terminal of the batteries to the case, which in turn is connected to the negative terminal by the spring **21**. Since the high resistance element is in parallel with the lower resistance bulb circuit, when the switch **18** is closed most of the current from the batteries will flow through the bulb circuit and provide the desired light.

The flashlight can be modified to incorporate a high resistance element which would be connected across the terminals of the batteries to provide a low current drain only when the switch is open. With such a modification, when the switch is closed all current from the batteries would be used to operate the bulb. One such modification could be the use of a single pole double throw switch. This three-terminal switch would alternately connect the batteries with either the bulb circuit to light the flashlight for use or with the high resistance element when not in use, thus increasing the shelf life of the batteries.

The following examples illustrate the improvement in the shelf life of 1.5 volt Leclanche-type dry cell batteries when stored under a high resistance load. In these examples the cell capacity is measured by discharging a dry cell battery through a 3-ohm resistance load to determine the total milliamp-hours (mAh) delivered to the load to a terminal voltage of 1 volt. In each example the results are referenced to a control set of batteries held at $23°C$ without a resistance load.

Example 1: A series of 11 dry cell batteries type D were stored for 70 days under various loaded and unloaded conditions. The initial and final cell capacities were determined as described above. The following table illustrates the effect of providing an external shunting resistor on the battery.

Condition	Cell Capacity (11 battery average)
Reference batteries	1237 mAh
Batteries under no load after 70 days aging ($50°C$)	936 mAh
Batteries under 10,000 ohm load after 70 days aging ($50°C$)	1067 mAh
Batteries under 43,000 ohm load after 70 days aging ($50°C$)	890 mAh

The results show that the 10,000 ohm load provided sufficient current drain to improve the shelf life of the batteries at $50°C$ whereas the 43,000 ohm load provided an insufficient external drain.

Example 2: A series of six dry cell batteries type D were stored at $75°C$ for 20 days under loaded and unloaded conditions. The initial and final capacities were determined as described above. The following illustrates the results.

Condition	Cell Capacity (6 battery average)
Reference batteries	1237 mAh
Batteries under no load after 20 days heat aging ($75°C$)	<200 mAh
Batteries under 10,000 ohm load after 20 days heat aging ($75°C$)	946 mAh

Four of the batteries stored at $75°C$ without load had no usable capacity. All of the batteries stored under a 10,000 ohm load retained 77% of their initial capacity.

Current Take-Off Connector for Wrapped Electrodes

According to a process described by *G. Schenk and H. Haake; U.S. Patent 4,009,053; February 22, 1977; assigned to Varta Batterie AG, Germany* a wrapped battery electrode is connected to the current take-off conductor by projections from radial slots in the conductor which engage the rim of the electrode. The projections are formed in the course of the punching operation which forms the slots.

FIGURE 1.16: WRAPPED-ELECTRODE BATTERY

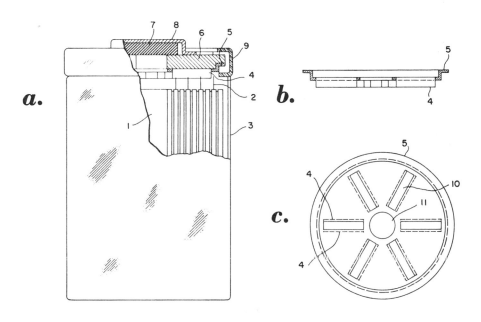

(a)   View in elevation, partly in section, of battery.
(b)   Cross-sectional elevation of the current take-off conductor of the
         battery of Figure 1.16a.
(c)   Top view of conductor.

Source: U.S. Patent 4,009,053

Referring to Figure 1.16a, this shows a cell casing **3** in which the wrapped electrode **1** is positioned. The opposite poled electrodes of wrap **1** are axially displaced with respect to each other, in such a manner that the mass-free rims of the wrapped coils protrude in both directions from the wrap. The upper mass-free rim **2** is attached at several points in its spiral path to the projecting edges **4** of slots **10** of current take-off conductor **5**. The attachment preferably takes place through electrical welding, using a spot welding technique. Lid **6** and cell casing **3**, which are at different potentials, are insulated from each other by gasket 9. Lid 6 rests on current take-off conductor 5. The conductive connection between these elements is provided by the closing pressure which is created when the cell is sealed. If desired, lid 6 and current take-off conductor 5 may be electrically welded together to further reduce the junction resistance.

Cover 6 has an opening equipped, in customary manner, with an elastic seal 7 retained by cell connector 8. Figures 1.16b and 1.16c show the current take-off conductor 5, which is seen to have the form of a dish with an offset rim forming a seat for lid 6. The underside of current take-off conductor 5 is provided with several radially extending slots 10, each of which has two projecting edges 4. These edges 4 are preferably produced by means of a punching tool which is narrower than the slot in the female die by about twice the thickness of the material of which conductor 5 is formed. This punch cuts the metal and turns up the edges of the cut so that they project as shown. In its center, the current take-off conductor has an opening 11 which, together with slots 10, permits electrolyte exchange.

The large number of slots and projecting edges makes possible many electrical contacts to the rim of the spiral electrodes. Moreover, the current take-off conductor can be connected to the electrodes principally by pressure, which considerably simplifies the assembly of such a cell. Preferably, however, the projecting edges are connected to the electrode rim which they engage through spot welds. Cells of this type have considerably better properties under high current loads than comparable cells with wrapped electrodes. This manifests itself as a 5 to 10% higher terminal voltage for equal discharge currents.

Cathode Contact Member

K.V. Kordesch; U.S. Patent 4,011,103; March 8, 1977; assigned to Union Carbide Corporation describes an alkaline round cell construction which uses a molded tubular cathode shell wherein provision is made to accommodate possible shrinkage of the cathode shell thereby preventing loss of contact between the cathode shell and the metallic can.

In accordance with the process, a cathode contact member is positioned at the interface between the molded tubular cathode shell and the interior side walls of the metallic can, the cathode contact member making permanent contact with the metallic can and being at least partially embedded in the cathode shell. In this manner, permanent electronic connection with the cathode shell is always maintained even though shrinkage of the cathode shell may occur after molding which might otherwise result in loss of contact with the metallic can, rendering the alkaline cell useless.

Elimination of the problem of cathode shrinkage, i.e., loss of contact with the metallic can without the need to reconstruct the cathode member is particularly advantageous since the most practical and economic way to assemble the cell is to extrude the tubular cathode shell inside the metallic can. Furthermore, the use of the molded tubular cathode shell together with the cathode contact member of the process does not involve extensive and costly modification of cell manufacturing methods and equipment. During cathode molding, the cathode contact member is first placed inside the metallic can adjacent to its interior side walls and permanent electronic connection with the metallic can is then established. Next, the tubular cathode shell is extruded inside the metallic can using the same manufacturing techniques and equipment as heretofore employed in the assembly of cells. In this instance, however, the cathode shell is also extruded in contact with the cathode contact member such that the latter is at least partially embedded in the finished cathode shell. Should any shrinkage of the cathode shell occur, permanent contact with the metallic can is still maintained by the provision of the cathode contact member in a permanent position throughout the useful life of the cell.

The cathode contact member may be a solid or foraminous structure. In a preferred case, the cathode contact member is made from an open metal mesh or metal screen which is formed into a cylinder having an external diameter approximately the same or slightly less than the internal diameter of the metallic can. During cathode molding, the mesh or screen cylinder is first placed inside the metallic can and is secured in electrical connection therewith, for example, by means of spot welding or simply by providing a tight force-fit. The tubular cathode shell is then formed by extruding the cathode mixture over and through the interstices of the mesh or screen cylinder so that the latter is at least partially embedded in the body of the formed cathode shell.

Snap Terminal and Connector

W.K. Nailor, III; U.S. Patent 4,024,953; May 24, 1977; assigned to E.I. duPont de Nemours and Company describes a battery terminal which includes a ring and a plurality of upstanding fingers extending inwardly of the ring. Each finger includes an inwardly and outwardly directed shoulder. The inwardly directed shoulders are engageable with a male battery terminal and the outwardly directed shoulders are engageable with a female battery terminal. Also, a polarizing shroud is provided having an opening to the inwardly directed shoulders of the terminal or an opening to the outwardly directed shoulders to provide the required polarization.

Referring to Figures 1.17a and 1.17b, a snap terminal **10** according to the process comprises a ring **12**, a plurality of inwardly directed inclined fingers **14** and a crimp barrel **16** for attaching the terminal **10** to a wire. The ring **12** includes an annular, embossed rib **18** and the fingers **14** are integrally formed with the ring **12**. Each finger **14** includes a tapered segment **20** extending inwardly toward a central axis of the ring inclined at a 45° angle, a transition segment **22** including a first inwardly directed shoulder **24** and a second outwardly directed shoulder **26**, and an end segment **28** terminating in a free end **30**.

The inwardly directed shoulders **24** are annularly disposed about a central axis of the ring **12** a first radial distance from the central axis, and the outwardly directed shoulders **26** are annularly disposed about the central axis a radial distance greater than the inwardly directed shoulders. The crimp barrel **16** includes a wire crimp barrel **32** and an insulating crimp barrel **34**.

The snap terminals **10** of the process may be formed from metal strip stock, e.g., bronze, by die-stamping and may include a conventional carrier strip for transporting the terminals from the die and for reeling and feeding the terminals. With a particular reference to Figure 1.17c, a package according to the process comprises a plurality of terminals disposed between first and second strips **40, 42** of the insulating material.

FIGURE 1.17: BATTERY SNAP TERMINAL

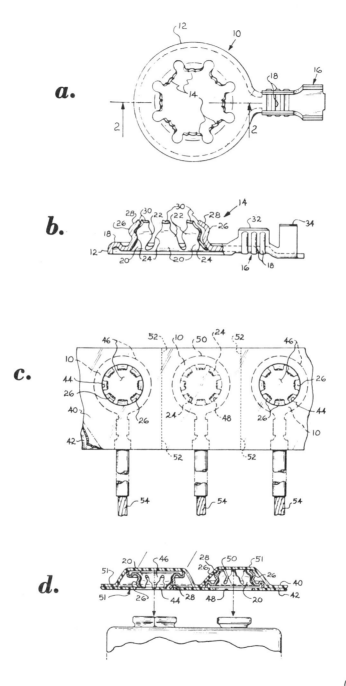

(continued)

FIGURE 1.17: (continued)

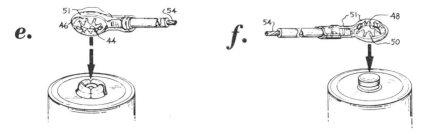

(a) Plan view of a snap terminal.
(b) Side view in partial section along line 2—2 of the terminal of Figure 1.17a.
(c) Plan view of a package of snap terminals.
(d) Sectional view through a pair of snap terminals shown opposite a battery having adjacent male and female terminals.
(e) Side view illustrating the use of a polarized snap terminal with a female battery terminal.
(f) Side view illustrating the use of a polarized snap terminal with a male battery terminal.

Source: U.S. Patent 4,024,953

The first strip of insulating material **40** includes alternately spaced openings **44** which open onto the outwardly directed shoulder **26** of alternately spaced terminals **10** to provide male polarized snap terminals **46**. The second strip of insulating material **42** includes alternately spaced openings **48** (one shown) which open into the inwardly directed shoulders **24** to provide female polarized snap terminals **50**. The insulating strips **40, 42** include transverse perforations **52** for readily separating the male polarized terminals **46** and female polarized terminals **50**.

The insulating strips **40, 42** may be any suitable insulating material, e.g., polypropylene, and may be either heat-sealed or sealed by use of an adhesive. The package including the terminals **10** may be preloaded with wires **54** in the crimp barrels **16**, as illustrated, or may be supplied with the crimp barrels ready for receiving a wire.

A pair of male and female terminals **46, 50** may be encased in a suitable insulated shroud **51**, as illustrated in Figure 1.17d, for use with a battery having male and female terminals at one end. In use, the male polarized terminal **46** is engaged with a female battery terminal with the free ends **30** of fingers **14** inside the rim of the female terminal and pressed to snap the outwardly directed shoulders **26** on the fingers **14** in engagement with the interior rim of the female terminal. Similarly, the female polarized terminal **50** is positioned over a male terminal on a battery with tapered segment **20** over the male terminal and pressed to snap the inwardly directed shoulders **24** of fingers **14** over the rim of the male terminal.

The male polarized snap terminal **46** or female polarized terminal **50** may be used for connecting a load to a battery as illustrated in Figures 1.17e and 1.17f respectively. Alternatively, a terminal **46** can be mated with a terminal **50**.

A process described by *J. Fafa; U.S. Patent 4,002,808; January 11, 1977; assigned to Cipel–Compagnie Industrielle des Piles Electriques, France* relates to electrochemical cells intended to be electrically connected up in series. Such a connection is obtained by means of a mechanical link between the terminals of opposite polarities of two adjacent cells. That mechanical link may be formed by a spring-catch system, screwing or locking.

The process is applied, to great advantage, to a cylindrical cell in which the terminal having the first polarity is constituted by a metal cap covering, for example, the end of a graphite rod belonging to the positive electrode; the second terminal may be constituted by a metal cup or a bottom disc in electrical contact with the negative electrode, comprising, for example, a zinc cup.

Ejector Mechanism

A.W. Rigazio; U.S. Patent 4,031,295; June 21, 1977; assigned to General Time Corporation describes a housing which includes a compartment into which a battery is received and an ejector mechanism. The ejector mechanism is formed by a pair of lever arms extending from a hinge in one wall of the compartment. One lever arm extends from the compartment wall while the other lever arm follows the compartment wall, terminating at a location below the battery.

Finger pressure on the former lever arm results in the ejector mechanism pivoting about the hinge so that the latter lever arm is engaged to partially eject the battery supported by the housing between a pair of terminals from the compartment. The terminals each carry a raised rounded portion on a spring arm contacting the battery. The raised rounded portions act as obstructions so that the battery, once ejected, is not permitted to return to the supported position.

Protective Cap

M.M. Garcin; U.S. Patent 3,948,683; April 6, 1976 describes a protective cap which is affixed to a dry cell and can be removed before use, which includes a disc-like cap covering the exposed electrode at one end of the cell. Axial protrusions on one face of the disc extend through holes in the end face surrounding the electrode, the protrusions being deformed while in a plastic state so as to be not removable without breakage.

The cap is removable by fracturing the protrusions with a rotary motion. In another case, a single annular flange fits in an annular groove around the electrode. A further example includes a two-part cap with one part externally threaded and molded onto the electrode, and the other internally threaded and screwed onto the first. Removal is effected by tightening the second which breaks the first loose.

Copper Sheet as Visual Charge Indicator

P. Depoix; U.S. Patent 3,992,228; November 16, 1976; assigned to Saft–Société des Accumulateurs Fixes et de Traction and Cipel–Compagnie Industrielle des Piles Electriques, both of France describes an alkaline electrolyte electric primary cell comprising a visual indicator of its state of charge. The primary cell has a negative electrode basically containing zinc powder in contact with its outside casing. The charge indicator is constituted by a perforated copper or copper alloy sheet which is visible from the outside of the primary cell and is in contact with the negative mass. On assembly, the copper sheet becomes coated with a film of zinc and appears to have a grey color. During cell discharge the zinc of this film oxidizes and the original coloration of the sheet becomes visible through a transparency provided in at least a portion of the casing. The sheet may be located at any selected depth in the thickness of the negative electrode and the time of restoration of original coloration is a function of the location depth and thus indicative of the residual state of charge.

Mercury Trap

F. Ciliberti; U.S. Patent 3,970,477; July 20, 1976; assigned to P.R. Mallory & Co., Inc. describes an electrochemical cell which is provided with means for isolating and segregating the mercury droplets formed during the electrochemical consumption of amalgamated anodic materials. The segregating means comprises an amalgamable metal sheet, preferably of expanded surface area as provided by grids of screening or expanded metal lath, in contact and adjacent to the inactive face of the anode. While zinc is the exemplified anode, the grid will protect other amalgamated anodes including cadmium and lead in aqueous systems, and lithium, sodium and calcium in nonaqueous systems.

Intermetallic Compounds for Tamping Tools

A process described by *M. Watanabe, Y. Fujii and K. Takayanagi; U.S. Patent 4,011,075; March 8, 1977; assigned to The Furukawa Electric Co., Ltd., Japan* relates to materials for tamping battery mix, composed of an intermetallic compound comprising about 60 to 40 mol percent of titanium, about 0.5 to 5 mol percent of molybdenum, with the balance being cobalt, in which part of the constituents are replaced by other elements and having excellent wear resistance as well as corrosion resistance against the battery mix.

More particularly, the process relates to tamping materials for tamping a battery mix of the Leclanche-type and a manganese-alkaline dry battery composed mainly of an intermetallic compound of CoTi comprising 60 to 40 mol percent of Ti, preferably 48 to 52 mol percent of Ti, about 0.5 to 5 mol percent of Mo, and the balance Co. As a modified article of the process, less than 50% of the cobalt atom is replaced by one or more of Ni, Cr, Fe, and/or less than 10%, preferably less than 5% of the cobalt atom is replaced by one or more of Nb, Ta, W, Al, Zr, V, Mn, Be and Mg and/or less than 5%, preferably less than 2.5%, of the cobalt atom is replaced by one or more of elements from each of the platinum group and rare earth metals.

The materials are very useful for the dry battery industry, particularly for tamping battery mix, when used as powder boxes, tamping molds, holders, sleeves, push-bars, tips, stoppers, plates, screws, rods, bars, extrusion dies for small batteries, tablets, pin, and other parts of machines for production of packed powders.

SILVER OXIDE-ZINC, AIR-ZINC AND OTHER BATTERIES

DIVALENT SILVER OXIDE-ZINC

Gold Layer Between Electrode and Separator

R.A. Langan, N.J. Smilanich and A. Kozawa; U.S. Patent 4,015,055; March 29, 1977; assigned to Union Carbide Corporation describe a metal oxide alkaline cell, such as a silver oxide-zinc cell, having a negative electrode, an alkaline electrolyte, a positive electrode comprising, for example, divalent silver oxide housed in a positive terminal container with a separator disposed between the negative electrode and the positive electrode.

To substantially eliminate the voltage variations of metal oxide cells during the silver discharge period, a thin layer of an electronically conductive material, such as gold, is disposed at the interface of the separator and the active cathode electrode and is extended to contact the positive terminal (cathode collector) of the cell. Therefore since the initial cathode reaction will proceed from the cathode-separator-anode interfaces back through the body of the cathode to the cathode collector terminal, the conductive layer will greatly decrease the internal resistance at the separator-cathode-cathode collector interfaces so as to result in effectively eliminating the voltage variations usually associated with the initial discharge of metal oxide cells such as silver oxide-zinc cells.

Thus by disposing the conductive layer only at the interface of the cathode and separator, the initial discharge reaction can be effectively localized to this area, thereby effectively utilizing the minimum ionic path length so as to result in lowering the internal resistance of the cell. This lower internal cell resistance is primarily beneficial during the initial discharge period and has the effect that the cell will discharge immediately at the intended voltage level.

It is not necessary for the conductive layer to cover the entire cathode-separator interface or the entire wall area of the container. All that is necessary is that sufficient electronically conductive material be present to make electronic contact between the surface of the cathode adjacent the separator and the wall of the cell container. Thus the conductive layer should be disposed on a portion

74

of the area of the cathode surface facing the separator and extend over the peripheral side wall of the cathode so as to contact the wall of the cathode container. Since the conductive layer can be confined to the area specified above, the total weight of the conductive material to be employed to substantially decrease the internal resistance of the cell during initial discharge can vary between about 130 $\mu g/in^2$ to about 3,000 $\mu g/in^2$ preferably between about 500 $\mu g/in^2$ and about 2,000 $\mu g/in^2$ and more preferably about 1,800 $\mu g/in^2$, based on the surface area of the cathode facing the separator.

Referring to Figure 2.1a, there is shown a sectional elevation of a metal oxide cell having a negative electrode 2, separator 3, and positive electrode 4 housed within a two-part container comprising a cathode container 5 and anode cup 6. As shown, cathode container 5 has a flange 7 which is crimped inwardly against a U-shaped flange 11 on anode cup 6 via grommet 8 during assembly to seal the cell as described, for example, in U.S. Patent 3,069,489. The cathode container may be of nickel-plated steel, nickel, nickel alloys, stainless steel, or the like, while the anode cup 6 may be made of tin-plated steel, copper-clad stainless steel, or the like. The grommet 8 may be made of a suitable resilient electrolyte-resistant material such as neoprene, nylon, or the like.

FIGURE 2.1: METAL OXIDE CELLS

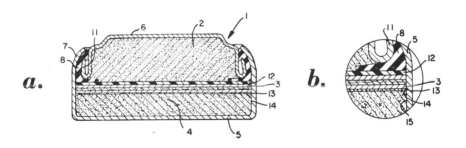

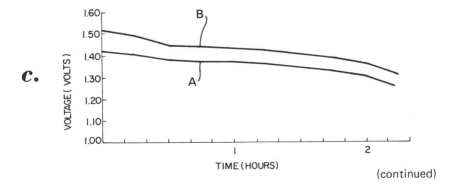

(continued)

FIGURE 2.1 (continued)

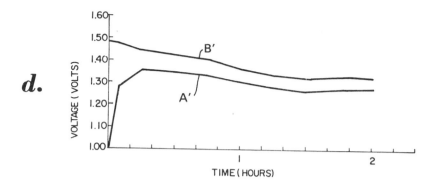

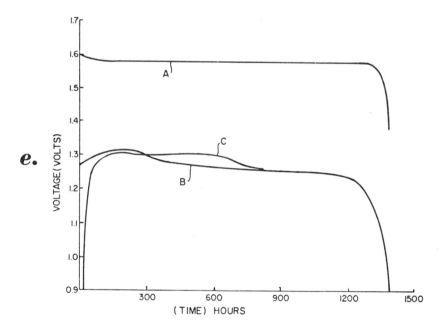

(a) Cross-sectional view of a miniature button size metal oxide cell
     having an electronically conductive layer at the interface of the
     separator and cathode and extending along the side wall of the
     cathode thereby contacting the cathode collector or container
(b) Enlarged view of a section of Figure 2.1a showing the loca-
     tion of the conductive layer in greater detail
(c) Curves of the discharge voltages of a silver oxide-zinc cell em-
     ploying an electronically conductive layer
(d) Curves of the discharge voltages of a silver oxide-zinc cell of the
     prior art
(e) Curves comparing the discharge voltage of two miniature silver
     oxide-zinc cells, one cell employing a conductive layer in ac-
     cordance with this process and the other cell made without such
     a conductive layer

Source: U.S. Patent 4,015,055

The separator **3** may be a three-layer laminate consisting of two outer layers of radiation-grafted polyethylene and an inner cellophane layer or the like. Disposed between anode **2** and separator **3** is a layer of electrolyte-absorbent material **12** which may consist of various cellulosic fibers. The anode (negative) electrode can comprise a lightly compressed pellet **2** of finely divided amalgamated zinc containing, if desired, a gelling agent. Cadmium may also be used as the anode material. The cathode (positive) electrode can comprise a rather densely compressed pellet **4** of a metal oxide powder such as divalent silver oxide powder which could comprise a mixture of divalent silver oxide powder and monovalent silver oxide powder.

The cell electrolyte may be an aqueous solution of potassium hydroxide, sodium hydroxide, or mixtures thereof. As shown in Figures 2.1a and 2.1b, a porous electronically conductive layer **13** is disposed at the interface of the separator **3** and cathode **4** and extends partially downward over the side wall **14** of cathode **4** thereby contacting the inner upstanding wall **15** of container **5**. The positioning of the conductive layer as shown in Figures 2.1a and 2.1b as being at the interface of cathode **4** and separator **3** and extending along the side wall **14** in contact with upstanding wall **15** of container **5** will effectively reduce the internal cell resistance during initial discharge of the cell.

Although not shown, the conductive layer could have been disposed on the surface of the separator facing the cathode and extended onto the side wall of the separator where it would contact the upstanding wall of the container. The following examples illustrate the process.

Example 1: Two miniature button cells of the general type shown in Figure 2.1a were made except that a zinc screen was interposed between the cathode and the inner bottom surface of cathode container. The cells, having a diameter of 0.450 inch (1.14 cm) and an overall height of approximately 0.160 inch (0.40 cm), were produced using a gelled zinc powder anode, a pellet of active cathode material of 50/50 by weight AgO/Ag_2O molded at about a 2-ton pressure and a three-layer separator consisting of two outer layers of cellophane and an inner layer of radiation grafted polyethylene. An additional electrolyte-absorbent separator composed of two layers of a rayon material was employed adjacent to the anode.

These components, along with a 33% KOH electrolyte (7.7 M KOH) which was employed in an excessive amount so as to effectively fill the voids of the cathode material, were assembled in a nickel-plated cathode container and a gold-plated copper-clad stainless steel anode cup and then the cell was sealed by crimping the top annular section of the cathode container inwardly against the anode cup via a grommet of nylon as described in U.S. Patent 3,069,489.

Each cell was made identically except that in one of the cells a gold layer was vacuum deposited on the top surface of the cathode facing the separator and extended over the side wall as shown in Figure 2.1a. The amount of gold employed was about 600 $\mu g/in^2$.

Both of the cells were tested on a 20 mA drain for over 2 hours. The voltage data for the cell employing the gold layer in accordance with this process (Cell A) are shown plotted as curve A in Figure 2.1c. The data of the corresponding discharge voltages for the cell not having the gold layer (Cell A') are shown plotted as curve A' in Figure 2.1d. Using the test procedure as described in the *Journal of Electrochemical Society*, Vol. 107, No. 6 June 1960 and Vol. 119 No. 8 August 1972, the resistance-free discharge voltage was observed for the cell with the gold layer and the cell without the gold layer and the data obtained were plotted as curves B and B' in Figures 2.1c and 2.1d, respectively.

As is apparent from Figures 2.1c and 2.1d, the cell without the gold layer exhibited a higher internal resistance during initial discharge and took over 15 min before the voltage exhibited the normal or expected voltage level. The normal or expected voltage level for cells of the type tested was found to be about 0.060 to 0.085 volt lower than the resistance-free discharge voltage. The cell employing a gold layer exhibited a substantially lower internal resistance than the cell without a gold layer and therefore was able to exhibit a discharge voltage curve substantially parallel to the resistance-free discharge voltage curve. Thus the cell containing the gold layer displayed the expected discharge voltage immediately on discharge.

To further confirm the above observations, the ac impedance of the cells was measured both before and after discharge at 40 and 1,000 Hz. The data observed are shown in the following table and confirm that the internal cell impedance is lower before discharge for a cell employing a gold layer as opposed to a cell not having a gold layer.

Cell Sample	Cell Impedance (ohms)			
	. . . . . 40 Hz		. . . . .1,000 Hz	
	Before Discharge	After Discharge	Before Discharge	After Discharge
Cell A	10.5	2.8	4.0	2.7
Cell A'	32.0	2.3	19.0	2.1

Example 2: Two cells, identical to the miniature button cells of Example 1 had an additional amount of the electrolyte added as in Example 1 to insure that the voids of the cathode were effectively filled with the electrolyte. Each cell was discharged at 70°F across a 140,000-ohm continuous load and for 1.25 sec of every 10-min period the cell was discharged across a 30-ohm load. This pulse discharge regime represents simulated operating conditions of battery-powered watches which require high current pulses for proper operation. The discharge voltage data for both the cell without the gold layer and the cell with the gold layer on a 140,000-ohm continuous load were observed to be substantially the same and thus are shown as a single curve in Figure 2.1e identified as curve A.

The data obtained from the pulsed 30-ohm load test were plotted as points on the graph of Figure 2.1e and then the points were connected to yield curve B for the cell employing a gold layer and curve C for the cell without a gold layer. As is apparent from the curves in Figure 2.1e the initial voltage discharge level for the cell without the gold layer (curve C) took over 100 hours before it reached the 1.2 V immediately upon discharge and continued to exhibit such a voltage level far in excess of 1,200 hours.

This comparison of curves B and C clearly demonstrates that using the process the internal resistance or impedance of a silver oxide-zinc cell can be greatly reduced during initial discharge so as to make it suitable for various battery-powered devices such as watches.

In related work *A. Kozawa; U.S. Patent 3,920,478; November 18, 1975; assigned to Union Carbide Corporation* describes an alkaline silver oxide-zinc cell having a negative electrode, an alkaline electrolyte, a positive electrode comprising divalent silver oxide housed in a positive terminal container, a separator disposed between the negative electrode and the positive electrode. A discontinuous oxidizable metal, such as a zinc screen, is interposed between the positive electrode and the inner wall of the positive terminal and/or between the positive electrode and the separator so as to achieve a unipotential discharge level on low drain conditions.

A. Kozawa; U.S. Patent 3,925,102; December 9, 1975; assigned to Union Carbide Corporation describes an alkaline silver oxide cell having a negative electrode, an alkaline electrolyte, a positive electrode comprising divalent silver oxide housed in a positive terminal cylindrical container having an upstanding wall and a closed end. An oxidizable metal strip, such as a zinc ring, is interposed between the positive electrode and the inner upstanding wall of the container so as to achieve a cell having a unipotential discharge level on low drain conditions.

The outer surface area of the metal strip that contacts the inner upstanding wall of the cylindrical section of the container housing the positive electrode should be at least about 5% of the area of the inner upstanding wall that would be contacted by the positive electrode if the ring were not present. This minimum outer surface area of the metal strip is preferred so as to insure sufficient electrical and physical contact of the metal with both the cathode container and the positive electrode since, without sufficient contact with the container, the metal ring will be isolated and will not function for the purpose of this process.

The projected area of the metal strip perpendicular to the inner upstanding wall of the cylindrical section of the cathode container should be at least about 30% of the area of the plane perpendicular to the upstanding wall and defined by the upstanding wall. This minimum projected surface area of the metal strip perpendicular to the inner upstanding wall of the cathode container is necessary to insure that sufficient oxidizable metal material will be available to react with the positive electrode in the presence of the cell's electrolyte to form metallic silver and possibly a layer of monovalent silver oxide with or without the oxide of the oxidizable metal, which, in turn, will initiate the formation of metallic silver between the interface of the positive electrode and the cathode container.

The amount of the oxidizable metal strip used, as based on the capacity of the total active cathode material, should be at least about 0.5%, with an upper limit being less than that which would completely reduce the divalent silver oxide to the monovalent level. A practical range of oxidizable metal strip should be between about 5% and about 15% of the divalent silver oxide capacity. The use of less than the lower limit of 0.5% would provide insufficient oxidizable metal to effectively react with the cathode to produce the unipotential discharge.

As used in this process, oxidizable metal means a metal that will electrochemically react with divalent silver oxide in the presence of the electrolyte of the cell during storage or during the initial discharge of the cell so as to produce metallic silver with possibly a minor amount of monovalent silver oxide with or without the oxide of the oxidizable metal which will effectively produce a unipotential discharge.

A suitable metal can be selected from the group consisting of zinc, copper, silver, tin, cadmium and lead. Of the above metals, zinc is preferable in zinc anode systems because it introduces no foreign ions into the cell and will easily form zinc oxide in the presence of an alkaline electrolyte. Furthermore, since zinc oxide has a low electrical resistance, it will provide a good electrical path between the silver oxide and the cathode container. Similarly, when using a cadmium anode system, cadmium would be preferred as the oxidizable metal.

Surface Reduction of Tablet

A process described by *H.-M. Lippold and D. Spahrbier; U.S. Patent 4,038,467; July 26, 1977; assigned to Varta Batterie AG, Germany* relates to a galvanic cell with negative zinc electrode and positive electrode of bivalent silver oxide (AgO) as well as alkaline electrolyte. The bivalent silver oxide has its surface so surface-reduced that the cell possesses the discharge voltage of the Ag_2O/Zn System.

Figure 2.2a shows various discharge characteristics of different button cells of conventional size (diameter equals 11.6 mm; height equals 5.4 mm). Curve **a** shows the discharge characteristic for a load resistance of 6 K Ω of such a cell utilizing univalent silver oxide and zinc as the active material. Curve **b** shows the discharge characteristic under the same conditions if bivalent silver oxide is used instead of univalent. This curve **b** shows the voltage step of about 250 mV which is characteristic of bivalent silver oxide, and also the about 40% higher capacity. Curve **c** corresponds to a cell construction according to the process.

Figure 2.2b illustrates diagramatically in cross-section such a button cell. It includes a cell cup **1**, the AgO mass **2**, the separator **3**, the electrolyte absorption layer **4**, negative mass **5**, cell cover **6**, and sealing ring **7**. Compared to a conventional button cell the process $AgO/Ag_2O/Zn$ cell has additional characteristics important for its functioning. These involve layer **8**, as well as layers **10** and **11** and contact ring **9** which serves to insure good electrical contact.

Layer **8** serves to provide electrical insulation of the AgO mass **2** from metallic cup **1**. Preferably, this layer **8** is so arranged that (see enlarged fragmentary view in Figure 2.2d) it terminates between contact ring **9** and the outer edge of layers **2**, **10** and **11**. It therefore protects tablet **2** completely from housing portion **1**, i.e., cup **1**, both on the bottom and on its circumference. Preferably layer **8** consists of polytetrafluoroethylene, polyethylene or other oxidation and acid resistant plastic material.

Layer **10** consists of silver(I) oxide (Ag_2O) and layer **11** of porous metallic silver. The method of producing these layers is described further below. Electrochemical contact between porous silver layer **11** and cup **1** is provided by a contact ring **9** which preferably consists of silver. Porous silver layer **11** is an intimate electrical contact with silver(I) oxide layer **10**. This provides the sole electrical contact with AgO mass **2**.

Figure 2.2c shows a top view of the positive half-cell portion. These are visible cell cup **1**, contact ring **9**, and porous silver layer **11**. In this figure the diameter of the inner opening of contact ring **9** is designated by dimension symbol **d**. As is apparent from Figure 2.2d, which shows in diagrammatic form an enlarged fragment in the region of contact ring **9**, the porous silver layer **11** is thicker in the vicinity of contact ring **9** than in the middle of the electrode.

FIGURE 2.2: GALVANIC CELL

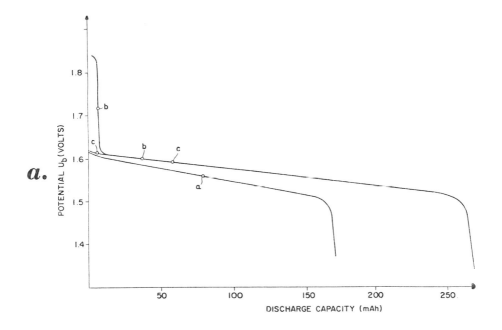

a.

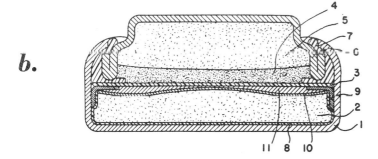

b.

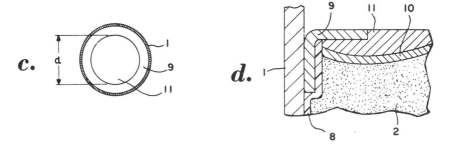

c.   *d.*

(continued)

FIGURE 2.2: (continued)

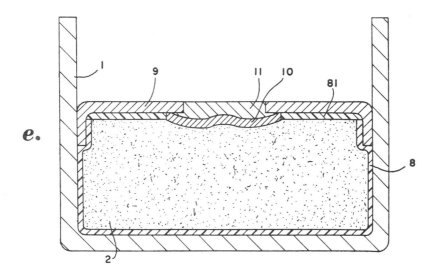

e.

    (a)   Discharge characteristics of various silver oxide cells
(b)-(d)   Views, in diagrammatic form, of the construction of
           a AgO/Ag$_2$O/Zn button cell
    (e)   Details of another example of a button cell

Source: U.S. Patent 4,038,467

This profile of layer **11** results from electrochemical surface reduction. This automatically produces Ag$_2$O layer **10**. The load capacity of the cell is determined in part by the degree of coverage of the surface of layer **11**. The less this coverage, the greater the current loading capacity, but the greater also the self-discharge of the cell. Contact ring **9** may cover between 10 and 98% of electrode surface **11**. A construction with little coverage, e.g., between 10 and 40%, is suitable for applications in which relatively rapid discharge of the cell takes place.

In applications which involve very slow discharge, as, for example, in electronic watches (discharge time in excess of 1 year), it is desirable to reduce surface **11**. This can be accomplished, for example, by reducing the inner diameter **d** of contact ring **9** until the coverage is in the range of 50 to 90%. In this connection it is important to have electrical insulation between the top end of AgO tablet **2** and widened contact ring **9** by means of insulating layer **81**. A diagrammatic illustration of such a construction is shown in Figure 2.2e which is a cross section of the positive cell portion in the region of contact ring **9**. This shows that insulating layer **8**, by means of horizontal extending portion **81**, also protects a considerable portion of the top surface of electrode **2** below contact ring **9**. Here, too, it will be seen that porous silver layer **11** has greater thickness in the vicinity in the edge of contact ring **9** than in the middle.

In producing cells in accordance with the process one starts with an AgO powder of grain size between 2 μ and 20 μ. This can be produced, for example, by oxidation of metallic Ag powder with the aid of peroxide sulfate. Utilizing the teaching in an article by S. Yoshizawa and Z. Takehara (*Journal of the Electrochemical Society of Japan* 31, 3, pages 91 to 104, 1963), this oxidation is carried out in known manner by addition of Au_2O_3 to enhance the AgO formation rate. Preferably between 2 g and 6 g of Au_2O_3 are used per kg of silver powder. To achieve good compression characteristics for the AgO powder, between 1 and 4% of Teflon powder may be added if desired. To improve the electrical conductivity, a mixture of 2 to 10% of silver powder may also be made.

The button cell housing may have conventional dimensions—diameter equals 11.4 mm, height equals 5.4 mm. About 800 mg AgO powder of the above-described mixture are compressed into a tablet (dimensions 10 mm x 2 mm). This is then wrapped in a 50 μ to a 100 μ thick TPFE foil 8. This may be done by placing a foil disc concentrically upon the cell cup and pressing the tablet down into the cup carrying the central portion of the foil with it. The foil is then folded up and over the top of the tablet. The silver contact ring 9 may simultaneously be pressed into cell cup 1 at high pressure, e.g., about 4 tons. For high current conditions according to Figures 2.2b through 2.2d a contact ring with an internal diameter of 8 mm may be used. For low current conditions according to Figure 2.2e a contact ring with an internal diameter of 2 mm may be used. This has placed beneath it a PTFE foil 81 with a central aperture of about 4 mm diameter.

In accordance with the process the silver layer 11, as well as the silver(I) oxide layer beneath it (Ag_2O) is only then produced by surface reduction, i.e., after insertion of the AgO tablet. The surface reduction may take place chemically or electrochemically. For electrochemical surface reduction the prepared positive portion of the cell is inserted in a container with sodium or potash lye. The cup is electrically energized and a counterelectrode, e.g., of nickel, is placed above the cell. The electrochemical reduction then proceeds. It is carried out on the not-yet-closed positive cell portion consisting of elements 1, 2, 8 and 9. Layers 10 and 11 are automatically formed from the AgO body 2 through the surface reduction. Electrochemical surface reduction is preferred, being carried out with a current density of between 300 and 600 mA/cm^2.

A suitable electrolyte is that which is used as the cell electrolyte in the finished product, e.g., 20% NaOH + 2% ZnO. The electrochemical reduction is complete once 80 to 100 mAh/cm^2 are derived. The chemical surface reduction may, for example, be carried out using 80% hydrazine monohydrate, sodium boronate, or other suitable reducing agents. The reduction active mass, and the reduction period is to be so proportioned that an equivalent of 80 to 100 mAh/cm^2 is reduced. For chemical reduction, the unreduced cell portion is simply inserted in the reducing medium.

The advantage of the electrochemical process consists particularly in the evolution of a thickened silver lay 11 in the vicinity of contact ring 9 (see Figure 2.2d and Figure 2.2e). This enhances the voltage stability of the cell. This profile of layer 11 is automatically obtained because the reduction begins at the perimeter of the contact ring and then proceeds progressively toward the center. After washing and drying of the surface reduced portion, its assembly into a finished cell takes place in a conventional manner. Membrane 3 should be oxidation- and acid-resistant and also should adequately inhibit the Ag ion diffusion toward the zinc electrode.

This membrane may be of known form, e.g., a porous foil based on PTFE. Preferred is a three-layer construction, the middle layer being cellophane, and the outer layers porous polypropylene. Button cells in accordance with the process have capacities between 250 mAh and 280 mAh. The high current example exhibits this capacity at discharge currents of 250 μA. They tolerate impulse loads up to 80 mA and are well suited, for example, for the operation of wrist watches with light emitting diode display.

Oxidation of Silver Powder

A process described by *W.K. Lux and T. Chobanov; U.S. Patent 4,003,757; January 18, 1977; assigned by Varta Batterie AG, Germany* involves treating metallic silver powder in an alkaline solution with an oxidant, thereby oxidizing it to silver(II) oxide. Preferably, potassium peroxidisulfate is used as the oxidant and the treatment takes place in caustic soda solution. The foregoing oxidizing agent is particularly characterized by a high resulting level of oxidation and by convenience of usage.

In the process, silver powder is suspended in an alkaline medium, such as an aqueous NaOH solution or KOH solution. The solution is heated to speed up the reaction, for example, to the range of about 50° to 95°C. After reaching a predetermined temperature, the oxidant is added, preferably in incremental portions, the total quantity being such that it provides at least complete transformation of the silver powder to silver(II) oxide. Preferably an excess of oxidant is used. Up to twice the minimum quantity of oxidant necessary to achieve complete transformation may be utilized. As an example, 28 g of the above mentioned oxidant would be the minimum quantity for the transformation of 10 g of silver powder. The following example illustrates the process.

Example: In 1.5 liters of aqueous solution containing 150 g of NaOH, 65 g of silver powder are suspended with continuous stirring. The silver powder has a density of approximately 1.6 g/cc. Its grain size distribution is: 52% under 10 μ, 33% 10 to 30 μ, 15% above 30 μ.

The liquid is then heated to about 85°C. Upon reaching this temperature, a total of 200 g of potassium peroxidisulfate ($K_2S_2O_8$) in portions of about 40 g each is added at intervals of, for example, 1 hour. After addition of the final portion of oxidant, stirring is continued for 3 hours. The product is then filtered, washed to free it of alkali substances, dried at a temperature of approximately 80°C and reduced to particle form.

The foregoing yields approximately 73 g of silver(II) oxide with more than 95% content of pure silver(II) oxide. The silver oxide produced is characterized by high thermodynamic stability, low internal discharge and consequent long shelf life. The rate of gas evolution of these products in 18% NaOH is below 1 μl/g-hr at room temperature. This stability is attributable to the fact that the process produces single crystals of exceptionally regular shape and monoclinic form.

It is believed that the monoclinic crystalline structure is particularly advantageous for use in an electrode mass because it is capable of bearing high current loads. Large surface area and low internal resistance are conducive to this. The oxidant, of which a specific example is provided above, must be active in alkaline media, inasmuch as the AgO is a basic oxide. Only those oxidants are suitable which have a higher reduction potential than the Ag_2O/AgO system. A particular advantage of the process, as compared with conventional processes, such as those using ozone to oxidize AgO, is the low cost of the technique.

Silver Oxide-Sulfur Mixture

P.L. Howard; U.S. Patent 3,935,026; January 27, 1976; assigned to Timex Corporation describes an energy cell for a wrist watch comprising a silver-zinc button-type cell having a bottom can for the cathode material and a top cap for the anode material mounted thereover with separating means between the anode and cathode materials. The cathode material in the bottom can is divalent silver oxide mixed with sulfur. This cathode material provides a lower voltage similar to that of the conventional monovalent silver oxide material but has a higher cell capacity.

A second example of the process comprises a cathode material of silver sulfide mixed with divalent silver oxide. This system also retains the higher capacity of the divalent silver oxide while supplying a lower voltage similar to that of the monovalent silver oxide. The anode material in both cases is a predetermined zinc amalgam.

Referring to Figure 2.3b, the process comprises an energy cell 9 for a wrist watch comprising a top cap 10 and a bottom can 11. The top cap 10, a descending flange 18 and a skirt 19 at the lower end thereof while the bottom can 11 includes a top open end and upwardly extending sides 2. The top cap 10 provides one terminal of the cell, the other terminal being provided by the bottom can 11. An annular grommet 12 is positioned between the top cap 10 and the bottom can 11 and electrically insulates the two terminals of the cell.

The top cap 10 contains the anode material 13, which may for example, be a zinc amalgam which is compressed within the top cap 10. The bottom can 11 contains a depolarizing silver cathode material 15. The anode and cathode materials, 13 and 15 respectively, are separated by one or more barrier layers 31 of a suitable plastic microporous membrane material and separators 14 of absorbent material which also extends under the annular grommet 12. Bibulous separators 14 are positioned between the anode and the cathode. An auxilliary cap 22 of plastic material is later assembled over the top cap 10 to lengthen the leakage path and to serve as an additional seal.

FIGURE 2.3: ENERGY CELL FOR WATCH

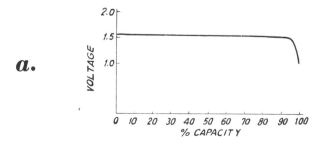

a.

(continued)

FIGURE 2.3: (continued)

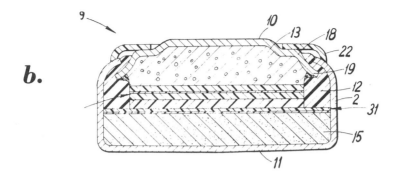

(a)   Diagrammatic illustration of the voltage profile for energy cell
(b)   Side cross section view of a typical energy cell

Source: U.S. Patent 3,935,026

In the first example of the process, the cathode material **15** in the bottom can **11**
is a mixture of divalent silver oxide and sulfur. This cathode material provides
a lower voltage, see Figure 2.3a, similar to that of monovalent silver oxide material with a higher cell capacity similar to that of the divalent silver oxide. Thus,
by adding a predetermined amount of sulfur to the divalent silver oxide in the
cathode, it is possible to obtain an output voltage suitable for highly accurate
timepieces.

The addition of sulfur affects the lattice arrangement of the divalent silver oxide
which contains a loose oxygen and adds resistance thereby providing a metastable
compound with the prime advantages of the monovalent silver oxide voltage while
using divalent silver compounds of a smaller amount for the same capacities as
the monovalent cell or increased capacities as required. Since the functioning
of silver-zinc cells is well known, the internal reactions of the cell are not described.

More specifically, the divalent silver oxide, AgO, and sulfur are intimately mixed
either by a pestle-type arrangement or in an agate mill using only agate balls.
The AgO is mixed with sulfur in proportions ranging up to 20% sulfur by weight
depending on how much the voltage must be initially depressed to maintain a
voltage between 1.68 and 1.48 V throughout the useful life of the battery. The
container with the mixture should be vented to prevent a pressure build-up and
care must also be taken to keep the temperature from rising over 100°F. Mixing
is timed to insure uniform distribution of the sulfur and a slight reduction of the
AgO particle size.

The mixture is then compacted into pellets which are subsequently inserted into
the bottom can **11** of the energy cell **9** or the mixture is compacted directly into
the can **11** with sufficient pressure to maintain a density between 4.0 and 4.8 g/cc
depending on the desired performance of the energy cell **9**. The quantity of the

mixture **15** is predetermined to provide the proper can height and a cell capacity which commonly ranges from 150 to 250 mAh. The improved cell **9** is intended for use with highly accurate timepieces particularly quartz crystal or solid state type watches where a constant low voltage power source is required. Furthermore, the subject cell **9** meets the design requirement of relatively long life due to its higher cell capacity.

Figure 2.3a gives a representative voltage profile over 100% capacity discharge for the cell **9**. It is noted that the voltage remains constant at approximately 1.54 V over about 95% of the cell life and then drops off slowly at first and then very rapidly as it reaches 100% of capacity. The cell **9** thus permits accurate functioning of the watch until the cell reaches almost the very end of its useful life.

In an alternative example, the cathode material may comprise a mixture of silver sulfide mixed with divalent silver oxide. This also retains the higher capacity of the divalent silver oxide while supplying voltage similar to that of the monovalent silver oxide.

As explained above, since divalent silver oxide, AgO, includes a loose oxygen in its makeup, it is possible by means of the process to disturb the lattice arrangement by evenly distributing sulfur throughout the divalent silver oxide to obtain the lower constant voltage of the more stable monovalent silver oxide, Ag_2O. The cathode mixture, nevertheless, still provides the higher capacity of the divalent silver oxide which is approximately double that of the monovalent silver oxide. It is therefore, possible to obtain a metastable compound without the loss of the higher cell capacity of the divalent silver oxide compound.

Metal Hydroxide Electrolyte

A process described by E.S. Megahed; U.S. Patent 3,907,599; September 23, 1975; assigned to ESB Incorporated relates to a sealed low drain rate dry cell having a stable divalent silver oxide depolarizer and more particularly to a cell for low drain rate applications which utilizes a metal hydroxide electrolyte in molar concentrations of from about 0.5 M to about 6.0 M. This cell will have particular utility as energy source in areas such as the electric or electronic watch industry.

Figure 2.4 is a cross-sectional view of a primary silver-zinc cell according to this process. The cell has a two part container comprising an upper section or cap **1** which houses the negative electrode or anode, and a lower section or cup **2** which houses the positive electrode or cathode. The bottom cup **2** may be made of any suitable material such as nickel plated steel and the cap **1** may likewise be made of any suitable material such as tin plated steel. Cap **1** is insulated from the cup **2** by means of an insulating and sealing collar **3** which may be made of any suitable resilient electrolyte-resistant material such as high density polyethylene or neoprene and it may be integrally molded around the edges of the cap **1** for insulating the cap from the can **2** and also to constitute an airtight enclosure therewith.

The negative electrode **4** of the cell comprises for example, a gelled or semigelled zinc. The zinc electrode **4** is separated from the positive electrode by means of an electrolyte-absorbent layer **5** and a membrane barrier **6**.

FIGURE 2.4: PRIMARY SILVER-ZINC CELL

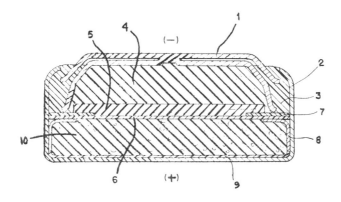

Source: U.S. Patent 3,907,598

The electrolyte-absorbent layer **5** may be made of electrolyte-resistant highly absorbent substances such as matted cotton fibers. Such a material is available commercially as Webril. The barrier layer **6** may be any suitable semipermeable material such as cellophane in combination with polyethylene grafted by chemical initiators with methacrylic acid or cellophane in combination with Permion 2291 (extruded polyethylene grafted with methacrylic acid by atomic radiation). Such material is described by V. D'Agostino, J. Lee and G. Orban, "Grafted Membranes" pages 271 to 281 in *Zinc-Silver Oxide Batteries* edited by A. Fleischer and J.J. Lander, John Wiley & Sons, Inc.

A guard ring **7** made of any suitable plastic material such as polystyrene, polyethylene and polypropylene is situated between sealing collar **3** and barrier **6** to prevent the top assembly **1** through **5** from cutting through barrier **6** when the cell is closed under pressure.

The positive electrode **10** of the cell is divalent silver oxide and is isolated from the can **2** by means of a zinc plated mix sleeve **8** and a liner **9** at the bottom of the can which may be of zinc or a suitable material such as polyethylene. In the event that the liner **9** at the bottom of can **2** is zinc, the cell is discharged at the monovalent silver oxide voltage level through the zinc liner **9** at the bottom of the can and through the zinc plated mix sleeve **8**, with the divalent silver oxide being reduced to monovalent silver oxide upon contact with the zinc sleeve and/or liner.

In the event that the liner **9** on the bottom of can **2** is of a material such as polyethylene, the cell is discharged at the monovalent silver oxide voltage level through the zinc plated mix sleeve only and in this case the capacity which would be exhausted by the zinc liner is saved. The following examples illustrate the process. Unless otherwise indicated, all quantities are by weight.

Example 1: Cells (75 size; 0.455 inch diameter, 0.210 inch height, and 0.034 in³ volume) were constructed according to Figure 2.4 utilizing divalent silver oxide with an average gassing rate of approximately 100 μl/g/hr in 18% sodium hydroxide solution at 165°F. The cell electrolyte was sodium hydroxide solution in concentrations from 2 to 42% by weight as seen below and saturated with zinc oxide. Cell height was measured initially and then after 9 months of storage at room temperature (70°F) and humidity. Some cells were placed at elevated temperature of 113°F and 50% RH (relative humidity) for 12 weeks. The following cell expansion was recorded:

Sodium Hydroxide Concentration (Weight %)	. . . Cell Expansion (inch) After. . .	
	9 Months at 70°F	12 Weeks at 113°F/50% RH
2	0.0010	0.002
6	0.0020	0.003
12	0.0025	0.005
18	0.0030	0.007
24	0.0060	0.013
30	0.0070	0.016
36	0.0090	0.018
42	0.0105	0.025

Results show that cells stored for 12 weeks at 113°F/50% RH made with sodium hydroxide concentration lower than 24% bulged below the cell expansion of 0.010 inch while those above 24% bulged more than 0.010 inch.

Since a watch cell should not expand beyond 0.010 inch in 2 years of storage at room temperature, and since elevated temperature of 113°F/50% RH is equivalent to 2 years of room temperature storage, it is clear that cells employing the electrolyte concentration range of this process, i.e., from about 2 to about 20% of sodium hydroxide, satisfy this requirement while cells with concentrations outside the scope of the process do not. Cells made with sodium hydroxide concentrations greater than 20% by weight can therefore be expected to expand beyond 0.010 inch after 2 years of storage at room temperature.

Example 2: Cells of the process such as those of Example 1 were constructed with 18% sodium hydroxide plus 1.25% zinc oxide as the cell electrolytes were discharged at 6.5K, 10K, 50K and 100K ohm load at room temperature (equivalent to approximately 240, 157, 316 and 159 μA per cell respectively) and were checked for capacity and voltage of operation.

Cells discharged at 6.5K and 10K delivered their full capacity at 1.56 and 1.57 V per cell while cells discharged at 50K and 100K delivered more than 75% of their capacity at 1.58 V per cell without failing. It is expected that these cells will continue to deliver their full capacity at the same voltage level (1.58 V/cell).

Example 3: Cells (41 size; 0.455 inch diameter, 0.160 inch height, and 0.026 in^3 volume) were constructed according to Figure 2.4 utilizing divalent silver oxide with an average gassing rate of 120 μl/g/hr in 18% sodium hydroxide solution at 165°F. The cell electrolyte was potassium hydroxide in concentrations from 5 to 46% by weight as seen below and saturated with zinc oxide. Cell height was measured initially and then after 6 weeks at room temperature and humidity. Also, some cells were placed at elevated temperatures of 113°F/50% RH, 130°F/50% RH, and 145°F/50% RH for 6 weeks. The following cell expansion was recorded.

Potassium Hydroxide Concentration (Weight %)	...... Cell Expansion (inch) After 6 Weeks at			
	Room Temp.	113°F/50%	130°F/50%	145°F/50%
5	0	0	0	0.0005
10	0	0	0.0005	0.0015
15	0	0.0005	0.0010	0.0025
20	0	0.0015	0.0025	0.0040
25	0.0005	0.0025	0.0050	0.0100
30	0.0005	0.0050	0.0070	0.0180
35	0.0010	0.0075	0.0095	cell rupture
40	0.0030	0.0090	0.0125	cell rupture
46	0.0050	0.0160	0.0185	cell rupture

Storing cells at elevated temperatures is one way of speeding up the elevation of shelf life expectancy from a battery. With divalent silver oxide-zinc batteries storing cells at 113°F/50%, 130°F/50% and 145°F/50% for 6 weeks is equivalent to storing the same cells for 1 year, 1.5 years and 2 years at room temperature, respectively. Keeping this relationship in mind, the data in the table indicate clearly that cells made with potassium hydroxide concentration below 25%, i.e., cells of this process, by weight will bulge below the maximum allowable expansion of 0.010 inch after 6 weeks at 145°F/50% or the equivalent of 2 years of room temperature storage. Cells made with potassium hydroxide concentration higher than 25% by weight will bulge above the maximum allowable expansion of 0.010 inch after 2 years of room temperature storage.

Reducing Agent Treatment of Depolarizer Mix

E.S. Megahed, C.R. Buelow and P.J. Spellman; U.S. Patent 4,009,056; Feb. 22, 1977; assigned to ESB Incorporated describe a primary alkaline cell having a stable divalent silver oxide depolarizer mix comprising a negative electrode (anode), a divalent silver oxide (AgO) depolarizer mix, a separator between the negative electrode and depolarizer mix, and an alkaline electrolyte. The surface of the depolarizer mix is treated with a mild reducing solution to form a reduced layer surrounding the mix, and the surface of the reduced layer adjacent to the separator is coated with a layer of silver.

The reduced layer surrounding the depolarizer mix in combination with the layer of silver provides improved stability of the depolarizer mix in the alkaline electrolyte and a single voltage plateau during discharge of the cell. The primary alkaline cell is characterized by a maximum open circuit voltage of about 1.75 V.

The silver layer on the surface of the reduced layer surrounding the depolarizer mix can be formed by treating the reduced layer with a strong reducing solution to form a substantially continuous and electrolyte permeable silver layer, or alternatively, the silver layer may be formed by placing a silver screen, expanded silver metal, perforated silver foil or porous silver powder layer on top of the reduced layer surrounding the depolarizer mix and adjacent to the separator.

It is preferred that the layer of silver be substantially continuous and electrolyte permeable which may be accomplished by treating the reduced layer surrounding the mix with a strong reducing solution.

When forming the layer of silver by treating the reduced layer with a strong reducing solution, it is preferred that the depolarizer mix surrounded by the reduced layer be consolidated in a cathode container prior to treating it with the strong reducing solution. The depolarizer mix may contain from about 50% by weight to about 100% by weight of divalent silver oxide based on the total silver oxide content and still provide a cell with a single voltage plateau discharge.

In closely related work *E.S. Megahed and P.J. Spellman; U.S. Patent 4,015,056; March 29, 1977; assigned ESB Incorporated* describe a method for manufacturing a stable divalent silver oxide depolarizer mix wherein the mix is treated with a mild reducing solution of a reducing agent such as methanol followed by a treatment with a strong reducing solution of a reducing agent such as hydrazine to form a layer of silver on the surface of the depolarizer mix. The depolarizer mix is specially prepared for use in a primary alkaline cell which is discharged at a single voltage plateau, and the preliminary treatment of the mix with the mild reducing solution permits the incorporation of greater amounts of divalent silver oxide in the mix, thereby increasing the electrochemical capacity while still providing a single voltage plateau discharge.

The treatment with the mild reducing solution generally requires soaking the mix in a solution containing a reducing agent for up to about 10 minutes and may be carried out at either room or elevated temperature. The subsequent treatment with a strong reducing solution may also require up to about 10 minutes at either room or elevated temperature, with the purpose to provide a substantially continuous and electrolyte permeable layer of silver on the surface of the depolarizer mix. The depolarizer mix is particularly adapted for use in a button cell construction, and it may contain from about 50% by weight to about 100% by weight of divalent silver oxide and still provide a cell with a single voltage plateau discharge.

SILVER PEROXIDE-ZINC

Silver Peroxide and Manganese Oxide

H. Nishimura and Y. Nomura; U.S. Patent 4,041,219; August 9, 1977; assigned to Citizen Watch Company Limited, Japan describe a silver peroxide-zinc cell, wherein the feature of high cell capacity of a silver peroxide is retained, while solving the problem of initial higher discharge potential. In accordance with an essential feature of the process, the positive electrode comprises silver peroxide and a proportion of an oxide of manganese as a positive electrode active material. It is preferred that the oxide of manganese content of the positive active material is selected within a range from about 0.3 to about 10% by weight.

Thus, the positive electrode of the battery comprises a layer of active material comprising a mixture of a silver peroxide and an oxide of manganese such as manganese dioxide, manganese oxide (MnO), manganese trioxide (Mn_2O_3), and trimanganese tetroxide (Mn_3O_4). More specifically, a very small proportion of an oxide of manganese is added to the silver peroxide so that significantly more silver peroxide can be included in the electrode, thereby increasing the capacity of the cell. In accordance with the process, a silver peroxide-zinc cell 11.6 mm in diameter and 5.4 mm thick was prepared using as the positive electrode a 96:4 mixture of silver peroxide and manganese dioxide.

Figure 2.5a shows a discharge characteristic of this cell obtained with a load of 5 kohms. The initial voltage of the 1.8 V cell fell to 1.5 V or less within 30 to 40 min, and the cell output became stabilized at 1.5 V within an hour.

FIGURE 2.5: PERFORMANCE CHARACTERISTICS FOR SILVER PEROXIDE-ZINC CELL

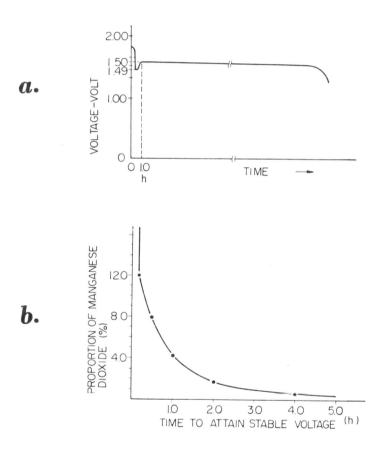

Source: U.S. Patent 4,041,219

Tests were run with cells of this process having the same size but with varying proportions of silver peroxide to manganese dioxide. Tests results are given in Table 1. Based on these results the proportions of manganese dioxide added were plotted in Figure 2.5b against the time required to stabilize the output voltage. In addition to manganese dioxide, other manganese oxides, such as manganese oxide, manganese trioxide, and trimanganese tetroxide were also used as a component of positive electrode active materials to obtain similar test results. These are given in Table 2.

TABLE 1

Sample Number	Mixing Proportions of Positive Electrode Materials		Time Required to Stabilize Output Voltage at 1.55 V with a 5 k-ohm Load (hr)
	AgO (%)	MnO$_2$ (%)	
1	99.7	0.3	50.0
2	99.5	0.5	5.0
3	99	1	4.0
4	98	2	2.0
5	96	4	1.0
6	92	8	0.5
7	88	12	0.3
8	84	16	0.3

TABLE 2

Types of Manganese Oxides	Sample Number	Mixing Ratios of Positive Electrode Materials		Time Required to Stabilize Voltage at 1.5 V with a 5 k-ohm Load (hr)
		AgO (%)	Manganese Oxides (%)	
MnO	9	99	1	3
	10	96	4	0.5
	11	92	8	0.5
Mn$_2$O$_3$	12	99	1	5
	13	96	4	1.0
	14	92	8	0.5
Mn$_3$O$_4$	15	99	1	6
	16	96	4	1.5
	17	92	8	1.0

As obvious from the above results, the silver peroxide-zinc cell of this process gives the following advantages:

(1)   The period of discharge at an initial high voltage level can be shortened considerably. This initial period is so short that it would create no problem in practical use.

(2)   The cell eliminates the necessity of forming a composite positive electrode of two-layer construction, thus making production steps shorter and production easier.

(3)   Because a comparatively small proportion of an oxide of manganese is added, the use of silver peroxide enables a cell having a discharge capacity of 3.22 AH/cc to be constructed. Thus, the production of small-size high capacity cells is made possible.

One example of this process is a silver peroxide-zinc cell 11.56 mm in diameter and 2 mm thick. This cell has a current capacity of 48 mAH, higher than the minimum level of 40 mAH required for use in wrist watches.

Coated Surface

T. Sakai and T. Harada; U.S. Patent 3,994,746; November 30, 1976; assigned to Kabushiki Kaisha Daini Seikosha, Japan describe a silver peroxide cell having a substantially constant single stage discharge voltage from an economic, efficient and useful cathode of silver peroxide.

The silver peroxide cell comprises in combination: a cathode member which is composed of a powder of silver peroxide as the cathode active material, an alkali resistant metal mesh member which is coated with at least one metal which deoxidizes the silver peroxide in the presence of electrolyte and constitutes a local cell with the silver peroxide, at least a part of the alkali resistant metal mesh member being embedded into a part or all of the surface of the cathode member, whereby a long life and a flat single stage discharge voltage are easily obtained.

Figure 2.6a shows the sectional view of the cell comprising anode can **1**, cathode can **2**, the anode active material **3** which may be composed of amalgamated zinc, the cathode **4** which is composed of the cathode-active powder of silver peroxide, the separator which is composed of the diaphragm and an absorbent layer including and containing electrolyte, an insulating packing member **6**, an alkali resistant metal mesh **7** which is coated with the deoxidizing Zn or Cd, and the silver metal layer **8** formed by the reaction of the local cell.

FIGURE 2.6: SILVER PEROXIDE CELL

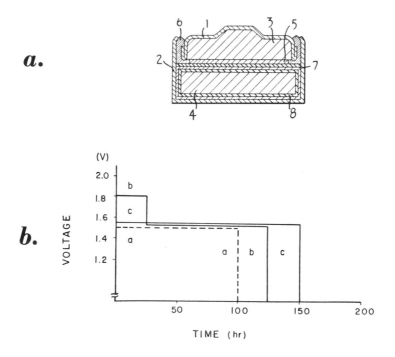

(a)   Sectional view of cell
(b)   Graph of the data of the continuous discharge testing at a
       constant discharge resistance of 1.5 K Ω between cells of
       the process and conventional type cells

Source: U.S. Patent 3,994,746

The cathode member **4** is obtained by mixing cathodic powder mix including silver peroxide and 3 wt % of graphite for conductivity. The mixed powder is pressed at a pressure of 4 tons/cm^2 to form the cathode pellets which are cathode member **4**.

The cathode member **4** is inserted into cathode can **2** with a pressure of 8 tons/cm^2. Before pressing cathode member **4** into cathode can **2**, an alkali resistant stainless steel mesh **7** (200 mesh), wherein the diameter of the openings of the stainless steel mesh correspond to holes in which a wire having diameter of 74 μ can pass, coated preferably by vacuum evaporation, with Zn or Cd metal of 2 or 3 μ thickness, is positioned on cathode member **4**. The pressure of 8 tons/cm^2 is then applied to form the cathode assembly in cathode can **2**. As a result at least one part of the metal mesh **7** is embedded to the surface of the cathode member **4**.

Then, the separator **5** is positioned on top of cathode can **2** which is filled with the cathode member **4**, and alkali-electrolyte is inserted into the absorbent layer of separator **5** on the cathode can **2**, and the cell assembly is completed by caulking and sealing the anode can **1**, via the packing **6**, to cathode can **2**. Anode can **2** had been previously filled by the anode material **3** of zinc (Zn) or cadmium (Cd).

The discharge characteristic of the cell by the above construction is indicated in c of Figure 2.6b; it indicates that the voltage is the flat voltage of 1.5 V with one stage and further the capacity is almost 1.5 times that of the monovalent silver oxide cell of the conventional type as indicated in a.

A current collecting effect is also obtained by embedding the metal mesh **6** into the cathode member and a significant reduction of the inner resistance of the cell is obtained. Comparison data of the average inner resistance based upon measurements of 100 cells of the process and the conventional cells are as follows:

The cell of this process type	4.7 Ω
The cell of conventional silver peroxide type	19.2 Ω

According to the process, the alkali metal mesh is coated by vacuum evaporation with the deoxidizing metal such as zinc, and at least 1 part of the metal mesh **7** is embedded to the circumference of the cathode member, whereby the metal mesh with zinc constitutes the local cell, and the zinc is oxidized to the zinc oxide, the silver peroxide of the surface layer is deoxidized to the silver metal and/or the monovalent silver oxide.

Therefore, the circumference region of the silver peroxide cathode member adjacent or near the coated mesh member is substantially coated by the silver metal or the monovalent silver oxide, then first discharge voltage of 1.5 V of the monovalent silver oxide is obtained, further the metal mesh acts also as a current collector and this action results in the reduction of the internal resistance of the cell.

According to another experiment, it is confirmed that the same effect as with zinc is obtained by the metal mesh which is coated with metallic Cd, Pb, Cu and Ag. Suitable methods for coating the alkali resistant mesh include vacuum evaporating, sputtering and plating. The useful thickness of the coating can range from 2 to 15 μ with 2 to 5 μ being preferred.

WATCH BATTERY DESIGN

Battery Construction

H. Nishimura and Y. Nomura; U.S. Patent 4,025,702; May 24, 1977; assigned to Citizen Watch Co., Ltd., Japan describe a battery having a two-part container comprised of a battery cap and a bottom cup with the bottom cup normally being used as a part of an electronic device, such as the back cover of an electronic timepiece.

In the design of an electronic wristwatch, it has been usual practice to consider a battery having a cylindrical shape as being indispensable, and to design the arrangement or positioning of other electronic components and mechanical parts to accommodate the shape of such a battery. As shown in Figure 2.7a, it is common practice to place a battery **10** in a watch case **12** near the side wall. Under this circumstance, spaces **14** of the watch case **12** are wasted, as shown by the hatched areas. These areas cannot be effectively utilized for accommodating electronic components or mechanical parts of the wristwatch, resulting in increased thickness of the wristwatch.

To solve this problem, it has previously been proposed to provide a cylindrical battery having a container which includes a part common to the back cover of the wristwatch. Such a battery has a thickness greater than 2.5 mm, from manufacturing considerations and, therefore, makes it difficult to manufacture a wristwatch of reduced thickness. In order to overcome the above shortcomings, it has also been proposed to provide a battery **16** having a semicircular periphery as shown in Figure 2.7b. This battery has a drawback in that difficulty is encountered in achieving complete sealing of the battery container by crimping operations during assembly. Thus, leakage of the electrolyte from the battery container can frequently take place. Such a battery is therefore not suitable for use in electronic devices in which long battery life is required.

FIGURE 2.7: WATCH BATTERY CONSTRUCTION

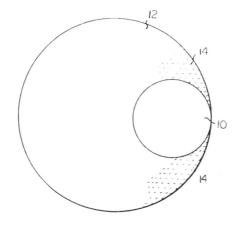

a.

(continued)

FIGURE 2.7: (continued)

b.

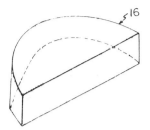

c.

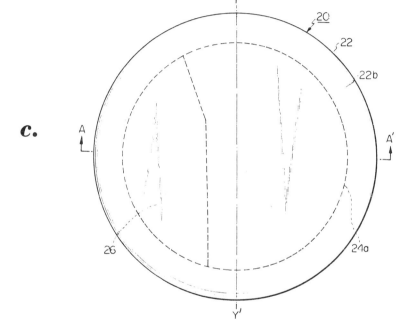

d.

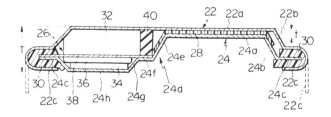

(continued)

FIGURE 2.7: (continued)

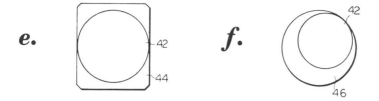

e. *f.*

(a) View illustrating an example of an electronic timepiece in-
     corporating a conventional battery
(b) Perspective view of another conventional battery having a
     semicircular periphery
(c) Plan view of a preferred example of a battery according to
     the process
(d) Cross-sectional view taken along line A–A' of Figure 2.7c
(e) Plan view of a battery incorporated into the back cover of
     an electronic timepiece
(f) View similar to Figure 2.7e but shows another example of
     the battery

Source: U.S. Patent 4,025,702

This process contemplates the provision of an improved battery having reduced
thickness yet ensuring long life. According to the process, the battery has a battery chamber with an outer periphery differing from the outer peripheries of the
joining flanges of the battery cap and the bottom cup. More specifically, the
joining flanges of the battery cap and the bottom cup are formed in a substantially annular shape to provide ease of sealing during crimping operation of the
battery assembly, whereas the battery chamber has a contour at its periphery
other than circular, to provide increased space for the electronic components
and mechanical parts of the electronic device. With this arrangement, the battery
has a high discharge capacity and long life.

Figures 2.7c and 2.7d illustrate a preferred example of a battery according to the
process, achieving the concept mentioned above. In this illustrated case, the battery **20** generally comprises a battery cup **22**, and a battery cap **24**. As best
shown in Figure 2.7d, the battery cup **22** is used as a back cover of an electronic
wristwatch and has a circular flat wall **22a**, a slanted wall **22b** and an annular
flange **22c** which will be crimped inward during battery assembly.

The battery cap **24** has a flat wall **24a** of partially circular periphery and a slanted
wall **24b** contiguous with an annular flange **24c** forming an outer periphery of
the battery cap **24**. The battery cap **24** also has a ridge portion **24d** comprising
a first slanted wall **24e**, a shoulder **24f**, a second slanted wall **24g** and a flat wall
24h. A battery chamber **26** is thus formed by the battery cup **22** and the
ridge portion **24d** of the battery cap **24**. The electrode construction of the battery is housed in the battery chamber **26**. The cap **24** is insulated from the cup
flat wall **22a** and the flange **22b** by means of an insulating layer **28** disposed

between the flat portion of the cap **24** and the flat wall **22a** of the cup **22**, and a grommet **30** which is compressed between the flange **24c** of the cap **24** and the flange **22b** of the cup **22** by a crimping operation during battery assembly, providing a compression seal between these parts. Since the grommet is compressed near the peripheries of the cap and cup, the battery is easily and reliably sealed. As shown in Figure 2.7c, the outer peripheries of the battery chamber **26** and the flat wall **24a** are asymmetric with respect to the axis **Y—Y'**.

The electrode construction is comprised of a positive electrode **32** disposed in the cup **22**, and a negative electrode **34** housed by the cap **24**. The negative electrode **34** is separated from the positive electrode **32** by means of an electrolyte absorbent layer **36** and a membrane barrier or separator **38**. The separator **38** is disposed at its periphery between the grommet **30** and the flange **24c** of the cap **24** and between the shoulder **24f** of the ridge portion **24d** and a separator support **40**. The separator support **40** is placed between the flat wall **22a** of the cup **22** and the separator **38** to hold the separator in place.

The battery cup **22** and the battery cap **24** are made of sheet metal (stainless steel, carbon steel, clad steel, etc.) by press working into the shapes shown in Figure 2.7d, and the sealing is performed by crimping the flange **22** inward to compress the grommet between the flange **22c** and the flange **24**. In this manner, the sealing problem for the battery is overcome in the same way as in the prior art.

The battery cup mentioned above will normally be used as the back cover of an electronic wristwatch. In this application, the battery is manufactured such that the battery chamber **26** has a thickness **T** approximately equal to the thickness of the usual back cover of the wristwatch. If, in this case, the electronic components and mechanical parts are designed to be accommodated within the space remaining in the watch case after the back cover has been attached thereto, the battery can be manufactured to have a significantly increased capacity. Thus, it is possible to manufacture a wristwatch of compact construction having a battery of longer life.

In another practical case, the battery chamber can be formed into various shapes or thicknesses, from design considerations of electronic components or mechanical parts of the electronic wristwatch. In this case, since the sealing surfaces of the battery are formed in a substantially annular shape, no problem is presented in crimping operations during battery assembly, and complete sealing can be reliably obtained even in cases where the battery chamber has contours other than circular. In all cases, the capacity of the battery will be proportional to the volume of the battery chamber.

In a thin electronic wristwatch which has been produced incorporating a battery of the process, the thickness **T** between the cup **22** and the cap **24** is less than 0.7 mm, and the battery chamber **26** has an area parallel to the watch face larger than that of a conventional battery of the cylindrical type. Thus the thickness **T** of the battery chamber **26** may have a smaller value. It is therefore possible to manufacture an electronic wristwatch of remarkably reduced thickness by employing a battery of the process.

The battery cup forming part of the battery can be utilized as the back cover of a wristwatch. Where the wristwatch has a rectangular shape as shown in Figure 2.7e, a battery of the process may be such that a battery chamber **42** is combined with the back cover **44** of the rectangular wristwatch. It is also to be noted that the battery chamber **42** may be of cylindrical type, and, in this case, the battery chamber **42** can be combined with a round back cover **46**, in a manner shown in Figure 2.7f.

Power Cell Compartment

H. Schneider; U.S. Patent 4,041,213; August 9, 1977; assigned to Societe Suisse Pour l'Industrie Horlogere Management Services S.A., Switzerland describes a power cell compartment for an electrically energized timepiece wherein an envelope formed from an anisotropically conductive elastomer partially encloses the power cell and is fixed to the timepiece casing so as to provide a mechanical seal between the movement compartment and the power cell compartment and to assure electrical connection between the movement contacts and the power cell electrodes.

Referring to Figure 2.8, there is shown a support plate **6** for an electrically energized watch movement. Within such support plate is to be found a cut-out portion **7** adapted so as to be suitable to accommodate a power cell **4** of the dimensions suitable for the watch movement in question. Within the cut-out portion are mounted electrical contacts **3** which are placed so as to be proximate the electrodes of the power cell **4**. Normally, with the power cell in place the electrodes thereof would be contacted by contacts **3** thereby to provide an electrically conductive path to the movement.

Between contacts **3**, the surface of the power cell **4** and its electrodes is placed an envelope **1**. Such envelope will partially enclose the power cell except for the portion by which it is retained within the compartment by means of the cover **5**. Envelope **1** is formed of a conductive elastomeric material such as that set forth in U.S. Patent 3,883,213. In addition to presenting good mechanical impermeability property, it is electrically conductive in a highly anisotropic manner.

FIGURE 2.8: POWER CELL COMPARTMENT FOR TIMEPIECE

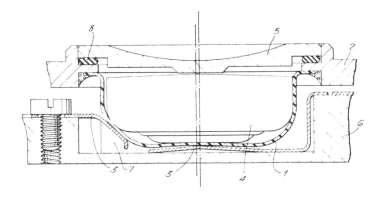

Source: U.S. Patent 4,041,213

Thus when placed between a pair of facing contacts the electrical resistance of thicknesses ranging between 0.010 and 0.020 inches will be in the order of less than 1 ohm as soon as slight pressure is applied. On the other hand resistance between two points on the same surface, even when placed together a distance almost as small as the thickness, will exceed hundreds of megaohms.

Accordingly, the placing of the power cell 4 within such an envelope 1 will enable the proper electrical contact to exist between such power cell and the circuits of the timepiece while assuring that none the less the power cell electrodes continue to be electrically insulated from one another. The envelope 1 may be fastened to the interior of the watch case at any location suitable in respect to the location of the power cell compartment. The fastening may be effected by means of glueing or by any other suitable manner such as through use of a press fitted retaining ring or a screw threaded retaining ring 2. When the power cell is in place the compartment may be sealed by means of a cover member 5 and if desired a sealing ring 8.

It will thus be clear that removal of the cover member 5 merely permits access to the power cell itself and that no access to the movement compartment is provided. Thus to replace the power cell it is merely necessary to remove cover 5, shake out the old power cell and replace it with a new one thereafter replacing the cover. Since no seals have been broken there is no need to be concerned about replacing seals in order to assure continuing impermeability. At the same time this power cell compartment provides an added measure of security in the event that there should be any leakage in the power cell itself. It is clear that even should it be necessary in such event to replace the elastomer element 1 this will be far less costly than the replacement of a partially destroyed movement.

In practice it has been found advantageous from a manufacturing standpoint that the useful anisotropic properties of the elastomer element 1 be confined substantially to the bottom portion. Thus it is preferred that the movement contacts be arranged as shown in the drawing wherein one such contact is adjacent the center contact of the power cell and the other is adjacent the curved wall of the power cell in proximity to the center contact.

Battery Retainer Spring

According to a process described by *W.D. Hart; U.S. Patent 4,043,113; Aug. 23, 1977; assigned to Hughes Aircraft Company* an electric watch has its battery held in the battery block in the watch module by means of a retainer spring which resiliently engages the batteries to urge them into proper installed position. Fingers on the spring resiliently engage in the spacer block to provide the spring retaining force. The fingers can be manually disengaged from the back of the spacer block for retainer spring and battery removal.

Battery retainer 10 shown in Figures 2.9b, 2.9c, and 2.9d is part of watch module 12 and acts with the rest of the components of the module to form the module into unitary construction which is complete, including batteries. Watch module 12 comprises battery block 14 which is generally circular or is otherwise shaped to fit within a watch case. Battery block 14 has battery openings 16 and 18 therein which respectively contain batteries 20 and 22. Opening 24 is for one of the larger components of the watch module such as the time reference crystal.

FIGURE 2.9: ELECTRIC WATCH BATTERY RETAINER

a.

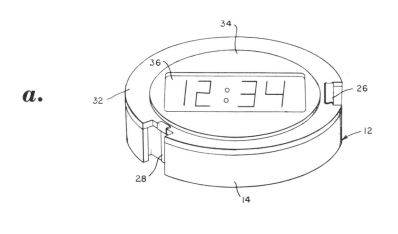

b.

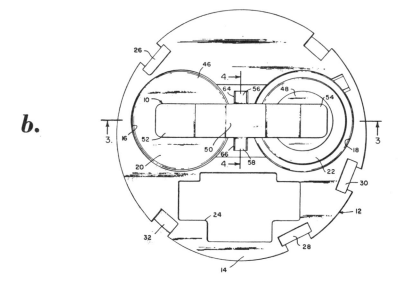

c.

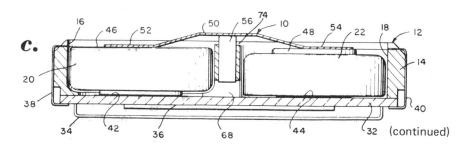

(continued)

FIGURE 2.9: (continued)

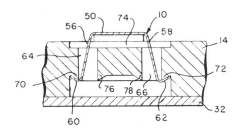

d.

(a)   Isometric view of an electric watch module which incor-
      porates the battery retainer
(b)   Bottom view of the module showing the battery retainer
      in place
(c)   Section through the watch module, taken generally along
      the line **3—3** of Figure 2.9b
(d)   Section through the center of the module, taken generally
      along the line **4—4** of Figure 2.9b, with parts broken
      away, showing the manner in which the battery retainer
      engages in the battery block

Source: U.S. Patent 4,043,113

Tee slots **26, 28**, and **30** are for the reception of J-shaped springs which depend
from substrate **32.** Such J-shaped springs are for control of the watch electronics
by exterior signal such as illustrated in U.S. Patent 3,846,971. The tee slots position the free legs of the contact springs.

Substrate **32** carries cover plate **34** thereon to protect electronic components
mounted on the substrate. These components include digital display devices **36**
as well as chips for timekeeping and display energization, printed circuitry and
interconnections. As seen in Figure 2.9a, a window in the cover plate permits
viewing of the digital display devices from the front of the watch module. Clamp
springs **38** and **40** engage over the top spacer and battery block to clamp them
together to make a unitary structure of the module, which can be handled, packaged, and shipped.

Contacts **42** and **44** are formed on the underside of the substrate, that is the side
away from the display. These contacts are beneath and in line with battery openings **16** and **18** so that batteries **20** and **22** lie against these contacts when the
batteries are installed. When thus installed in a watch case the usual battery contact spring in the back of the watch case contacts the batteries and urges them
against the contacts on the substrate. Also they make an electrical connection
between the battery contacts **46** and **48** which are away from the substrate.
However, in order to retain the batteries in place when there is no case around
the module and to make the connection between battery contacts **46** and **48**,
battery retainer **10** is provided.

Battery retainer **10** is stamped from resilient sheet metal, for example, 0.005 inch thick austenitic stainless steel or equivalent. It is formed into shape to resiliently engage the batteries and the battery block. As seen in Figures 2.9b, 2.9c and 2.9d, battery retainer **10** comprises bridge **50** which has battery contact legs **52** and **54** at the ends. Legs **52** and **54** are in contact with the respective batteries. As is seen in Figure 2.9c, the contact legs are bent below the plane of the bridge, and in the position shown, are already deflected from the free state to provide battery contact pressure. In view of the fact that retainer **10** is metallic, it provides electrical connection between contacts **46** and **48**.

To hold battery retainer **10** in place with respect to battery block **14** and the rest of the module, arms **56** and **58** are bent downwardly from bridge **50**. Retainer arms **56** and **58** terminate in retainer fingers **60** and **62**, see Figure 2.9d, which extend outwardly for hook engagement with the battery block.

Openings **64** and **66** extend through the battery block to pocket **68** on the substrate side of the battery block. Pocket **68** provides hook walls **70** and **72** on which fingers **60** and **62** engage to hold down battery retainer **10**. Thus, when the battery retainer is pressed in place, retainer arms **56** and **58** with their fingers **60** and **62** are squeezed together to enter down through openings **64** and **66** for resilient deflection of legs **52** and **54** until fingers **60** and **62** are engaged. There is sufficient opening to permit the use of tweezers to squeeze arms **56** and **58** together to release the fingers from the hook walls to permit nondestructive removal of battery retainer **10**. For those installations in which it is necessary to save space by turning fingers **60** and **62** toward each other instead of away from each other as shown in Figure 2.9d, alternate hook walls **76** and **78** are also provided.

The completed watch module **12** with its battery retainer allows setting and testing of the watch electronics without breaking the supply of power during storage, shipping, or assembly of the module into a case. When the module is installed into a case, battery retainer **10** can remain on the module and act against the watch case so that it acts as the battery springs during normal usage.

Recess **74**, see Figures 2.9c and 2.9d, is provided in the top of battery block **14** so that when the module is installed in a case the case back can push bridge **50** into the recess level with the back of battery block **14**. When stressed in that manner fingers **60** and **62** move downward away from walls **70** and **72** into pocket **68** so that now battery retainer **10** thrusts the entire module toward the front of the watch case. Thus battery retainer **10** does not require any additional case thickness for its accommodation. In this way the battery springs in the case can be eliminated. This same action occurs whether or not there are battery springs in the back of the watch case.

J.C. Salin; U.S. Patent 4,041,214; August 9, 1977; assigned to Timex Corporation describes an improved button cell configuration for wristwatches which can be snapped into the case back. The button cell has an outer can configuration or shape having major and minor circumferential retaining flanges adapted to hold a gasket. The outer can flanges and gasket are arranged to mate with the case back to provide a snug snap-in and snap-out fit therewith.

The button cell configuration of the process, when inserted into a wristwatch, provides a dustproof and watertight seal between the button cell and watch case.

Self-Tightening Sealing Arrangement

A process described by *F. Sperandio and M. Guglieri; U.S. Patent 3,907,602; September 23, 1975; assigned to Saft-Societe des Accumulateurs Fixes et de Traction, France* concerns a self-tightening sealing arrangement for an enclosure constituted by two parts in the form of cups defining a bottom part and a lid part, the first of which is metallic and is turned down or crimped over the second. According to the process, the second part is made of a plastic material and comprises, over its whole periphery, a lip providing in the vicinity of its rim a space located inside the enclosure. The lip is in contact with a folded back portion of the first part and cooperates with the first part to form a self-tightening seal. The arrangement is particularly applicable to the containers of electrochemical cells and provides self-tightening of the seal with increases of internal pressure in such cells.

In Figure 2.10a, reference **1** designates the cup-like metal part forming the bottom of the casing of an electrochemical cell **C**, the casing defining a sealed container having the self-sealing arrangement closing according to the process. The bottom **1** which is cup-shaped constitutes the first part.

The lid part **2** made of plastic material with some elasticity, for example, a polyamide, is also of cup-like shape and constitutes the second part. The active components of the electrochemical cell are arranged inside the casing formed by the metal bottom **1** and the plastic lid **2**. A positive electrode **7** is placed between two negative electrodes **5**, the separators **6** soaked with electrolyte being inserted between them. The negative electrodes **5** are electrically connected together and one of them is in electrical contact with the bottom of part **1**, which latter forms the negative terminal of the cell. The positive electrode **7** is assembled, for example, by welding to the rivet **3** traversing the lid part **2** in a fluid-tight manner. The rivet **3**, therefore, constitutes the positive terminal of the cell.

Referring to Figure 2.10b, it will be observed that the peripheral rim portion of the plastic lid part **2** comprises a first lip **8** which is separated by an annular groove or space **10** from the rim portion **9** of the lid. A second lip **11** is provided in the vicinity of the upper periphery of the outermost face of lid part **2**. As seen in Figure 2.10a, it is onto the second lip **11** that the rim portion **13** of the bottom **1** is turned down as by crimping. In Figure 2.10a, too, it will be observed that the lip **8** then closely engages the inside curve of the fold **12** joining the cylindrical portion of the metal bottom part **1** to its plane portion.

FIGURE 2.10: SEALING ARRANGEMENT FOR CELL

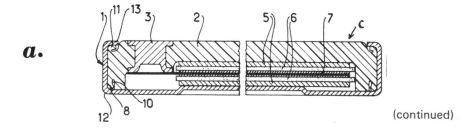

a.

(continued)

FIGURE 2.10: (continued)

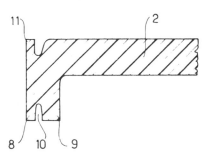

b.

(a)   Sectional view of an electrochemical cell whose casing
        constitutes a sealed container having the self-sealing
        arrangement
(b)   Partial sectional view, the portion adjacent the periphery
        of the lid of the cell, before its assembly in a cell

Source: U.S. Patent 3,907,602

The result of this arrangement is as follows: The crimping which has caused the turning down of the rim **13** of the metal bottom **1** onto the lip **11** of the lid **12** presses the rim portion **9** of the lid against the plane portion of the metal bottom part **1** in an elastic manner due to the elasticity of the lip **11**. A similar result which would, however, not be as good would be obtained by crimping directly the edge **13** of the bottom **1** onto the lid **2** not, however, provided with a lip **11**.

In either case, when the pressure increases inside the completed casing **C** thus formed, it increases more particularly in the space or groove **10** and tends to press the lip **8** all the more closely and tightly against the inner fold **12** of the bottom part **1**. Thus, the effectiveness of the seal increases with any rise in internal cell pressure and a self-tightening seal is indeed obtained. The process is particularly advantageous for shallow casings. The horizontal cross section of the casing of the cell may, moreover, have various shapes: circular, rectangular, square, oval, etc.

Fatty Polyamide Seal

A process described by *J. Winger; U.S. Patent 3,922,178; November 25, 1975; assigned to Union Carbide Corporation* relates to the use of fatty polyamide as a seal or protective coating to prevent alkaline electrolyte from wetting surfaces within the alkaline galvanic cell.

As a result, the process provides a means for improving the leak resistance of alkaline galvanic cells, and also, it provides a means for protecting certain surfaces in an alkaline galvanic cell from being wet by and attacked by the alkaline electrolyte.

The fatty polyamide can be employed in any physical shape or configuration that is appropriate for the particular protective application to be performed. For instance, the fatty polyamide can be applied as a coating over a surface to be protected. Alternatively, the fatty polyamide can be employed as a seal, gasket, or other configuration.

In general, it is preferred that the fatty polyamide have an amine number of above about 9. The amine number is the number of milligrams of KOH equivalent to 1 g of fatty polyamide, and is determined by known procedures such as ASTM D 2074-62T. The fatty polyamide can be applied by known procedures, as from a hot melt or from solution in a solvent such as an alcohol/aromatic hydrocarbon mixture. Among the specific commercially available fatty polyamides that are particularly effective in the process is Swift's Z-610 hot melt adhesive, available from Swift & Company, Adhesive Products Department, Chicago, Illinois. Swift's Z-610 has an amine number of about 70. The following example illustrates the process.

Example: In order to compare the performance of cells made in accordance with the process with the performance of conventional cells, a series of silver oxide/zinc cells of the type used for hearing aids and watches were made. The cells were sealed with a nylon gasket, which in one case was left uncoated, and in the other cases was coated with either Swift's Z-610 or with a hydrocarbon wax. The cells contained a conventional quantity of KOH electrolyte (15 mg) or a higher quantity of KOH electrolyte (20 mg).

The cells were stored under room temperature conditions for 381 days, and then visually examined for salting (appearance of potassium carbonate on the exterior of the cell formed by reaction of KOH from the electrolyte with CO_2 from the atmosphere) and wetness (caused by absorption of moisture from the atmosphere by potassium carbonate).

The nylon gaskets were coated with the hydrocarbon wax by dipping once or twice (as indicated) in a heated solution of 135 g of wax per liter of trichloroethylene. The Z-610 fatty polyamide was applied from a solution of 20% by weight fatty polyamide in 50/50 (volume) isopropyl alcohol/toluene. The table below displays the results of these experiments. 35 to 40 cells were evaluated in each series. The superior performance of the cells having the nylon gasket coated with Z-610 is obvious.

Series	Gasket	KOH Level (mg)	% OK	% Salted	% Wet
1	Uncoated	15	2.6	97.4	66.7
2	One coat wax	15	43	57	53
3	One coat Z-610	15	95	5	0
4	Two coats wax	15	90	10	0
5	Two coats Z-610	15	98	2	0
6	One coat wax	20	13	87	80
7	One coat Z-610	20	84	16	0

Casing Design

A process described by *F. Sperandio and M. Guglieri; U.S. Patent 3,928,077; December 23, 1975; assigned to Saft-Societe des Accumulateurs Fixes et de Traction, France* concerns an electrochemical cell which is provided with a casing which comprises a metallic sheet, one portion of which is partly in contact with an electrode of one polarity. Such a cell is characterized more particularly in that the portion of the sheet in contact with the electrode is offset inwardly towards the inside of the generator in relation to the remainder of the portion. The process is applicable to electrochemical cells having small bulk in at least one dimension.

Figures 2.11a and 2.11b show a metal sheet intended to form a part of the casing according to the process and to which the form of a cup **1** is imparted. The bottom of that cup comprises an inwardly offset part **2** defining a peripheral edge **3** as is shown in Figure 2.11b.

Figure 2.11c shows an electrochemical cell **C** whose casing comprises on the one hand, a metal cup **11** similar to the cup **1** shown in Figures 2.11a and 2.11b and, on the other hand, a lid **17** made of plastic material, also in the form of a cup. The active portion of the cell is constituted by two interconnected negative electrodes **15** surrounding a positive electrode **14** from which they are separated by the separators **16** carrying imbibed electrolyte.

As will be seen in Figure 2.11c, the metal cup **11** comprises, at its bottom, an inwardly offset part **12** which is in contact with one of the negative electrodes **15**. The positive electrode **14** is electrically connected, for example, by welding to a metal member such as a rivet **18** traversing the lid **17** in a fluid-tight manner and constituting the positive terminal.

FIGURE 2.11: CASING DESIGN

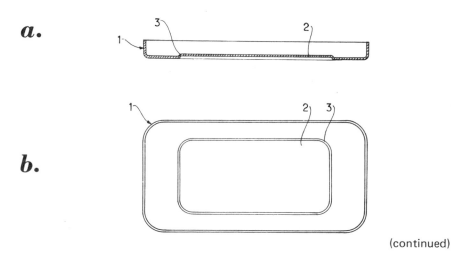

(continued)

FIGURE 2.11: (continued)

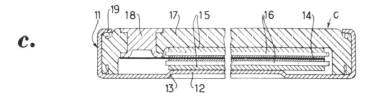

c.

    (a)   Sectional view of a cell cup
    (b)   Plan view of the cup in Figure 2.11a
    (c)   Sectional view of a cell comprising a casing

Source: U.S. Patent 3,928,077

The sealed closing of the casing **C** is effected by the crimping of the rim **19** of the cup **11** onto the bottom of the plastic lid **17**. The assembly formed by the electrodes **14** and **15** and the separators **16** is tightly maintained between the inwardly offset part **12** and the inner surface of the lid **17**. If an overpressure occurs inside the casing, for example, at the end of the charge in the case of a storage cell, the part **12** will bend outwardly, but its defining edges **13** will remain in contact with the negative electrode **15**, thus transmitting current from the electrode **13** to casing **11**. The swelling of the depressed or inwardly offset part **12** will not exceed the outer level of the remainder of the bottom of the cup **11**.

By way of an example, the extent of the offsetting may be of about the order of thickness of the metallic sheet, that is, a few tenths of a millimeter. The assembly of the cup **11** and of the plastic lid **17** whose structure is like that of the lid in U.S. Patent 3,907,602 may be effected in the manner described in the patent.

Plastic Cased Cell

R.W. Lewis; U.S. Patent 3,945,850; March 23, 1976; assigned to Timex Corporation describes a plastic-cased alkaline cell which may be of indeterminate or noncircular or circular shape. The cell includes opposed plastic casing members sheathed in a split metal case, anode and depolarizing cathode materials in the respective casings with suitable separators and electrolyte, metallic spacers electrically connecting the respective anode and cathode materials with the conductive terminal portions, and an insulating seal of nonconductive plastic forming part of a peripheral flange junction between the two casing members.

Referring to Figure 2.12a, an alkaline energy cell is shown which includes a first plastic casing member **1**, a second opposed plastic casing member **2** joined together at respective juxtaposed peripheral flanges **3, 4** with each plastic casing member **1, 2** being sheathed in a metal case **5, 6**.

The internal active materials of the cell, which may be more or less conventional are described as follows with reference to Figure 2.12a.

FIGURE 2.12: PLASTIC CASED CELL

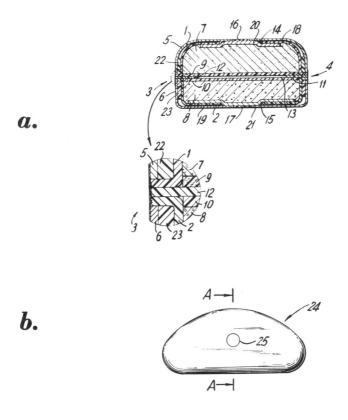

a.

b.

(a)   Horizontal cross section of energy cell
(b)   Plane view of an energy cell especially suited for electric
       watches

Source: U.S. Patent 3,945,850

Casing **1** contains an anode **7** which may be of granulated zinc, cadmium or indium. The lower casing **2** contains a depolarizing cathode **8** which can be a mixture of mercuric oxide or monovalent silver oxide mixed with graphite or a mixture of monovalent or divalent silver oxide mixed with silver powder. Separating
the anode and cathode materials is a cellulosic absorbent separator **9** and a barrier
membrane **10** made, for example, of a crosslinked high molecular weight polyethylene methacrylic acid graft which is commercially available as a polymer called
Permion 2291.

In order to provide an insulating and electrolyte seal between the juxtaposed peripheral flange area **3, 4** a nonconductive plastic ring **12**, made for example from
a semirigid plastic, with central aperture **13** is used as an intermediate member
between the flanges **3, 4** and membrane members **9** and **11**. The casing mem-

bers **1, 2** are formed, for example, from an electrically nonconductive plastic such as a rigid PVC, each with a central aperture **14, 15**. A metal terminal **20, 21** made from, for example, Phos Bronze or nickel 205 is disposed in the central apertures **14, 15** providing electrical connection to its respective anode or cathode.

The plastic casing members **1, 2** are each sheathed in a metal case **5, 6** such as stainless steel. The metal cases **5, 6** are contoured to mate with, i.e., be substantially aligned with or to, the flange portions **3, 4** to provide a smooth exterior surfaced cell so that the cell does not have exposed or protruding flanges and to provide provision for contact to be made to the side of the cell. The outer metal casings **5, 6** are electrically insulated from each other by the nonconductive flange portions **3, 4** which separate the split metal casings **5, 6**.

An adhesive bond or seal **18, 19** is formed between the plastic and the metal cases to form a unified cell case. The metal casings **5, 6** can have a central aperture or hole into which a silver filled epoxy resin portion **16, 17** is formed so that each half cell is sealed at the terminal (hole) portion, and electrical connection is provided between a respective terminal portion and metal case.

It should be recognized, however, that the metal casings **5, 6** could alternatively be formed without such holes and the metal casings electrically coupled to the terminals by disposing a suitable electrically conductive epoxy resin over the terminals and then sealing the metal casing over the plastic casings such that its resin forms a seal and conductor between the metal and plastic casings.

A tube or grommet-like ring **22, 23** made from an absorbent material or plastic is provided between the plastic and metal cases of each half-cell. The tubes **22, 23** may be bonded to the plastic and metal cases or so formed therebetween to prevent electrolyte leakage. And, it was discovered, by sealing the outer metal case to the inner plastic case and providing a tube-like ring and sealed terminal portion as described above or in similar manner, the electrolyte leakage path at both the anode and cathode are lenghtened substantially such that virtually the only possible leak area is confined to the flanges **3, 4**.

The assembled cell comprises the components indicated above with a suitable electrolyte such as potassium hydroxide or sodium hydroxide added, and with the peripheral flange portions **3, 4** then being fused together by pressure and heat or by a high frequency weld applied around the periphery. The energy cell above described is particularly useful in electric devices, such as electric watches, where in the case of the energy cell and electric device, when they are inserted, are in electrical contact providing a ground or reference potential path. Thus, if metal case **6**, which is connected to the cathode of the energy cell shown in Figure 2.12a, is provided with mating screw threads to the watch case, reliable electrical coupling will be provided between the terminal portion **17** of the energy cell and the watch by mere insertion of the battery thereinto.

Reference to Figure 2.12b shows a plane view of a noncircular cell shape suitable for an electric watch. The sheathed plastic cell casing is shown by phantom outline **24**. Also, a filled conductive terminal portion **25** is shown which is similar to the terminal portions **16, 17** shown in Figure 2.12a and a cross section taken along lines A–A would appear as shown in Figure 2.12a.

Metallic Silver Layer

According to a process described by *T. Naruishi, Y. Kataoka, H. Sasabe and S. Tutiya; U.S. Patent 4,021,598; May 3, 1977; assigned to Toshiba Ray-O-Vac Co., Ltd., Japan* a zinc-silver oxide dry cell comprises a zinc anode, a silver oxide cathode, an immobilized body interposed between the anode and the cathode, and a metallic silver layer formed at least on the surface of the anode side of the cathode in which the metallic silver layer is formed through a reduction reaction resulting from a light exposed to the surface of the cathode. The cathode of the dry cell requires no graphite for an electroconductive ingredient, making it possible to increase the amount of silver oxide which is an active ingredient for the cathode and to increase the discharge capacity correspondingly.

The following table indicates a comparison in properties between the alkaline dry cell according to this process and a conventional alkaline dry cell in which a cathode pellet contains graphite and has no metallic silver at the surface. As will be evident from the table, the cathode pellet of this process shows a smaller electrical resistance than that in the conventional alkaline dry cell, and, in consequence, the electroconductivity is excellent.

It is also evident that, as compared with the conventional alkaline dry cell, the alkaline dry cell of this process provides increased short-circuit current and increased discharge capacity of about 25 to 30%.

Comparison item		This process (A)	prior art (B)
Composition of cathode pellet	silver oxide	99.5 parts	95 parts
	electroconductive ingredient (graphite)	0	4.5 parts
	binder (polystyrene resin)	0.5 part	0.5 part
Weight of cathode pellet		1.2 g	1.0 g
Weight of silver oxide in cathode pellet composition		1.194 g	0.95 g
Theoretical electrical capacity		275.8 mAH	219.5 mAH
Electrical resistance from the top surface of the cathode pellet up to the outer surface of the cathode vessel*		6 to 8 Ω	8 to 10 Ω
Open circuit voltage		1.58 V	1.58 V
Short-circuit current		0.71 A	0.64 A
Impedance (1K Z AC frequency)		2.0 Ω	2.3 Ω
Discharge capacity (Total End Voltage 1.2 V	20° C 5000 Ω continuous discharge	253 mAH	191 mAH
	20° C 500 Ω continuous discharge	201 MAH	155 mAH
	20° C 150 Ω continuous discharge	175 mAH	140mAH

*A nickel-plated steel rod with a diameter of 3 mm was contacted under a pressure of 5 kg/cm² with the central top surface portion of a cathode pellet with a diameter of 11 mm and a thickness of 2 mm and electrical resistance between the bottom of a cathode vessel and the steel rod was measured using a DC resistance meter.

Impregnated Sintered Silver Heat Sink

D.H. Fritts; U.S. Patent 4,022,952; May 10, 1977; assigned to U.S. Secretary of the Air Force describes a bipolar electrode having, between active materials, a heat sink fabricated from a porous electrical and heat conductive material, such as sintered silver. Impregnating the porous material with a phase-change heat absorbent material having a high heat of fusion, such as beeswax, provides an electrode having a very large thermal capacity for a high discharge current density battery.

The method of preparation of an electrode assembly will depend on the particular battery system that is involved. A typical assembly technique for forming the heat sink module for a AgO-Zn-KOH system is to start with a relatively thin sheet of porous metal matrix, such as silver, cut to the proper shape for the cell. The porosity should be as high as possible and still maintain adequate strength and thermal and electrical conductivity. Suitable commercially available materials of this type are known as Feltmetal or Foam metal. Suitable matrix material may also be conventionally prepared by sintering a sheet prepared from a mixture of metal and organic material such as silver and polyethylene.

The porous metal sheet is then wrapped or covered, on at least two sides including the two faces, with a single layer of a thin sheet of metal which will not interfere with the electrochemical reactions of the battery. This layer of grid metal on each side of the matrix is sealed to the matrix in a conventional manner such as by sweat soldering, sintering or welding. The preferred manner of attaching will depend upon the quantity of units being manufactured and the facilities available.

The resulting plate of porous metal covered on the two faces may, if desired, be indented or grooved, which can be accomplished by pressing in a die, so that the active material will adhere better. This is a conventional practice to achieve better adhesion of the active materials. The negative and the positive active materials are applied to their respective faces by electrodeposition or pasting or sintering in the conventional manner. The method of application depends in part on the particular grid metal and active material involved and the porosity of active materials that is desired.

The porous metal matrix is filled with molten heat sink material, such as beeswax, and usually cooled to room temperature, before further assembly, to solidify the wax. This may be done either before or after the active materials are applied to the grid surfaces. An alternative method of preparation of the bipolar electrode module is to fill the porous plate with the heat sink material, scrape the surfaces and paint them with a conventional conducting paint, such as silver, and then electrodeposit the grid metal on the surfaces.

AIR-ZINC SYSTEMS

Air Access Design

F.K. Nabiullin, Z.M. Buzova, E.M. Gertsik, I.I. Koval, V.M. Maslov and L.N. Khamits; U.S. Patent 3,920,475; November 18, 1975; and F.K. Nabiullin; U.S. Patents 4,022,949; May 10, 1977; 4,042,761; August 16, 1977 describe an alkaline galvanic cell comprising coaxially arranged positive and negative electrodes separated by an ion-permeable membrane; the positive electrode has a terminal which forms a housing, and the negative electrode has a bar terminal. There is a washer and a spacer, both of an electrically insulating material, mounted adjacent the end faces of the electrodes; and a seal assembly including a metal cap in contact with the bar terminal, and a gasket with a hub which is tightly fitted on the bar terminal. The gasket is made of an electrically insulating plastic material; the washer is located on the side of the metal cap, and is also provided with a hub tightly fitted on the bar terminal; the cap is provided with holes filled with the material of the gasket to form passages in the gasket body by punching so as to provide for air access to the positive electrode.

Referring to Figure 2.13a, the alkaline galvanic cell comprises a positive electrode **1** having a terminal **2** which forms a housing of the cell, and a negative electrode **3** preferably of pasty zinc, having a bar terminal **4**.

FIGURE 2.13: ALKALINE GALVANIC CELL

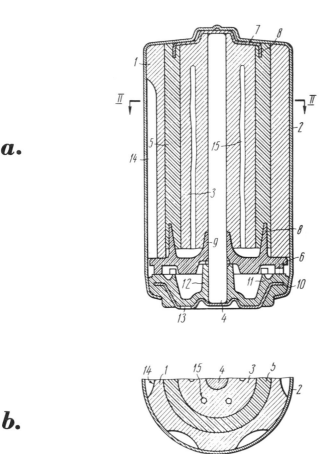

a.

b.

(continued)

FIGURE 2.13: (continued)

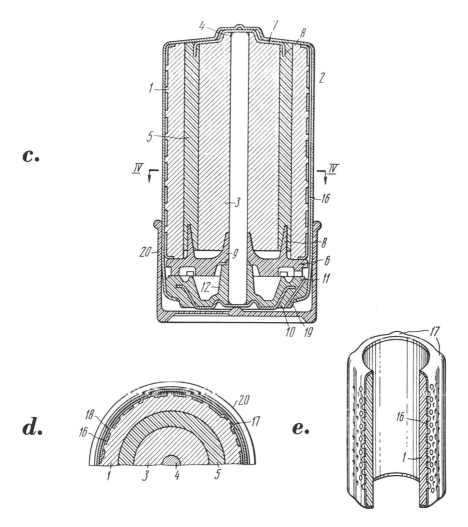

c.

d.

e.

(a) Longitudinal sectional view of an alkaline galvanic cell wherein a seal assembly cap is provided with holes, a washer mounted adjacent the cap is provided with a hub, and positive and negative electrodes are provided with passages

(b) Sectional view taken along the line II—II in Figure 2.13a

(c) Longitudinal sectional view of a modified alkaline galvanic cell wherein a seal assembly gasket is provided with passages extending in the body thereof, the mass of the positive electrode being pressed into a shell, and the cell is provided with a cap

(d) Sectional view taken along the line IV—IV in Figure 2.13c

(e) General view of the shell with the positive electrode shown in Figures 2.13c and 2.13d

Source: U.S. Patent 3,920,475

As shown in Figure 2.13a, the positive electrode **1** and the negative electrode **3** are arranged coaxially and are separated from each other by an ion-permeable membrane **5**. Mounted on one side adjacent the end faces of the electrodes **1** and **3** is a washer **6**, and a spacer **7** of an electrically insulating material is located on the other side adjacent to the end faces of the electrodes, both the washer and the spacer having an annular rib **8** which partly protrudes into the ion-permeable membrane **5** along the entire periphery of the end face.

The washer **6** is provided with a hub **9** which is tightly fitted on the bar terminal **4** of the negative electrode **3**. Located adjacent the washer **6** is the seal assembly of the cell which comprises a metal cap **10** and a gasket **11** having a hub **12**, which is also tightly fitted on the terminal **4**, the gasket being formed by reinforcing an electrically insulating plastic material.

The inside diameter of the hub **9** and the washer **6**, as well as the inside diameter of the hub **12** of the gasket **11** are smaller than the diameter of the bar terminal **4**, whereby the above-mentioned parts can be tightly fitted on the terminal. The outer surface of the metal cap **10** constitutes the negative contact of the cell. The gasket **11** with the hub **12** is made of a plastic which is nonwettable with alkali, such as polyethylene.

The central part of the cap **10** is supported by the bar terminal **4** extending through the hub **12**. The hub **12**, which is compressed around the terminal **4**, prevents alkali from penetrating to the cap **10**. The cap **10** is pressed against the terminal **4** through the interposition of the gasket **11** by a peripherally rolled edge of the cell housing, which in this case is the terminal **2**. Holes **13** are spaced along the periphery of the cap **10** and filled with the plastic material of the gasket **11**, the holes being adapted to form passages in this material to ensure air access inside the cell by punching, which is effected prior to the beginning of the operation of the cell.

In order to provide for free access of air oxygen to the positive electrode **1** along the height and cross section thereof, as well as for hydrogen escape from the negative electrode **3** without deformation thereof, axially extending passages **14** (Figures 2.13a and 2.13b) are made in the body of the positive electrode **1**; they are preferably of arcuated shape in the cross section, and the body of the negative electrode is also provided with axially extending passages **15**.

Figure 2.13c shows a modified example of the alkaline galvanic cell, in which the mass of the positive electrode **1** is pressed into a shell **16** (Figures 2.13c and 2.13d) mounted in the housing **2** of the cell. The shell **16** comprises a perforated metallic component and is galvanically plated with nickel. The shell is externally provided with projections 17 (Figures 2.13d and 2.13e) extending in parallel with the shell axis and adapted to define a uniform gap **18** between the cell housing **2** and the shell **16**. The gap **18** provides for air oxygen access to the positive electrode **1** along the height and cross section thereof.

The shell **16** is made of a steel band or sheet which is galvanically plated with nickel on the side in contact with the mass of the positive electrode **1**. The outer surface of the shell **16** may have no coating. The edges of the shell **16** are rolled. Electric contact between the shell **16** and the housing **2** is ensured by means of the projections **17** and the rolled edge of the shell which rests against the bottom wall of the housing **2**.

In this alkaline galvanic cell air access inside the cell is ensured by means of passages 19 made in the body of the gasket 11 of electrically insulating plastic material. In order to ensure long-term storage of the cell (to prevent air from penetrating inside the cell during the storage period), the cell is provided with a cap 20 of a plastic material, such as of polyethylene, which is tightly fitted on the cell housing on the side of the metal cap 10. The central position of the cap 20 is made thinner so as to ensure hydrogen escape from the cell. The cap may be used repeatedly, that is, it may be removed for operation of the cell and then put on the cell again for a long-term storage period.

The ion-conductive diaphragm 5 is manufactured by the introduction of a particulated bibulous moisture-retaining material, for instance wood flour as an additive to the gelling starch. The approximate composition of the ion conductive diaphragm is 65 g starch and 130 g wood flour per liter of electrolyte. Other materials, for instance sawdust, paper or cotton dust and the like may be used as the additive.

Helix Collector

G. Gerbier and P. Depoix; U.S. Patent 3,928,072; December 23, 1975; assigned to Saft-Societe des Accumulateurs Fixes et de Traction, France describe an air depolarized electric cell comprising a negative electrode occupying a peripheral position, an electrolyte which is preferably gelled and a positive electrode mass occupying a central position within the negative electrode. The positive electrode is provided with a funnel in the center of the positive mass and with a metallic current collector disposed in the funnel in contact with walls of the funnel. The collector comprises a metallic wire in the form of a helix at least partially incrusted or embedded in the wall of the funnel.

Figure 2.14a shows a collector 1 in the shape of a helix, according to the process. The rectilinear part 2 at one end of the collector is intended to be connected to the positive terminal of the cell. This part 2 is in the axis of the helix and the end turn 3 has a smaller diameter than that of the other turns of the helix. According to other variants, the rectilinear part lies in a plane tangent to the helix and all the turns are identical.

Figure 2.14b shows a cutaway view of a cell C incorporating the described current collector 1. It comprises a casing 4 of suitable nonconductive material closed by a cover 5 provided with the two aeration holes 6 and 7 which normally are sealed off before the cell is put into service. The cover 5 also bears terminals 8 and 9, respectively positive and negative. Inside the casing 4 a tabular negative electrode 10, constituted for example by zinc powder in suspension in a potassium hydroxide and/or sodium hydroxide gel, occupies a peripheral position.

The outer longitudinal surface and lower end surface of the electrode are in surface contact with the inner surface of the casing 4. An electrolyte 11 covers completely the inner tubular surface and upper end surface of the negative electrode to protect it against an undesirable intake of air.

This electrolyte is constituted, for example, by a gelled sodium and/or potassium hydroxide solution and is itself protected by a directly overlying layer of gas-impermeable material such as pitch 12.

FIGURE 2.14: AIR DEPOLARIZED CELL

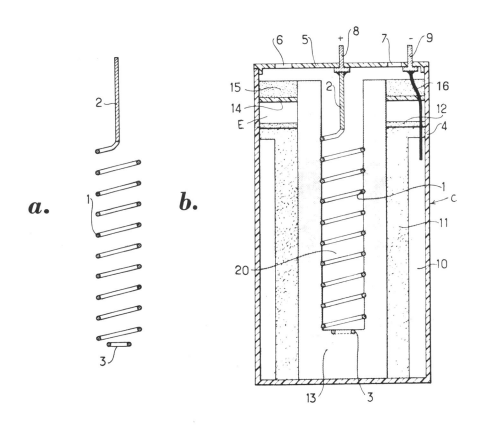

a.   *b.*

(a)   Cutaway view of a current collector
(b)   Cutaway view of an air depolarized electric cell provided
       with a positive electrode

Source: U.S. Patent 3,928,072

A second protective means is constituted, for example, by a washer **14** made of plastic material bearing an upper sealing layer of pitch **15**. This means **14, 15** is situated above the electrolyte and spaced from layer **12** so as to provide an expansion space **E** for the negative electrode **10**

The cell **C** comprises, moreover, a positive electrode occupying an axially extending central position whose positive mass **13** has a funnel or vent **20** which extends axially but not completely the length of the electrode thus remaining closed adjacent its lowermost end. A metallic current collector **1** having a helical configuration is arranged in intimate contact with the internal wall of the funnel 20 and its end turn 3 is sunk into, i.e., embedded in the positive mass of the closed or bottom end of the funnel. The individual turns of the metallic helix are partly embedded in the inner wall of the funnel 20. Such embedding may, for example, be effected during the molding of the positive mass 13. The metal helix 1 may, for example, be of nickel, nickel-plated copper, nickel-plated iron or an alloy, such as stainless steel.

The other end of helix 1 is provided with an axially extending or rectilinear part 2 joined to the terminal of the cell to which it may be welded. Likewise, a current collector 16 connected to the negative electrode 10 is joined to the terminal 9 to which it may be welded. Such an arrangement enables the inner surface of the funnel 20 to be exposed completely to contact with the air flowing into the cell via holes or openings 6; the result of this is a better supply of air to the active catalytic sites of the positive mass 13. In consequence, the cell is capable of discharging higher currents than conventional cells without any substantial lowering of its voltage.

The helical shape imparted to the collector 1 enables it to accommodate itself or follow the shrinkage or expansion of the positive mass without exerting stresses on the latter. It has been observed, moreover, that the ohmic resistance of the positive electrode is decreased due to the closer contact of the collector turns with the mass of the conglomerate of positive mass 13 in which they are sunk.

Button Cell

M. Wiacek; U.S. Patent 4,041,211; August 9, 1977; assigned to Unican Electrochemical Products Ltd., Canada describes a substantially leak-proof button cell, particularly adapted for use in hearing aids.

The cell comprises an outer cathode can and an inner anode can containing a zinc electrode material, preferably a zinc paste, the inner can being positioned within the outer can, with the outer wall of the inner can spaced from the inner wall of the outer can, a polymeric seal, preferably a polypropylene seal, positioned between the outer wall of the inner can and the inner wall of the outer can.

The outer wall of the inner can has a coating of an alkali-resistant elastomer or elastomeric adhesive, e.g., a self-curing butyl elastomer modified by a phenolic resin, which is pressure sensitive and has the property of deforming when subjected to pressure without cracking or forming voids.

The catalyst electrode is on a flexible and permeable, preferably polytetrafluoroethylene, membrane and includes a current collector, positioned between the polymeric seal and the inner wall of the outer can.

A separator is used between the zinc electrode material and the catalyst electrode. The outer wall of the inner can preferably is provided with a locking means such as a flange, to lock the seal onto such wall, and the seal, also coated with the above elastomeric adhesive, is preferably tapered downward and forms a leak-proof seal with the inner wall of the outer can.

The collector mesh or wire from the catalyst electrode is exposed in contact with the inner wall of the outer cathode can.

Electrode Assembly

R.W. Kelm; U.S. Patent 3,966,496; June 29, 1976; assigned to Gould Inc. describes an electrode assembly for an air depolarized cell in which an electrical conducting tab is mechanically attached to the electrode assembly to produce a low resistance electrical connection. The tab includes a crown which has sharp tines thereon for piercing the electrode assembly which generally comprises a separator, a current collecting screen, an electrode material and an outer hydrophobic member. The projections on the conducting tab are wide enough to insure that each of the tines will engage a section of current collecting screens.

This process, without any further application of fusing, cleaning, etc., produces a low resistance electrical connection between the screen and the tab. In addition, the use of more than one set of crowns provides torsional resistance to the tab thus making it sturdier for handling in the unassembled state.

Vent Tube

According to a process described by *J.E. Schmidt; U.S. Patent 3,970,480; July 20, 1976; assigned to McGraw-Edison Company* an air-vented battery cell containing liquid electrolyte is rendered leakproof against spillage during shipping and during usages when the cell is tilted from a vertical position in any direction to a horizontal position, as when used on a marine buoy. The process involves providing the cell with a vent reservoir across the top having a liquid connection with the cell at only one corner and having a vent tube in the reservoir at the top part, leading through one end wall diagonally to nearly an opposite corner portion.

In three tilt positions of this cell at 90° intervals from a vertical to a horizontal position, a limited flow of electrolyte into the reservoir is well below the inner open end of the vent tube, and in the fourth such tilt position the electrolyte in the reservoir enters the vent tube through only a small length, typically 10 to 20% thereof, with the result that the cell is protected against any leakage from shaking thereof while tilted and is provided further with a reserve against leakage through the vent tube if an increase in gas pressure occurs within the cell while it is so tilted.

OTHER BATTERY PROCESSES

Nickel Oxide and Manganese Dioxide

A process described by *T. Takamura, T. Shirogami and Y. Kanada; U.S. Patent 3,961,985; June 8, 1976; assigned to Tokyo Shibaura Electric Co., Ltd., Japan* generally relates to an alkaline nickel oxide-zinc cell and more particularly, to a primary alkaline nickel oxide-zinc cell containing a mixture of nickel oxide and manganese dioxide as the positive electrode material, wherein, preferably, an electroconductive sheet containing many apertures such as a wire netting, tightly covers the surface of the positive electrode. The alkaline cells with this positive electrode configuration are characterized by improved stability during the storage or shelf life of the cells and an improved capacity even under heavy current discharge loads. The process involves using an improved positive electrode which is characterized by a positive electrode material consisting of 100 parts by weight of nickel oxide and 10 to 90 parts by weight of manganese dioxide, wherein the improvements are further enhanced by the addition of less than 10 parts by weight of at least one moiety selected from the group consisting of lithium cation, cobalt oxide and bismuth oxide. The positive electrode is further characterized by the presence of a wire netting, consisting of an expanded metal sheet or a metallic sheet containing many holes, which is placed tightly over the electrode surface opposite the negative electrode. In the following example the amounts of ingredients mixed are expressed in parts by weight.

Example: A 500 ml amount of 2 molar aqueous nickel nitrate was poured with stirring into 8 liters of a 10 molar KOH solution containing 10% sodium hypochlorite at 90°C. After mixing the reactants the temperature of the mixture was maintained at 90°C for 1 extra hour, and then the solution was allowed to stand for 24 hours. The resulting black precipitate was washed repeatedly with a large amount of 0.2 molar KOH solution, and afterwards the water was removed by vaporization. Finally, black nickel oxide electrode material was obtained by drying the black precipitate at 90°C overnight and by pulverizing it.

After mixing 50 g of the positive electrode material, 30 g of electrolytic MnO_2, 15 g of flake graphite and 5 g of polyethylene powder in a V-shaped mixer, 20 ml of 7 molar KOH was added to the mixture. A 0.75 g quantity of the mixed electrode material was weighed and pressed into the interior of a can of an H-C flat cell which is specified by the Japanese Industrial Standard (JIS) under a pressure of 2 tons/cm^2.

On the exposed face of the compressed composition, a sheet of Acropore was placed followed by a 0.3 mm thick sheet of a nonwoven cloth of cotton. On top of the cloth was placed 0.5 g of a negative electrode mixture followed by ZnO saturated with 210 μl of a 40% KOH solution. A sealed nickel-zinc cell was assembled by covering the can with a plate which acts as the negative electrode and a rubber packing which acts as an insulator. The negative electrode material was comprised of 100 parts of 100 mesh zinc particles which were amalgamated with 10% Hg, 2 parts of zinc oxide, 2 parts of magnesium oxide and 1.8 parts of carboxyvinyl polymer. A gel was formed by mixing 100 parts of the negative electrode material with 70 parts of 35% KOH.

As a reference example, the same cell was assembled as disclosed above except that nickel oxide was used in the same amount as that of γ-MnO_2 in the positive electrode material. The cells thus assembled were subjected to an intermittent 15 Ω discharge at 20°C, wherein the on and off periods were 10 and 20 seconds, respectively. The results in Table 1 are reported as the number of on times attained until the voltage of the cells was reduced to the specified minimum voltages indicated in the headings of Table 1. Each value shown in the table was obtained by averaging the voltage received from nine cells.

TABLE 1

Positive Electrode Material	Minimum Average Voltage (volts)					
	1.20	1.10	1.00	0.95	0.90	0.85
Reference electrode (no γ-MnO_2 present)	1	22	61	78	93	112
Process electrode (38% γ-MnO_2 present)	2	34	92	114	132	140

These results clearly indicate that the presence of 38% γ-MnO$_2$ in the positive electrode material improved the intermittent heavy load capacity of the cell. The same improvement was also found when acetylene black or silver powder was used as the electroconductive material instead of flake graphite. The same test as described above was conducted at -10°C. The reference example used was a silver-zinc cell, which was made by the same procedure with the same materials as the cell of this process except that the positive electrode material contained Ag$_2$O in the same amount as the MnO$_2$ used in the cell of the process.

Fifteen repetitive discharges were attained with the cell of this process until the minimum average voltage of 0.9 V was reached. On the other hand, only three discharges were obtained with the silver oxide reference cell. These data indicate that the cell of the process has better properties than the reference cells at lower temperatures.

The same nickel-zinc cell of this process and one of the reference cells described above were stored at 45°C in an atmosphere of 85% relative humidity for three months. The cells were then subjected to a 500 Ω continuous discharge and the capacities lost during storage of the cells were evaluated by comparing the capacity of each cell at the minimum voltages specified in Table 2 with the capacity of each cell before storage. The percent reduction of the capacity of each cell during storage at the minimum voltages shown in Table 2 is based on the comparison of the data obtained for the cells before and after storage. The data obtained at the minimum voltages shown are the average values received from ten cells.

The data below show that a large capacity reduction was obtained in the reference cells containing no γ-MnO$_2$ in the positive electrode material. The presence of 38% by weight γ-MnO$_2$ in the electrode material of the cells of this process substantially prevents capacity reduction at high temperatures.

TABLE 2

Positive Electrode Material	Reduced Capacity During Storage (%) at the Minimum Average Voltage Shown (volts)				
	1.40	1.30	1.20	1.10	1.00
Reference cell (no γ-MnO$_2$ present)	22.2	19.9	20.4	21.0	21.8
Cell of process (38% γ-MnO$_2$ present)	10.5	10.2	11.8	11.2	10.2

Silicon Metal Anodes

According to a process described by *J.P. McKaveney; U.S. Patent 4,024,322; May 17, 1977; assigned to Hooker Chemicals & Plastics Corporation* anodes which exhibit electrochemical potentials comparable to zinc are prepared from alloys formed from silicon together with one or more highly reactive elements which are unstable when introduced alone into water. The highly active element component of the alloy is preferably selected from the group consisting of calcium, barium, magnesium, cerium and strontium. Preferably, the active element or elements comprise from about 5 to about 65% of the alloy. The following examples illustrate the process.

Example 1: An alloy of calcium and silicon containing 32.4 calcium, 63.1 silicon, 4.1 iron and 0.4 barium in percent by weight was crushed to a powder having an average size of about 600 μ. A layer of the powdered alloy as coated onto a graphite rod painted with silver epoxy paste. The electrode was placed in a drying oven at 160°C for 15 hours to effect curing of the epoxy.

Following the curing period, the anode was cooled and placed in a dilute hydrochloric acid solution for a light pickle to remove any surface oxide films. Following the acid pickle, the anode was immediately water rinsed, and while still wet, attached to a copper wire at the graphite end, with the alloy-coated end placed in a beaker of test electrolyte. The opposing end of the copper was attached to a Beckman terminal connector inserted in the connection normally used for the glass electrode of a pH meter. The anode was measured in various electrolytes compared with iron and zinc, with the results shown on Table 1.

TABLE 1: POTENTIALS vs STANDARD CALOMEL ELECTRODE

| | .... Electrolyte .... | | |
Alloy	0.10N $CaCl_2$	0.10N $MgSO_4$	0.10N NH_4Cl
$CaSi_2$	−1.35	−1.20	−1.40
Zn	−0.98	−0.99	−1.07
Fe		−0.65	−

Example 2: Experiments essentially duplicating that of Example 1 were conducted utilizing the alloys set forth in Table 2. The electrochemical potentials for each compared with those of iron and zinc are set forth in Table 2.

TABLE 2: POTENTIALS vs STANDARD CALOMEL ELECTRODE

| | .... Electrolyte .... | | |
Alloy	0.10N $CaCl_2$	0.10N $MgSO_4$	0.10N NH_4Cl
CaSiBa	−0.65	−0.68	−0.63
Mg_2Si	−1.25	−1.15	−1.21
MgFeSi	−0.89	−1.07	−1.12
Zn	−0.98	−0.99	−1.07
Fe	−	−0.65	−

Mercury Oxide-Zinc Cell

H. Balters and G. Schneider; U.S. Patent 3,964,931; June 22, 1976; assigned to VARTA GmbH, Germany describes a cathode for alkaline primary batteries, which contain a mercury oxide depolarizing cathode and an addition of Ag_2O. An amount of Ag_2O in the cathode mass, between 2 and 10%, preferably between 4 and 6%, was found to be particularly preferable. The Ag_2O can simply be mixed to the cathode mass. However, the depolarization mass and a suitable Ag_2O mass may be processed in separate molds and be subsequently joined in the cell.

Alternatively, both masses may be processed while partially separated, in the same mold. The two last described methods are used for placing the Ag_2O mass on the side of the cathode that is turned away from the anode, which makes the diffusion of silver oxide (soluble in small concentrations in the electrolyte) more difficult in the direction of the anode. It is known that the solubility of the Ag_2O in alkaline hydroxide requires the use of narrow porous separators in cells that contain Ag_2O as depolarizers. A suitable placement of the Ag_2O masses in the cell has been shown to have the same beneficial effect as the use of a narrow porous separator. The following shows some examples of a cathode according to the process.

Example 1: A cathode mass of HgO, 87 parts by weight; Ag_2O, 5 parts by weight; graphite, 8 parts by weight; and binder, 0.75 part by weight, was pressed into tablets, with amalgamated zinc powder as the anode mass and potassium hydroxide with dissolved zincate as the electrolyte; cells were produced in a known manner—whose construction basically resembles that of conventional button cells. Four days after production, the open voltage of these cells was between 1.577 and 1.587 V. After a 1 year storage, one cell went completely dead; one cell showed the onset of self-discharge at 1.653 V, while the remaining, approximately 40 cells had voltages between 1.578 and 1.582 V and almost unchanged capacity.

Example 2: Small tablets consisting of a mass of 100 parts by weight Ag_2O, 1 part by weight graphite and 0.5 part by weight binder, which were first superficially reduced, were pressed into a cathode of a mass comprising 92 parts by weight HgO, 8 parts by weight graphite and 0.075 part by weight of an organic binder. The portion of Ag_2O in the cathode was about 5%. Some of these cathodes were so built into the cells that the Ag_2O tablet faced the anode; some were reversed so that the Ag_2O tablet lay upon the cup bottom.

In the first instance, following a 1 year storage at normal temperatures, a little silver had diffused through the separator into the anode chamber, while in the second embodiment, however, no trace of this was found. In both embodiments the open voltage was between 1.583 and 1.585 V, and after 1 year between 1.579 and 1.581 V. Cells that were opened following this period showed traces of incipient oxidation of the zinc; the Ag_2O tablets were partly reduced; and the HgO was still completely unused.

The practice of adding Ag_2O to cathodes of alkaline cells which contain HgO as depolarizer, provides a reliable indicator for the partial discharging of cells in batteries with series connection.

Beryllium Battery

H. Nishimura and M. Toda; U.S. Patent 4,008,357; February 15, 1977; assigned to Citizen Watch Co., Ltd., Japan describe a high efficiency, high capacity beryllium battery. The beryllium battery comprises an outer casing having an open end, an aperture plate closing the open end, a beryllium metal anode coupled to the aperture plate, a sulfide, sulfate or carbon fluoride composition cathodic material coupled to the outer casing, and an organic electrolyte.

Referring to Figure 2.15, shown therein is a cross-sectional view of a simplified beryllium battery in accordance with the process. The beryllium battery of Figure 2.15 comprises an outside casing 1 constructed in the shape of a battery and having one open end.

FIGURE 2.15: BERYLLIUM BATTERY

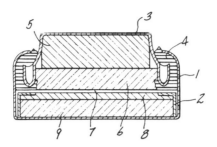

Source: U.S. Patent 4,008,357

Outside casing **1** also functions as the positive terminal of the battery. Retaining ring **2** is contained within the body of outside casing **1**, and may be omitted. Aperture sealing plate **3** closes the open end of outside casing **1** and forms the negative terminal of the battery. Beryllium metal **5** is the anodic active element and is coupled to aperture sealing plate **3** in the top of the battery. Gasket **4** couples aperture sealing plate **3** to and electrically insulates it from outside casing **1** and seals the electrolyte within the battery.

Furthermore, retaining ring **2**, when used, improves the tightness of gasket **4**. The organic electrolyte is contained within permeable materials **6** and **8**. Furthermore, permeable material **8** acts as an electrolyte chamber which is held in contact with the cathode **9** by a separator **7**. Since the electrolyte in permeable material **8** is held in contact with the cathode **9** by separator **7**, the contact between the cathode **9** and the electrolyte is enhanced thereby resulting in an increased electrical conductivity.

In practice, both the outside casing **1** and aperture sealing plate **3** may be made from nickel-plated or stainless steel sheets. Also, the electrolyte may be made of 1 mol of anhydrous lithium perchlorate dissolved in triacetin. Furthermore, the cathode **9** may be molded under high pressure from a composition which is 85% cupric sulfate, 14% lead, and 1% carboxymethylcellulose. Beryllium batteries have a high capacity with an output voltage of approximately 1.3 V. Such a battery has great practical value and great utility in devices such as electronic watches and the like.

LITHIUM NONAQUEOUS ELECTROLYTE SYSTEMS

LITHIUM-FLUORINATED CARBON

Sulfolane Electrolyte

G.W. Mellors; U.S. Patent 3,907,597; September 23, 1975; assigned to Union Carbide Corporation describes a nonaqueous cell utilizing a highly active metal anode, such as lithium, a solid cathode such as $(CF_x)_n$, copper sulfide or the like, and a liquid organic electrolyte consisting essentially of sulfolane or its liquid alkyl-substituted derivatives in combination with a cosolvent, preferably a low viscosity cosolvent such as dioxolane, and an ionizing solute, such as $LiClO_4$. Preferable nonaqueous cell systems according to this process are shown below.

Nonaqueous Cell Systems

Anode	Cathode	Solvent	Solute
Li	$(CF_x)_n$	Sulfolane-dioxolane	$KAsF_6$
Li	CuS	Sulfolane-dioxolane	$LiClO_4$
Li	CuS	Sulfolane-dioxolane	$LiAlCl_4$
Li	CuO	Sulfolane-dioxolane	$LiClO_4$

Example 1: Parallel plate test cells were made in airtight glass containers, the dimensions of which were 3.8 cm diameter and 0.7 cm in height. Electrical leads were passed through epoxy cement seals and connected to the electrodes within the container. The anode of each cell was a sheet of lithium with 4 cm^2 surface area; each cathode was of similar apparent area. The cathode contained 80 wt % active material ($\approx CF_{1.0}$), 10 wt % carbon black, 10 wt % hydroxyethylcellulose and a fibrous cellulosic binder of 10 wt % polytetrafluorethylene pressed onto an expanded nickel screen. About 30 ml of the selected electrolyte as indicated in the table below was employed in each cell.

The theoretical capacity of $(CF_x)_n$, where x = 1 was 0.864 Ah/g (calculated). The cells were made with about 0.23 to 0.28 g of active cathode material. The cells were tested by being discharged across a resistive load to a selected voltage cutoff. The results of the test are shown below.

Li/(CF$_x$)$_n$ Cells
(Electrolyte: 50/50 by Volume Sulfolane-Dioxolane + 1 M LiClO$_4$)

Load (ohms)	Cell Capacity (mAh)	Cutoff Voltage	Eff. of (CF$_x$)$_n$ (%)	Wt. (CF$_x$)$_n$ (g)	Wh/g (CF$_x$)$_n$	Wh/in^3*
1000	252	1.9	117	0.25	2.17	31.8
500	205	1.9	103	0.23	1.83	24.7
250	224	1.9	112	0.23	1.96	26.5
125	157	1.75	80	0.28	1.23	16.5

*Calculated from actual dimensions of cathode.

Cathode efficiencies of over 100% are believed due to the use of the active carbon black material which is a very active carbon and has been observed in previous nonaqueous cell systems to contribute to the capacity of the cell during discharge.

Example 2: Parallel plate cells of the type described in Example 1 were made using copper sulfide (CuS) cathodes. Fine mesh copper powder and sulfur in stoichiometric amounts were tumbled overnight to mix thoroughly, pressed onto a nickel screen, and then heat-treated under argon at 200°C for 16 hours. The electrode surface area used in the cells was 4 cm^2. The anode and the electrolyte were the same as that of Example 1. The cells were tested as in Example 1 and the results obtained are shown in the table below.

Li/50:50 Sulfolane-Dioxolane + 1 M LiClO$_4$/CuS System

Load (ohms)	Capacity (mAh)	Cutoff Voltage	Capacity (mAh/g CuS)	Eff. of CuS (Cu^{2+}→Cu^{1+}) (%)	Wh to 1.9V Cutoff	Wh/g CuS*	Wh/ in^3**
1000	408	1.86	272	97.1	0.775	0.516	19.62
500	303	1.74	266	95.0	0.576	0.505	15.13

*Using an average voltage of 1.9V.
**Calculated from actual dimensions of cathode.

It will be noticed that these results refer to the first step of the reduction of CuS, i.e., Cu^{2+} + 1e$^-$ → Cu^{1+} in this system; the second step, Cu^{1+} + 1e$^-$ → Cu^0, is absent. This is a distinct advantage in that substantially all of the cells' output is at the potential of the first step.

Crotonitrile and Propylene Carbonate Electrolyte

G.E. Blomgren and G.H. Newman; U.S. Patent 3,953,235; April 27, 1976; assigned to Union Carbide Corporation describe a nonaqueous cell which utilizes a highly active metal anode, such as lithium, a solid cathode such as (CF$_x$)$_n$, copper sulfide, copper oxide, nickel fluoride or silver chloride, and a liquid organic electrolyte comprising crotonitrile in combination with a protective cosolvent and a solute. It is an object of the process to provide an electrolyte

solvent system for nonaqueous solid cathode cells consisting essentially of crotonitrile in combination with a protective cosolvent, a low viscosity cosolvent and
a solute. It is a further object of this process to provide a nonaqueous cell which
utilizes a metal anode, a solid cathode and a liquid organic electrolyte based on
crotonitrile in combination with a protective cosolvent such as propylene carbonate and a solute such as $LiClO_4$. Thus, according to the process, crotonitrile can
be successfully employed in nonaqueous cells employing highly active metal anodes if a protective cosolvent is propylene carbonate.

Example 1: Nonaqueous button cells having a diameter of 1.7 inches (4.32 cm)
and height of 0.36 inch (0.91 cm) were fabricated using either a CuS or $(CF_x)_n$
solid cathode, an electrolyte of 1 M solution of $LiClO_4$ in crotonitrile, and a lithium anode. Specifically, each cell was constructed by placing a solid cathode in
a nickel container followed by superimposing thereon a separator containing one
milliliter of the electrolyte and composed of two layers of fibrous glass and a
layer of "Celgard 2400" material, the latter material being a wettable microporous polypropylene.

A lithium electrode composed of a lithium sheet pressed onto and into an expanded nickel screen was placed on top of the separator followed by a fibrous
glass pad, the pad being employed to contain additional electrolyte, if needed.
The cell was closed at the top by a nickel lid which was seated upon an annular
polypropylene gasket having an L-shaped cross section, which in turn rested on
the peripheral surface of the container. A peripheral crimp seal was then made
between the lid, gasket and container, thereby sealing the cell.

On different current drains to a specific cutoff voltage, the discharge capacity of
the cathode, cathode efficiency and average discharge voltage to cutoff were obtained for each cell and are shown in Table 1. Since the cells were cathode-limited, the cathode efficiency was calculated as a percentage based on the theoretical capacity of the cathode material available in each cell. For example, the
theoretical efficiency of CF (x = 1) as a cathode material in a lithium anode cell
discharging at a 1 milliampere per square centimeter drain to a 1.5 volt cutoff,
is calculated as follows: Assuming the reaction:

$$6.94 \text{ g Li} + 31 \text{ g CF} \longrightarrow 25.94 \text{ g LiF} + 12 \text{ g C}$$
$$\text{(1 equiv} \quad \text{(1 equiv} \quad \quad \text{(1 equiv} \quad \text{(1 equiv}$$
$$\text{wt)} \quad \quad \text{wt)} \quad \quad \quad \text{wt)} \quad \quad \text{wt)}$$

then if 1 gram CF is used the fraction of the equivalent weight is $\frac{1}{31}$. Since
one Faraday of electricity is obtained from one equivalent weight, then the Ah
per equivalent weight is calculated as follows:

$$\frac{96,500 \text{ coulombs/Faraday}}{3,600 \text{ coulombs/Ah}} = 26.8 \text{ Ah/equivalent weight}$$

Therefore, $\frac{1}{31}$ equivalent weight x 26.8 Ah/equivalent weight = 0.864 Ah. This
0.864 Ah or 864 mAh is the theoretical capacity of 1 gram of CF material when
used as a cathode in a lithium anode cell and, by using this calculation technique,
the cathode efficiency of $(CF_x)_n$ material and the other cathode materials can be
calculated when such are used as cathodes in cells having various electrolytes.
As shown by the test data in Table 1, the discharge capacity and cathode efficiency of the cells containing crotonitrile as the sole solvent were very low

thus demonstrating that crotonitrile would be unacceptable for nonaqueous cell systems if used as the only solvent.

TABLE 1

Cell Sample	Cathode	Current Drain (mA/cm^2)	Cutoff Voltage (V)	Theoretical Capacity (mAh)	Discharge Capacity (mAh)	Percent Efficiency	Average Discharge Voltage (V)
1	CuS	5.0	1.0	1,025.0	120	11.4	—
2	CuS	5.0	1.0	1,014.0	51	5.0	—
3	CuS	1.0	1.0	1,008.0	212	21.0	1.6
4	CuS	1.0	1.0	1,014.0	251	24.7	1.6
5	CuS	1.0	1.0	1,014.0	310	30.5	1.8
6	CuS	0.1	1.0	1,014.0	82	8.1	1.8
7	CuS	0.1	1.0	1,014.0	68	6.7	1.8
8	CuS	0.1	1.0	997.0	99	9.9	1.8
9	$(CF_x)_n$	5.0	1.5	738.0	427	57.8	2.0
10	$(CF_x)_n$	5.0	1.5	718.3	352	49.0	2.0
11	$(CF_x)_n$	5.0	1.5	792.1	201	25.4	1.8
12	$(CF_x)_n$	1.0	1.5	733.1	486	66.3	2.1
13	$(CF_x)_n$	1.0	1.5	800.3	449	56.1	2.2
14	$(CF_x)_n$	1.0	1.5	839.7	430	51.2	2.2
15	$(CF_x)_n$	0.1	1.5	813.9	231	28.4	2.3
16	$(CF_x)_n$	0.1	1.5	780.6	298	38.2	2.3
17	$(CF_x)_n$	0.1	1.5	806.9	153	19.0	2.2

Example 2: Several similar type button cells were produced as described in Example 1 except that the electrolyte was a 1 M solution of $LiClO_4$ with different concentrations of crotonitrile and propylene carbonate. The cathode used in each of these cells was a solid $(CF_x)_n$ cathode wherein the x value varied between 0.85 to 1.0. On a discharge drain ranging from 0.1 mA/cm^2 to 5.0 mA/cm^2 to a 1.5 volt cutoff, the average discharge voltage, cathode efficiency and discharge capacity to cutoff were obtained for each cell and are shown in Table 2.

The test data in Table 2 clearly illustrate the high cathode utilization obtainable when using an electrolyte based on crotonitrile in combination with propylene carbonate and $LiClO_4$. The test data also show that the concentration of the propylene carbonate can vary between 10 to 30 volume percent of the solvent mixture and still provide a good electrolyte for use in nonaqueous lithium cells.

TABLE 2

Cell Sample	Electrolyte System	Current Drain (mA/cm^2)	Theoretical Capacity (mAh)	Discharge Capacity (mAh)	Percent Efficiency	Average Discharge Voltage (V)
1	10% PC-90% CN	5.0	656	120	18.3	1.9
2	10% PC-90% CN	5.0	701	90	12.8	2.0
3	10% PC-90% CN	5.0	676	200	29.5	2.0
4	10% PC-90% CN	1.0	781	609	78.0	2.3
5	10% PC-90% CN	1.0	787	692	87.9	2.3
6	10% PC-90% CN	1.0	767	703	91.6	2.3
7	10% PC-90% CN	0.1	761	630	82.8	2.3
8	10% PC-90% CN	0.1	800	560	70.0	2.2
9	10% PC-90% CN	0.1	748	440	58.8	2.6
10	20% PC-80% CN	1.0	827	661	80.0	2.2
11	20% PC-80% CN	1.0	872	779	89.3	2.2
12	20% PC-80% CN	1.0	735	531	72.3	2.1
13	30% PC-70% CN	1.0	781	580	74.4	2.2
14	30% PC-70% CN	1.0	833	690	82.8	2.2
15	30% PC-70% CN	1.0	807	540	66.9	2.3

Lewis Acid in Thionyl Chloride Electrolyte

J.J. Auborn and S.I. Lieberman; U.S. Patent 3,923,543; December 2, 1975; assigned to GTE Laboratories Incorporated describe a cell having an alkali metal anode, a cathode including as the active cathode material an intercalation compound of graphite and fluorine of the general formula $(C_4F)_n$ wherein n refers to the presence of a large, but indefinite, number of recurring (C_4F) groups in the intercalation compound; and an electrolyte containing an inorganic solvent selected from the group consisting of phosphorus oxychloride, monofluorophosphoryl dichloride, thionyl chloride, sulfuryl chloride, and mixtures thereof, and a solute dissolved in the inorganic solvent material.

Of particular interest are those cells which have a Lewis acid, or an excess of a Lewis acid, present in the electrolyte either by actual addition or by dissociation. A special feature of such cells is that the cathode material will catalyze the electrochemical decomposition of the solvent resulting in a cell having a coulombic cathode utilization efficiency greater than 100% of the theoretical attainable according to reduction of the active cathode material.

Coated Metallic Screen Cathode

N. Marincic; U.S. Patent 3,907,593; September 23, 1975; assigned to GTE Laboratories Incorporated describes an electrochemical cell, particularly of the flat, button-type, having an alkali metal anode, a carbon or $(C_4F)_n$ cathode, and an electrolyte, comprising a solute dissolved in an inorganic oxyhalide or thiohalide solvent, between and in contact with the anode and cathode. The cathode material, which catalyzes the electrochemical decomposition of the solvent, is present as a composite structure wherein a metallic screen is coated on both sides with the cathode material. The screen is so positioned that the sharp spikes at the ends cut into the walls of the cell housing (i.e., can) and maintain positive electrical and structural contact.

The anode is an oxidizable material and preferably is lithium metal. Other anode materials include sodium, potassium, etc. The anode may be constructed of the oxidizable material in contact with a suitable supporting metal grid. The grid for a lithium anode, for example, can be made of nickel, nickel alloys (such as Monel), stainless steel, silver or platinum. The complete cathode structure comprises a metallic screen coated with the cathode material [e.g., carbon or $(C_4F)_n$] on each side to thereby form a composite in which the cathode layers are firmly held to the screen and interconnected through the openings in the screen.

Individual composite cathode structures can be prepared; preferably, however, individual, round cathode discs are punched out of a larger flat sheet of the composite material. When placed in the bottom of the cell can, the sharp spikes at the ends of the cathode screen cut into the adjacent walls of the cell can, thereby maintaining positive mechanical and electrical contact by the spring action of the contracted screen. To achieve this result, it is necessary to properly select the diameter of the cathode disc with respect to the internal diameter of the adjacent portion of the cell can. For example, to achieve the desired spring-type contact, the cathode discs can be made of a diameter equal to, or just slightly larger than the internal diameter of the can, and then force fitted to the bottom of the can.

Example 1: A flat, button-type electrochemical cell is fabricated from an 8 mil thick stainless steel can, 0.450 inch in diameter and 0.135 inch high. A No. 5 Nil2-1/0 (5 mil thick) expanded nickel screen is coated on each side thereof with a 20 mil thick layer of cathode material comprising 85% acetylene black (50% compressed), 10% graphite and 5% polytetrafluorethylene binder. A cathode disc of 0.450 inch diameter is punched out from the preformed composite structure and placed in the bottom of the stainless steel can. A 5 mil thick glass mat separator, also of 0.450 inch diameter, is placed over the cathode disc. A gasket formed of Kel-F is placed on the separator adjacent the inside perimeter of the can.

The electrolyte comprising 1.8 M lithium tetrachloroaluminate in thionyl chloride is added to the cell. The anode, a 15 mil thick lithium foil pressed into a No. 5 Nil2-1/0 (5 mil thick) expanded nickel screen, previously welded to the inside surface of an 8 mil thick stainless steel cap, is positioned on the gasket, and the cap is secured to the cell can. When discharged, this cell exhibits about 63 milli-amp-hours at an average voltage of about 3.47 volts, resulting in about 216.5 milliwatt-hours of energy. This is contrasted with commercial mercuric oxide-zinc cells of the same size (e.g., Eveready 343) rated at 110 milliamp-hours which, at an average discharge voltage of 1.25 volts, deliver 137 milliwatt-hours of energy, or about one-third less energy than the cells of this example. Cathode surface area is about 1 cm^2.

Example 2: Example 1 is repeated using a cell can of the same diameter, but 0.200 inch high (i.e., a can having the physical size of the Mallory RM 675 primary mercury battery). In the cell of this example, the anode is 20 mils thick, the cathode material is 60 mils thick, and the cell, because of its larger size, includes more electrolyte than the cell of Example 1. The cell was discharged at three different discharge rates. When discharged at a 1 milliamp rate, cell discharge above 3 volts lasted about 110 hours at an average discharge voltage of 3.45 volts. The total energy delivered was 375.5 milliwatt-hours. In contrast, the rated energy output of the Mallory RM 675 is 234 milliwatt-hours; thus the cell of the example realizes a 47% improvement in energy output over the aforementioned cell of the same size.

When the cell of this example was discharged at a 1.5 milliamp rate, cell discharge above 3 volts lasted about 75 hours, and when discharged at a 0.75 milliamp rate, cell discharge above 3 volts lasted about 150 hours. This example additionally illustrates an advantageous property of the cells of this process, namely that the substantially majority (i.e., 80% or so) of the useful lives are above 3 volts at a relatively constant discharge voltage. For example, when discharged at the 1 milliamp rate, cell discharge voltage remains essentially constant at about 3.4 volts for about 95 hours of the total cell useful life of about 115 hours. This property is advantageous where essentially constant voltages of this magnitude are desired over long periods of actual use.

LITHIUM-CARBON

Graphite-CrO$_3$ Intercalation Compounds

A.N. Day; U.S. Patent 3,998,658; December 21, 1976; assigned to P.R. Mallory & Co. Inc. describes a rechargeable nonaqueous electrolyte electrochemical cell based upon the $Li/LiClO_4/CrO_3$ system. A method by which the solubility and the chemical reactivity of strong oxidizing agents such as CrO_3 can be greatly reduced so that the material can be used as depolarizing cathodic material for organic electrolyte cells and batteries has been discovered. This was accomplished by preheating the CrO_3 with graphite at a temperature near the melting point of the CrO_3, viz 198°C. Such a procedure renders the CrO_3 insoluble to solvents such as tetrahydrofuran without reducing its electrochemical activity. It is possible that CrO_3 forms interstitial compounds with graphite under the above conditions. Such interstitial compounds have been reported by R.C. Croft, *Australian J. Chem.* 9, 184–194 (1956). These curious chemicals are classified as intercalation compounds.

The intercalation compounds (called compounds for lack of more accurate characterization) can be formed by mixing CrO_3 and graphite in proportions from 10 to 1 to 1 to 3 parts by weight. These compounds are formed at temperatures in the range from 175° to 225°C but generally it is preferred to heat the mixed powders at just below or about the melting point of CrO_3 (198°C). The heating can take place in air but is preferably accomplished in sealed tubes and should be continued for at least 3 hours but preferably for longer periods. A test for the completion of the formation of the intercalation compound consists in the introduction of small portions of the heated mixture into tetrahydrofuran or water. The absence of leaching and thus coloration indicates the formation and segregation of all of the CrO_3 into the intercalation compound with the graphite.

Example 1: Powdered CrO_3 and graphite powder were mixed thoroughly in 2 to 1 weight ratio with a mortar and pestle. The mixture was contained in a porcelain crucible and heated in a tube furnace for 4 days at 195°C in air. At the end of 4 days, the furnace was cooled to room temperature and the above mixture was taken out of the furnace. A small sample of this material when poured either into water, or into tetrahydrofuran, gave no coloration of the liquid indicating the insolubility of the CrO_3 in the intercalation complex. Prior to the above heat treatment, the CrO_3 and graphite mixture produced instantaneous coloration of the hexavalent chromium ion (deep yellow-orange) when poured into either water or tetrahydrofuran.

Example 2: The heat treated intercalation compound material of Example 1 was ground to a powder using a mortar and pestle. Cathodes were constructed without any further addition of graphite to the above powder (as it was found to be adequately conductive) using the following procedure: The powder was mixed with an aqueous Teflon dispersion (commercially sold as colloidal Teflon) so that the amount of solid Teflon content was 5% by weight of the mix. This was then treated with isopropyl alcohol and mechanically kneaded to form a rubbery mass.

This rubbery material was then dried under vacuum at room temperature and the solids were converted into a fluffy powder using a blender. Rectangular cathodes were molded on an expanded nickel current collector using the above powder by pressing with a force of 5,000 lb in a rectangular (2.27 x 0.95 cm) die. The apparent area of the finished cathodes was 2.16 cm^2. The cathodes were then further dried under vacuum for 2 hours at room temperature.

Example 3: Li/CrO_3 graphite cells were constructed in a parallel plate configuration using two rectangular lithium anodes (lithium ribbon pressed on stainless steel) placed on both sides of the rectangular cathode prepared as in Example 2, using one layer of filter paper separator on each side. The cells were packaged in a foil laminate bag (aluminum foil laminated with polyethylene). The electrolyte, 1 M LiClO$_4$ in an equivolume mixture of propylene carbonate and tetrahydrofuran electrolyte, was introduced and the cells were heat sealed. The cells were cathode limited. The open circuit voltage of the cells was 3.8 V.

The cells were discharged across constant loads of 200, 500, 1000 and 1500 ohms. The electrical performance of the cells was excellent. Assuming a cell reaction of 2Li + CrO$_3$ → CrO$_2$ + Li$_2$O the cathode efficiency, the average cell voltage, and cathode current density to a 2.0 volt cutoff are given in the table below:

Cathode Weight (g)	Stoichiometric Capacity (mAh)	Load (ohms)	Average Cell Voltage (V)
0.51	173	1500	2.8
0.49	166	1000	2.8
0.49	166	500	2.7
0.53	180	200	2.5

Average C-D (mA/cm^2)	Capacity to 2.0 V Cutoff (mAh)	Cathode Efficiency (%)
0.43	109	63
0.65	81	49
1.26	79	47
2.90	31	17

The theoretical energy density of the cells from these data was calculated to be 809 W-hr/lb (with the open circuit voltage of 3.8 volts).

Slotted Cathode Collector Bobbin

A process described by *D.H. Johnson, D.M. Kubala, and R.J. Bennett; U.S. Patent 3,985,573; October 12, 1976; assigned to Union Carbide Corporation* relates to an electrochemical cell employing a liquid active cathode material in conjunction with an active metal anode and an elastically deformable carbonaceous cathode collector in the form of a slotted annular bobbin.

Referring in detail to Figure 3.1a, there is shown a cylindrical can 2 having partially disposed therein an anode liner 4 in contact with the inner upstanding circumference of the can and completely disposed therein a bottom anode disk 6, shown in broken lines, in contact with the base of the can thereby adapting the container as the first or anodic terminal for the cell. Partially disposed within and in contact with the inner circumference of the anode liner is a separator liner 8 while a bottom separator disk 10, also shown in broken lines, is in contact with the bottom anode disk. If desired, the anode material could be extruded within the can.

A slotted cathode collector bobbin 12 is shown in Figure 3.1b having a longitudinal slot 14 with opposing faces 16 and 18 and an axial opening 20. The width X of the slot can vary somewhat as long as it is sufficient so that the bobbin can be radially compressed to reduce its diameter an amount sufficient to enable it to be easily slid into the anode-separator-lined can whereupon it will then radially expand to exert a bias against the separator layer which in turn contacts the anode so as to effectively maintain good physical contact between these components during cell discharge.

FIGURE 3.1: SLOTTED CATHODE COLLECTOR BOBBIN

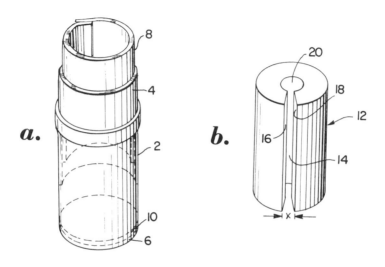

(a)   Perspective view of an anode and separator partially assembled
        in a cell can
(b)   Perspective view of a slotted cathode collector bobbin

Source: U.S. Patent 3,985,573

For conventional size cells, the width **X** of slot **14** for the bobbin could vary between about 0.05 inch (0.13 cm) and about 0.40 inch (1.02 cm), preferably about 0.075 inch (0.19 cm) for "AA" size cells, about 0.125 inch (0.32 cm) for "C" size cells, and about 0.20 inch (0.51 cm) for "D" size cells.

Example: Several elastically deformable carbonaceous cathode collector bobbins, as shown in Figure 3.1b were made using acetylene black and Teflon in the amounts shown in the table on the following page. The acethylene black of Mix A was wetted with the water-alcohol solution and mixed until the acetylene black was completely wetted and dispersed. The teflon emulsion was then added and thoroughly mixed with the solution after which the water content of the slurry was reduced to less than 5%.

The cake so formed was broken up into a powder form and then molded into a slotted annular bobbin. The bobbin, while still retained in a carrier, was heated for 30 minutes at 370°C. This sintering operation was found to impart elasticity or spring-like characteristics to the slotted bobbin which allowed the bobbin to be radially compressed so as to close the slot until the opposing faces touched and then, when the compression was released, the bobbin returned to its original geometry without breaking or splitting. This demonstrated the elastic deformability of a bobbin suitable for use in the subject process.

Mix Component	Weight or Volume	Dry Mix (percent by weight)
Mix A		
Acetylene black	10.0 g	75
Teflon Emulsion T-30B*	5.55 g	25**
Ethyl alcohol	100 ml	—
Water	420 ml	—
Mix B		
Acetylene black	10.0 g	90
Teflon Emulsion T-30B*	1.85 g	90**
Tergitol 1559***	0.45 g	—
Water	700 ml	—

*Obtained commercially—basically polytetrafluoroethylene, 60% solid
**Teflon
***Obtained comercially and contains mainly nonionic polyglycol ether

For Mix B, the acetylene black was slowly added to a well-stirred solution of water, Tergitol and Teflon emulsion until the acetylene black was thoroughly wetted and dispersed. The water in the slurry so formed was then substantially removed. The cake thus formed was then sintered in an inert atmosphere at 370°C for 30 minutes. The sintered cake was then broken up into fine particles of powder and molded into a slotted annular bobbin.

The bobbin so formed exhibited good elastic characteristics which permitted the bobbin to be radially compressed so as to close the slot until the opposing faces touched and then, when the compression was released, the bobbin expanded radially outward to its original geometry without breaking or splitting. This test again demonstrated the elastic deformability of a bobbin suitable for use in the process.

Several "C" size cells were constructed to demonstrate that an elastically deformable carbonaceous current collector in the form of a slotted annular bobbin can be used as a component part of an efficient liquid cathode cell system.

Globules of Cathode Material on Screen

F. Goebel; U.S. Patent 4,020,248; April 26, 1977; assigned to GTE Laboratories Incorporated describes a primary electrochemical cell which is capable of achieving high discharge currents. The primary electrochemical cell incorporates a special cathode structure. That cathode structure has a cathode current collector which is comprised with a plurality of electrically interconnected layers of a porous metallic material such as a nickel screen. Interposed between the layers of the cathode current collector are layers of globules of a cathode material.

This material has a composition of from about 40 to 99 weight percent of carbon black, at least 1 weight percent of a mechanical binder which is inert in the primary electrochemical cell and the remainder is graphite. When such a cathode structure is incorporated into a primary electrochemical cell, two features are obtained which contribute to the high discharge current capability of the cell. Firstly, because a multitude of globules of cathode materials are utilized in the cathode structure along with a porous current collector, large channels are maintained throughout the cathode structure thereby greatly facilitating the diffusion of the electrolytic solution of the cell throughout the cathode structure. Secondly, the conductive cathode current collector extends throughout the cathode structure and is in close contact with all of the cathode material.

Lithium Hexafluoride Electrolyte

D.J. Eustace and B.M.L. Rao; U.S. Patent 3,997,362; December 14, 1976; assigned to Exxon Research and Engineering Company describe an electrochemical cell which comprises an alkali metal anode, a cathode, an oxidant of at least one quaternary ammonium polyhalide salt, and a dipolar aprotic electrolyte containing an ionizable salt of an alkali metal.

A specific example includes a lithium anode, a carbon cathode, a tetraalkylammonium tribromide salt and an electrolyte of one mol lithium hexafluoride dissolved in propylene carbonate, which cell provides an open circuit voltage of about 3.6 volts. The oxidant can be incorporated in the cathode structure or be added independently to the cell, e.g., by use of a circulating electrolyte.

Example: A flat plate cell was prepared. The cathode consisted of a nickel grid, a carbon carrier and an oxidant. The nickel grid was one centimeter square and served as a current collector and as a support member. The cathode was prepared by dry pressing a mixture of 0.82 part of tetramethylammonium tribromide and 0.18 part of acetylene black on the nickel grid. The anode was prepared by pressing metallic lithium to another one centimeter square nickel grid. The electrolyte consisted of 0.6 mol of lithium hexafluorophosphate per liter of propylene carbonate.

The anode and cathode were immersed in the electrolyte and were separated from each other by a polypropylene separator. The resulting cell had an open circuit potential of 3.57 volts. The cell was discharged at 23.5°C. The cell was discharged at 20 milliamps per square centimeter and at 6.5 milliamps per square centimeter. The discharge curves show that at a discharge rate of 6.5 milliamps per square centimeter the cell provides a comparatively flat discharge curve. The potentials of the cells are significantly insensitive to the rate of discharge.

Concentric Design

F. Goebel and N. Marincic; U.S. Patent 4,042,756; August 16, 1977; assigned to GTE Laboratories Incorporated describe electrochemical cells having a first central anode, a second anode in the form of a concentric annular ring disposed around the central anode, a cathode, also in the form of a concentric annular ring, positioned between the first and second anodes, but not in mechanical contact therewith, and an electrolytic solution between and in contact with the anodes and cathode. The central anode can be in the form of a cylindrical rod or, preferably, a concentric annular ring.

Referring to Figures 3.2a and 3.2b, there is seen an electrochemical cell **10** having a housing or outer case **12**, commonly referred to as a "can." Adjacent **10**, and in mechanical and electrical contact with can **12**, is outer (second) anode **14** which is in the form of a hollow cylinder, the outer diameter of which is equal to the inner diameter of hollow can **12**.

FIGURE 3.2: ELECTROCHEMICAL CELL

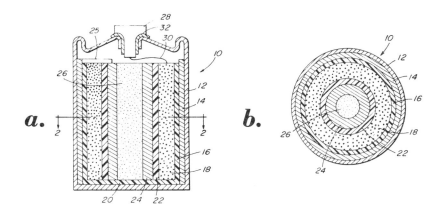

(a)   Vertical cross-sectional view of a symmetrical electrochemical cell

(b)   Cross-sectional view of the cell of Figure 3.2a taken along line 2-2 of Figure 3.2a

Source: U.S. Patent 4,042,756

Adjacent to and in contact with outer anode **14** is an insulating separator **16** which serves to prevent direct electrical and mechanical contact between anode **14** and cathode **18**, which is also in the form of a hollow cylinder. The cathode is prevented from contacting the can by a portion **20** of the insulator which is adjacent the bottom of the cell. Separator **22** prevents electrical and mechanical contact between cathode **18** and inner (first) anode **24** which, as with anode **14** and cathode **18**, is in the form of a hollow cylinder.

If desired, anode **24** can be in the form of a cylindrical rod, and suitable grids, for the electrodes can be provided if necessary or desired. Anodes **14** and **24** are connected in parallel by means of wire **25**. The operative terminals of the cell are the can **12** per se (generally the bottom thereof) which is in electrical contact with anode **14**, and contact **28** which is connected to cathode **18** via wire **30**. Contact **28** is separated from the can by an insulating separator **32**. The electrolytic solution (not numerically designated) fills the pores in cathode **18** and separators **16** and **22** so as to provide the necessary internal electrical circuit between the various electrodes. Central core **26** provides additional volume for the electrolytic solution thereby adding to the overall electrical capacity attainable with the cells described herein. Cells of the type described herein have been built in C- and D-size configurations with two lithium anodes and a carbon cathode. When tested under constant load and current conditions they exhibited an average of 15 to 20% capacity increase over corresponding high energy and high power cells at discharge rates of 1 to 2 mA/cm^2 of electrode surface area. C-size cells (lithium anodes and carbon cathode) when discharged at twice the rate recommended for the high energy C-size cell (e.g., 50 mA) exhibited capacities between 5.5 and 6 Ah above a 3.0 volt cutoff and energy densities of approximately 400 Wh/kg.

Example 1: A C-size cell is fabricated having a central lithium anode comprising a 1.4 inch x 1.4 inch x 0.035 inch lithium ribbon rolled about a ¼ inch diameter rod, an outer lithium anode (in the form of a concentric ring) comprising a 3.0 inch x 1.5 inch x 0.035 inch lithium sheet, and an intermediate carbon cathode (in the form of a concentric ring) 1.5 inches high and having an outside diameter of 0.85 inch and an inside diameter of 0.040 inch. A glass cloth separated each of the lithium anodes from the intermediate carbon cathode.

The cathode contained a blend of 97 g of Shawinigan carbon black (50% compressed) and 3 g of a tetrafluoroethylene binder. The electrodes were inserted into a C size nickel housing to which there is added 14 ml of an electrolyte comprising 1.8 M of lithium tetrachloroaluminate in thionyl chloride. Capacity to a cutoff voltage of 3.0 volts was 6 amp-hours at a 50 mA discharge rate.

Example 2: A D-size cell is fabricated having a central lithium anode comprising a 1.75 inch x 2.5 inch x 0.060 inch lithium ribbon rolled about a ¼ inch diameter rod, an inner lithium anode (in the form of a concentric ring) comprising a 3.75 inch x 1.75 inch x 0.070 inch lithium sheet, and an intermediate carbon cathode (in the form of a concentric ring) 1.75 inch high and having an outside diameter of 1.1 inch and an inside diameter of 0.062 inch. A piece of glass cloth separated each of the lithium anodes from the intermediate carbon cathode.

The cathode contained a blend of 97 g of Shawinigan carbon black (50% compressed) and 3 g of a polytetrafluoroethylene binder. The electrodes were inserted into a D-size nickel-plated steel housing to which there is added 32 ml of an electrolyte comprising 1.8 M of lithium tetrachloroaluminate in thionyl chloride. Capacity to a cutoff voltage of 3.0 volts was 11 amp hours at a 100 mA discharge rate.

Resilient Electrode Biasing Means

A process described by *L.F. Urry; U.S. Patent 4,032,696; June 28, 1977; assigned to Union Carbide Corporation* relates to a cylindrical type cell employing an outer cathode or cathode collector, a separator and an inner disposed anode in the form of at least two discrete bodies. The anode bodies have centrally disposed biasing means for continuously exerting an outward force against the anode bodies so as to provide good physical contact between the anode-separator-cathode or cathode collector interfaces.

The cell made in accordance with this process will have the advantage that as the inner anodic electrode is being consumed during discharge, the resilient biasing means will exert a force against this electrode thereby maintaining a good interface contact between the anodic electrode and the separator while the separator in turn is being maintained in a fixed position in contact with the cathode or collector of the cell.

Figure 3.3a shows a cross-sectional view of a cylindrical cell comprising a cylindrical container 2 having a cathode or cathode collector shell 4 in contact with the inner upstanding circumference of the container thereby adapting it as the cathodic terminal for the cell.

FIGURE 3.3: CYLINDRICAL CELL SYSTEM

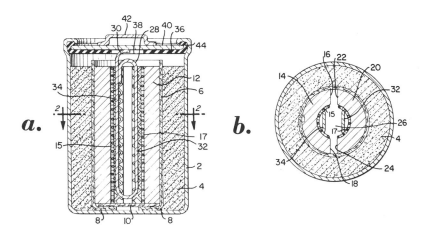

(a) Vertical cross-sectional view of a fully assembled electro-
       chemical cell
(b) Sectional view taken along line 2-2 of Figure 3.3a

Source: U.S. Patent 4,032,696

Disposed within and in contact with the inner circumference of cathode or cathode collector 4 is a separator liner 6 with its bottom end 8 radially folded inward and supporting a bottom separator or disc 10. If desired, the cathode or cathode collector material could be extruded within the container, rolled with the can material or composed of one or more segments to form a cylindrical tube and then placed in the can.

A two member anode 12 is shown in Figures 3.3a and 3.3b comprising a first half cylindrical annular member 14 having flat end faces 16 and 18 and a second half cylindrical annular member 20 having flat end faces 22 and 24. When the flat end faces of each cylindrical half member are arranged in an opposing fashion as shown in Figures 3.3a and 3.3b, an axial opening 26 is defined between the cylindrical half annular members 14 and 20. If desired, arcuate type backing sheets 15 and 17, such as inert electrically conductive metals screens or grids, could be disposed against the inner surface wall of the anode bodies 14 and 20 respectively, to provide uniform current distribution over the anode. This will result in a substantial uniform consumption or utilization of the anode

while also providing a substantial uniform spring pressure over the inner wall surface of the anode. An electrically conductive spring strip **28** is appropriately
bent into a flattened elliptically shaped member having an extending end **30**.
When inserting the spring strip **28** into a container the legs **32, 34** of the conductive strip **28** are squeezed together and forced into the axial opening between
the two screen backed anode members arranged in a container as shown in Figures 3.3a and 3.3b.

The inserted conductive spring strip resiliently biases the two anode members
14 and **20** via backing screens **15** and **17** so as to provide a substantially uniform
and continuous pressure contact over the inner wall of the anode members. The
extended end **30** of the spring strip is shown projected above the surface of anode members **14** and **20**. An insulator disc **36** has a central opening **38** through
which the projected end **30** of the spring strip **28** passes, whereupon the end is
then welded to a two part cover **40** and **42** thereby adapting the two-piece cover
40-42 as the anodic or negative terminal of the cell.

Before closing the cell, the electrolyte or cathode-electrolyte can be dispensed in
opening **26** whereupon it can permeate through the anode, separator and cathode
or cathode collector of the cell. In addition, the separator could be presoaked
with the electrolyte or cathode-electrolyte prior to being inserted into the cell.
The insulating disc has a peripheral depending skirt **44** disposed between cover
40 and the upper inner wall of container **2** for sealing the cell through conventional crimping techniques.

Example: Two C size cells were constructed similar to the cell construction
shown in Figures 3.3a and 3.3b except that the cathode collector was formed
of two half-cylindrical annular members and an anode backing sheet was not
used. The cathode collector half-cylindrical annular members were made using
acetylene black and Teflon in the proportions shown in Table 1.

Component	Amount	Percent
Acetylene black	10.0 g	90
Teflon Emulsion T-30B*	1.85 g	10**
Tergitol 1559***	0.45 g	—
Water	700 ml	—

  *Polytetrafluoroethylene
 **Teflon
***Nonionic polyglycol ether

The acetylene black was slowly added to a well-stirred solution of water, Tergitol and Teflon emulsion until the acetylene black was thoroughly wetted and
dispersed. The water in the slurry so formed was then substantially removed.
The cake thus formed was then sintered in a controlled atmosphere at 370°C
for 30 minutes. The sintered cake was then broken up into fine particles of
powder and molded into half-cylindrical annular members. Two of the annular
members were inserted in a 304 stainless steel container and then heated for 30
minutes at 370°F.

This heating operation expanded the members so that they became locked firmly
in the container. A tubular nonwoven glass separator was then inserted in the
container followed by the insertion of two half-cylindrical lithium anode members as generally shown in Figures 3.3a and 3.3b. A spring member as shown in

Figures 3.3a and 3.3b was compressed and inserted into the axial opening defined by the anode members. Upon release of the compressive force on the spring member, the anode was mechanically biased radially outward thereby imparting a good physical contact to the separator which in turn physically contacted the cathode collector of the cell.

An extended leg of the spring member was projected through an opening in an insulating disc and a Teflon gasket and then welded to a stainless steel cover. Before sealing the cover to the container, a 1.0 M solution of $LiAlCl_4$ in SO_2Cl_2 was fed into the container. The cell was then sealed in a conventional manner. Two such cells were discharged across an 88-ohm load and the data obtained are shown in the tables below.

Cell Sample	Impedance of fresh cell	Impedance after test	Hours to 2.5 volt cutoff	Ampere hours to 2.5 volt cutoff
1	1.83 ohms	1.58 ohms	205 hours	8.18 ampere hours
2	1.79 ohms	1.31 ohms	209 hours	8.41 ampere hours

Cell Sample	Watt hours per cubic inch to 2.5 volt	Average voltage to 2.5 volts	*Li utilization to 2.5 volts
1	19.3	3.51	81.3%
2	19.9	3.54	83.9%

$$*\text{Li Utilization} = \frac{\text{Amp. Hours out} \times 100\%}{\text{Total lithium in cell at start}}$$

As evidenced by the test data, the cells yielded high energy densities and high lithium utilization thereby demonstratively showing the benefits of the process.

Electrically Conductive Spacer

R. Okazaki, K. Aoki and T. Shinagawa; U.S. Patent 4,020,242; April 26, 1977; assigned to Matsushita Electric Industrial Co., Ltd., Japan describe a primary cell in which the expansion of the anode or cathode may be compensated by the contraction of a spacer so as to minimize the pressures acting upon the cell enclosure to expand it and in which a positive and efficient electrochemical reaction in the cell may be ensured by the smooth supply of the electrolyte to the cathode by the gradual contraction of the spacer.

Thus, the process provides a primary-cell wherein a cell element assembly consisting of a cathode, anode and separator interposed between the cathode and anode is enclosed in a primary-cell enclosure which serves as one of the terminals of the primary cell and which is sealed by and electrically insulated from a metal top closure which serves as the other terminals. Between the cathode or anode and the cell enclosure or top closure electrically connected to the cathode or anode is interposed a spacer which is chemically stable in the primary cell, electrically conductive and capable of holding the electrolyte and is adapted to reduce its apparent volume when pressure is applied.

FIGURE 3.4: BUTTON TYPE CELL

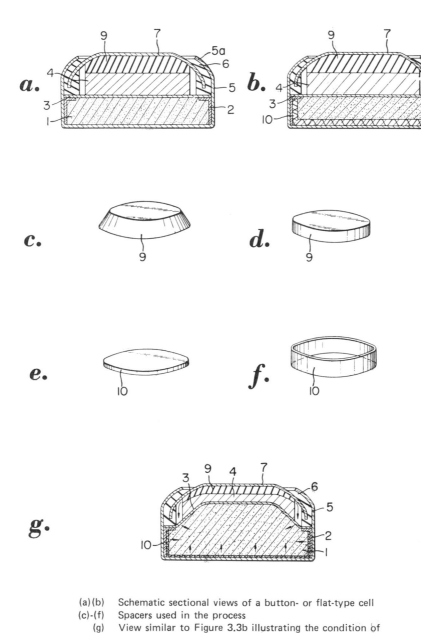

(a)(b)   Schematic sectional views of a button- or flat-type cell
(c)-(f)   Spacers used in the process
(g)   View similar to Figure 3.3b illustrating the condition of
          the primary-cell after some discharge time interval

Source: U.S. Patent 4,020,242

Referring to Figure 3.4a, reference numeral **1** denotes a disk-shaped cathode mainly consisting of fluorinated carbon (90% weight) mixed with carbon powder as a bonding agent; **2**, a stainless steel cathode ring surrounding the side wall of and peripheral portion of the upper surface of the cathode; **3**, a separator made of nonwoven fabric of polypropylene for covering the top of the cathode; **4**, an anode made of metallic lithium and placed above the cathode; **9**, a spacer made of an electrically conductive material and placed upon the anode, the spacer being made of a porous nickel sheet with the porosity of 90% and with the pore diameter of 100 μ; **5**, a stainless steel enclosure which serves as a cathode terminal; **6**, a polypropylene sealing and insulating gasket interposed between a metal top closure **7** placed over the spacer **9** and the top of the enclosure, the top end portion **5a** of the enclosure being bent inwardly toward the metal top closure to seal the cell.

The spacer, the cathode and the separator are impregnated with a suitable amount of electrolyte consisting of lithium perchlorate dissolved into propylene carbonate at the ratio of 1 mol per liter. The electrolyte may consist of a nonaqueous solvent such as dimethoxyethane, dioxolane, acetonitrile, tetrahydrofuran, dimethylsulfoxide, ethylenecarbonate, α-butyrolactone or the mixture thereof and a solute consisting of salt of light metal such as sodium perchlorate, aluminum chloride, lithium borofluoride.

The second example shown in Figure 3.4b is substantially similar in construction to the first embodiment shown in Figure 3.4a except that a cup-shaped spacer **10** made of carbon fiber nonwoven fabric is interposed between the cathode **1** and the enclosure **5** and the spacer **2**. The spacers used in the process may be made of a sponge-like porous metal such as stainless steel, nickel, titanium, copper, aluminum, iron, silver or alloy thereof. In addition, they may be provided by forming a woven or nonwoven fabric made of carbon, metal fibers or the mixtures thereof into a suitable shape.

The sponge-like metals may be provided by sintering of metal powder. It is preferable to use the sponge-like metals with the interconnecting channels or pores throughout the metals. The pore diameter is selected depending upon the construction of the cells and the viscosity of the electrolytes used, but in general the pore diameter is preferably less than 500 μ. The porosity is closely correlated with the electrolyte holding capacity and the resistance to the deformations by compression, and it is preferable that the materials have as high porosity as possible in that the high porosity will not adversely affect the machinability and handling. In general, materials with a porosity of 80 to 95% still have a satisfactory machinability.

The shapes of the spacers may be suitably selected depending upon the constructions of the cells. For instance, the spacers shown in Figures 3.4c and 3.4d may be used as the spacer **9** which is made into contact with the anode **4** while the spacers of the types shown in Figures 3.4e and 3.4f may be used as the spacer **10** on contact with the cathode **1**. Since these spacers are electrically conductive, the current flows through when they are made into contact with the metal top closure **7** and the cell enclosure **5**. The more reliable electrical and mechanical contact between the spacers and the top closure and the cell enclosure may be attained when they are joined by the spot welding or the like at the centers. The spacer is, of course, impregnated with the electrolyte.

Figure 3.4g is a cross-sectional view of the primary-cell shown in Figure 3.4b after it has been discharged for some time. It is clearly seen that the cathode **1** mainly consisting of fluorinated carbon has been considerably expanded upwardly because the lateral expansion is restricted by the stainless steel ring **2** while the lithium anode **4** has been consumed and reduced in volume. The expansion of the cathode is caused by the fixation or adhesion of the discharge product LiF to the cathode and by the absorption of the electrolyte by the cathode which is promoted as the discharge proceeds.

The expansion of the cathode is considerably greater as compared with the contraction of the anode so that the compressive forces are exerted to the spacers **9** and **10** thereby reducing their apparent volumes. However, the reduction in volume of the spacers **9** and **10** is cancelled by the increase in volume of the electrodes (that is, the cathode **1** and the anode **4**) so that neither the cell enclosure **5** nor the top closure **7** is subjected to the excessive forces. Thus, the expansion of the primary cell may be prevented.

Because of the reduction in apparent volume of the spacers **9** and **10**, the electrolyte is squeezed out of them in the directions indicated by the arrows in Figure 3.4g and absorbed by the cathode. Therefore, the electrolyte may be supplemented so that the discharge may be continued. The same is true for the primary cell shown in Figure 3.4a. Thus, the construction shown in Figure 3.4b is adapted for a relatively higher discharge rate while the construction shown in Figure 3.4a is adapted for a discharge of a relatively smaller rate because the smoother supply of the electrolyte may be ensured in the primary cell shown in Figure 3.4b than in the primary cell shown in Figure 3.4a.

Leak Neutralization Agent

A process described by *J.L. Hallet, P.H. Rollason and T.V. Rychlewski; U.S. Patent 4,011,371; March 8, 1977; assigned to GTE Sylvania Incorporated* relates to electrochemical cells and more particularly to leak neutralizing means associated with such cells which contain a corrosive electrolyte. Demands for smaller and more powerful energy sources have led to the development of batteries which employ very reactive materials. One such cell which is being developed employs an oxidizable lithium anode and an inert carbon cathode.

A proposed electrolyte for such a cell comprises a solvent of thionyl chloride and a solute of aluminum chloride. Both the thionyl chloride and aluminum chloride are corrosive materials. While cells employing such an electrolyte are theoretically hermetically sealed, the problem of a leaking cell always exists. It would be an advance in the art if some means were provided for neutralizing any of the corrosive electrolyte which might leak from such a cell.

This process provides an electrochemical cell which comprises a container or a body portion for containing positive and negative electrodes and a corrosive electrolyte therefor. The container is closed by means of a cap which is sealed to the container. Also provided are means associated with the seal which include a neutralizing agent for the corrosive electrolyte contained within the container.

Referring to Figure 3.5 an electrochemical cell **10** comprising a container **12** which has a substantially cup-shaped portion **14** of a suitable material, such as stainless steel, which is provided with a peripheral outstanding flange **16**.

FIGURE 3.5: LEAK DETECTION AND NEUTRALIZATION CELL DESIGN

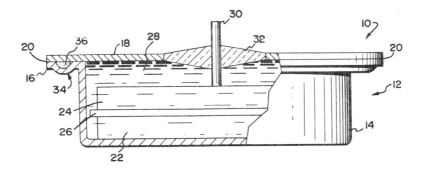

Source: U.S. Patent 4,011,371

A lid **18** of a suitable material, such as Kovar or Rodar, is provided to seal the opening of the can **14** and is peripherally sealed at **20**. Provided within the interior of container portion **14** are a positive electrode **22** and a negative electrode **24** which can be separated by a separator **26**. Filling the remainder of the interior and surrounding the electrodes is a corrosive electrolyte **28**. One of the electrodes, in this case the positive one, is in contact with the can and the negative electrode has an external connection **30** in the form of an electrically conductive pin which protrudes through the lid and is insulated therefrom by means of a glass seal **32**.

Means **34** is provided associated with the seal and contains a neutralizing agent **36** for neutralizing the corrosive electrolyte in the event of a leak. In the form shown in Figure 3.5 the means **34** comprises a depression or pocket formed within flange **16** and which contains the neutralizing agent. In a typical high energy cell having a lithium anode and a carbon cathode and a corrosive electrolyte of thionyl chloride and aluminum chloride, the neutralizing agent can be sodium carbonate. It will be obvious, however, that the neutralizing agent employed will be dependent upon the electrolyte used. The incorporation of a neutralizing agent at the seal area of an electrochemical cell is seen as a distinct advance by effectively neutralizing the leakage of a corrosive electrolyte.

Titanium Case

A process described by *A. Heller; U.S. Patent 3,922,174; November 25, 1975; assigned to GTE Laboratories Incorporated* is directed to an electrochemical cell having an alkali metal anode; a catalytic cathode upon which the solvent material is reduced, the cathode including a catalytic active cathode material selected from the group consisting of gold, carbon and $(C_4F)_n$; and an electrolyte having an inorganic solvent material selected from the group consisting of phosphorus oxychloride, thionyl chloride, sulfuryl chloride, and mixtures thereof, and a solute selected from a wide range of materials dissolved in the solvent.

The active elements of the cells are housed in a titanium or titanium alloy container which is inert to the electrolyte solvent materials. Since the material is inert, and specifically because it is inert electrochemically, the material can be utilized for the container, thereby reducing the overall weight of the cell while simultaneously increasing the overall energy density thereof (including the case or container material). The fact that titanium and its alloys, are compatible with the phosphorus oxychloride, thionyl chloride and sulfuryl chloride solvents utilized in this process is particularly significant in view of the corrosion of most transition and structural metals including iron, lead, copper and aluminum when in contact with the solvent materials or electrolytic solution.

Nonetheless, the titanium or titanium alloy cases lead to higher tolerances for internal pressure, impact, temperature, acceleration, etc., by the cell or battery than can be achieved by other cases (e.g., of iron or nickel) of equal weight. Additionally, for equal performance, the container walls can be thinner thus further reducing the weight of the cell or battery, and with less corrosion due to increased resistance to the solvent materials, the cells will have a longer shelf life.

Thermal Switch

J. Epstein; U.S. Patent, 4,035,552; July 12, 1977; assigned to GTE Laboratories Incorporated describes an electrochemical cell which comprises a hermetically sealed housing having an outwardly deformable wall section, an electrochemical system disposed within the housing and including a pair of electrodes respectively forming an anode and a cathode, and an electrolyte in contact therewith, a cover assembly affixed to the housing at least a portion of which serves as a first terminal associated with one of the electrodes, a second terminal associated with the second electrode. A normally closed thermal switch is disposed between the cover assembly and the housing.

The thermal switch is electrically connected between one of the electrodes and its associated terminal and is arranged to open-circuit the internal electrical circuit of the cell to prevent cell discharge when the cell temperature exceeds a preselected level and to close again when the temperature drops below the preselected value. The switch is additionally responsive to the wall section deflection to interrupt the internal cell circuit and prevent cell discharge when the internal cell pressure exceeds a predetermined level. Accordingly, the cell may be discharged at any rate so that the temperature and internal pressure levels are not excessive. Additionally, the cell is reactivated once the excessive level has dissipated.

ELECTROLYTES

3-Methyl-2-Oxazolidone

M.L. Kronenberg; U.S. Patent 3,996,069; December 7, 1976; assigned to Union Carbide Corporation describes a high energy density nonaqueous cell comprising a highly active metal anode, a solid cathode selected from the group consisting of FeS_2, Co_3O_4, V_2O_5, Pb_3O_4, In_2S_3 and CoS_2, and a liquid organic electrolyte consisting essentially of 3-methyl-2-oxazolidone in combination with at least one low viscosity cosolvent and a conductive solute. Highly active metal anodes include lithium, potassium, sodium, calcium, magnesium and their alloys. Of these active metals, lithium would be preferred because, in addition to being a ductile, soft metal that can easily be assembled in a cell, it possesses the highest energy-to-weight ratio of the group of suitable anode metals. Liquid organic 3-methyl-2-oxazolidone material (3Me2Ox) is an excellent nonaqueous solvent because of its high dielectric constant, chemical inertness to battery components, wide liquid range and low toxicity. However, it has been found that when metal salts are dissolved in liquid 3Me2Ox for the purpose of improving the conductivity of 3Me2Ox, the viscosity of the solution becomes too high for its efficient use as an electrolyte for nonaqueous cell applications other than those requiring very low current drains.

Thus, in accordance with this process, the addition of a low viscosity cosolvent is necessary if 3Me2Ox is to be used as an electrolyte for nonaqueous cells which can operate or perform at a high energy density level. Specifically, in order to obtain a high energy density level, it is essential to use a solid cathode along with a highly active metal anode. Thus this process is directed to a high energy density cell having a highly active metal anode, such as lithium, a solid cathode selected from the group consisting of FeS_2, Co_3O_4, V_2O_5, Pb_3O_4, In_2S_3 and CoS_2 and an electrolyte comprising 3Me2Ox in combination with at least one low viscosity cosolvent and a conductive solute.

The cathode efficiency of the cell, as based on the percentages of the theoretical capacity of the cathode material available in a cell operating on a drain of 1 milliampere per square centimeter to a 1.0 volt cutoff using a lithium anode, will be above about 50% and preferably above about 75% according to this process.

Example 1: The viscosity of several samples of 3Me2Ox, with and without a conductive solute and/or a low viscosity cosolvent, were obtained using a Cannon Fenske viscometer. The data obtained are shown in Table 1 and clearly demonstrate the high viscosity of a solution of 3Me2Ox containing a dissolved conductive solute.

TABLE 1

Sample	Solvent and Salt	Viscosity (Centistokes)
1	3Me2Ox; no salt	2.16
2	3Me2Ox; 1M $LiClO_4$	6.61
3	3Me2Ox; 1M LiBr	7.58
4	50-50 3Me2Ox, THF; no salt	1.05
5	50-50 3Me2Ox, THF; 1M $LiAsF_6$	3.59
6	50-50 3Me2Ox, THF; 1M $LiClO_4$	2.87
7	25-75 3Me2Ox, THF; 1M $LiAsF_6$	2.08
8	25-75 3Me2Ox, dioxolane; 1M $LiAsF_6$	1.83
9	25-75 3Me2Ox, THF; 1M $LiClO_4$	1.99

As shown in Sample 2, when one mol of $LiClO_4$ is added to one liter of 3-methyl-2-oxazolidone, the viscosity of the solution was found to be 6.61 centistokes. In Sample 6, when one mol of the same metal salt, $LiClO_4$, was added to one liter of equal parts of 3Me2Ox and tetrahydrofuran (THF), the viscosity of the solution was found to be only 2.87. Thus, it is clearly shown that the viscosity of a solution of 3Me2Ox and a metal salt can be decreased by the addition of a specifically selected low viscosity cosolvent.

Example 2: Each of six flat-type cells was constructed utilizing a nickel metal base having therein a shallow depression into which the cell contents were placed and over which a nickel metal cap was placed to close the cell. The contents of each sample cell consisted of a 1.0 inch diameter lithium disc consisting of five sheets of lithium foil having a total thickness of 0.10 inch, about 4 ml of a specific electrolyte as shown in Table 3, a 1.0 inch diameter porous non-woven polypropylene separator (0.01 inch thick) which absorbed some of the electrolyte, and a solid FeS_2 cathode mix compressed onto and into a porous 1.0 inch diameter cathode collector.

The FeS_2 electrodes were made of a mixture of FeS_2, acetylene black and a poly-tetrafluoroethylene binder compression-molded onto both sides of a nickel expanded mesh. The FeS_2 and acetylene black were first micromilled together, then blended with water, ethanol and a polytetrafluorethylene emulsion (obtained commercially from DuPont as Teflon emulsion designated T-30-B) in the proportions shown in Table 2 prior to draining off the excess liquid and compression molding (at 18,000 psi) onto the expanded metal carrier or mesh. Each finished FeS_2 electrode contained about 1.9 grams of the cathode mix and had a thickness of about 0.04 inch with a diameter of about 1.0 inch.

TABLE 2

Materials	Amount	Percent in Finished Electrode
FeS_2	20.0 g	87.5
Teflon emulsion	2.86 g	7.5
Acetylene black	1.14 g	5.0
Ethanol	20.0 ml	—
Water	110.0 ml	—

TABLE 3

Sample	Electrolyte Salt	Current Density (mA/cm^2)	Average Discharge Voltage (V)	Theoretical Capacity (mAh)	Average Discharge Capacity to 1.0 V Cutoff (mAh)	Cathode Efficiency (%)
1*	2 M LiBF$_4$	1.0	1.24	1,172	855	73.0
2**	2 M LiAsF$_6$	1.0	1.20	1,174	810	69.0
3**	1 M LiCF$_3$SO$_3$	0.8	1.27	1,168	835	71.5
4**	2 M LiCF$_3$SO$_3$	0.8	1.25	1,178	913	77.5
5**	2 M LiBF$_4$	0.2	1.44	1,194	1,071	89.7
6**	1 M LiBF$_4$	0.2	1.45	1,196	997	83.4

*The electrolyte solvent is 30 vol % 3Me2Ox and 70 vol % THF.
**The electrolyte solvent is 30 vol % 3Me2Ox, 40 vol % dioxolane, 30 vol % dimethoxyethane and a trace of dimethylisoxazole.

The total thickness of the anode, cathode plus cathode collector and separator for each cell measured about 0.15 inch. The average discharge voltage and discharge capacity on various current drains to a 1.0 volt cutoff were obtained for each cell and are shown in Table 3. Since the cells were cathode-limited, the cathode efficiency was calculated as a percentage based on the theoretical capacity of the cathode material available in each cell. For example, the theoretical efficiency of FeS_2 as a cathode material in a lithium anode cell discharging at a 1 milliampere per square centimeter drain to a 1.0 volt cutoff is calculated as follows. Assume the reaction:

$$4Li + FeS_2 \longrightarrow 2Li_2S + Fe$$
$$27.76 \text{ g Li} + 119.85 \text{ g FeS}_2 \longrightarrow 91.76 \text{ g Li}_2S + 55.85 \text{ g Fe}$$

Then if 1 gram FeS_2 is used, the fraction of the equivalent weight is 1/29.96. Since one Faraday of electricity is obtained from one equivalent weight, then the Ah per equivalent weight is calculated as follows:

$$\frac{96,500 \text{ coulombs/Faraday}}{3,600 \text{ coulombs/Ah}} = 26.8 \text{ Ah/equivalent weight}$$

Therefore, 1/29.96 equivalent weight x 26.8 Ah/equivalent weight = 0.894 Ah. This 0.894 Ah or 894 mAh is the theoretical capacity of 1 gram of FeS_2 material when used as a cathode in a lithium anode cell and by using this value as a reference, the cathode efficiency of FeS_2 material can be calculated when used as a cathode in a cell having various electrolytes.

As evidenced by the test data shown in Table 3, the cathode efficiency of the cells ranged from 69.0 to 89.7%, thus demonstrating that by using this process efficient, high energy density FeS_2 nonaqueous cells can be made.

Nitrobenzene

G.H. Newman; U.S. Patent 3,982,958; September 28, 1976; assigned to Union Carbide Corporation has found that nitrobenzene and substituted nitrobenzene compounds are excellent, electrolyte solvents for nonaqueous battery systems. Nitrobenzene is a liquid over the temperature range 5.7° to 210.9°C. It is stable in the presence of the materials of battery construction, even those which are normally considered to be highly reactive. Additionally, rechargeable battery systems containing nitrobenzene have displayed good efficiency on charge and have been capable of being charged at 10 volts with no gassing or apparent decomposition of the nitrobenzene.

The electrolytes comprise a solute dissolved in a solvent which is nitrobenzene or a substituted nitrobenzene compound. Preferred solutes are complexes of inorganic or organic Lewis acids and inorganic ionizable salts. The only requirements for utility are that the complex be compatible with the solvent being employed and that it yield a solution which is ionically conductive. Typical Lewis acids suitable for use in the process include boron bromide, aluminum chloride, aluminum bromide, boron fluoride and boron chloride. Ionizable salts useful in combination with these Lewis acids include lithium fluoride, lithium chloride, lithium bromide, sodium fluoride, sodium chloride, sodium bromide, potassium fluoride, potassium chloride, and potassium bromide.

Example: A test cell was constructed in a rectangular polytetrafluorethylene trough having two slots spaced one-half inch apart to accommodate an anode and a cathode. A sheet of lithium metal pressed into a nickel screen was the cell anode and the cathode, which consisted of silver chloride on a nickel screen, was charged to a total capacity of 440 milliampere-hours. The cathode had a nominal surface area of 10 square centimeters. The electrolyte consisted of a solution of about 11 weight percent lithium aluminum tetrachloride in nitrobenzene. The charge was begun at a rate of 2.0 milliamperes per square centimeter and a voltage of 3.5 volts. When the terminal voltage rose to 4 volts the charging rate was cut back to 0.25 milliampere per square centimeter and the last 60 milliampere-hours of charge were completed at a flat 3.5 volts. At a discharge rate of 1.0 milliampere per square centimeter this cell delivered 370 milliampere-hours at a voltage of 2.6 to 2.5 volts and 20 milliampere hours at a voltage of 2.5 to 2.0 volts.

Nonionic Surfactants

A process described by *R.G. Gunther; U.S. Patent 3,928,070; December 23, 1975; assigned to Yardney Electric Corporation* comprises the addition of a nonionic surfactant to cells which employ organic electrolytes. The surfactant employed is one which is soluble in the organic electrolyte which is employed in a particular organic electrolyte cell. It is added in an amount of at least about 0.01% by weight of the organic electrolyte.

Use of the surfactants results in a substantial improvement in the performance of organic electrolyte power cells without the attendant disadvantages of the prior art, such as the problems associated with the use of low-boiling organic solvents. Improved cell performance includes improved discharge performance and increased weight energy density (watt-hours/lb). Improvements in volume energy density may also be realized depending upon whether expansion of the carbonaceous cathode materials during cell operation completely negates the improvement in volume energy density which is theoretically possible in the absence of such expansion.

Additionally, the addition of the surfactant to the electrolyte is quite flexible in that it can be added directly to the electrolyte or it can be added indirectly, such as by including it in the cathode from which it can be dissolved by the organic electrolyte. Furthermore, substantial improvements are realized from the use of rather small amounts of the nonionic surfactants.

Examples of useful nonionic surfactants include: alkyl aryl polyethers such as nonyl phenoxy polyoxyethylene; alkyl aryl polyethylene glycol ethers such as nonyl phenyl polyethylene glycol ether and dodecylphenyl polyethylene glycol ether; and alkyl aryl polyether alcohols such as isooctylphenoxy polyethoxy ethanol and nonylphenoxy polyethoxy ethanol. Mixtures of these and other nonionic surfactants may also be employed.

Lithium Hexafluoroarsenate in Methyl Formate

S.G. Abens; U.S. Patent 3,918,988; November 11, 1975; assigned to Honeywell Inc. describes a current-producing cell system wherein the electrolyte comprises a solution of lithium hexafluoroarsenate in methyl formate. Lithium metal anode is used because of its high activity. If reduced activity is desired, the lithium may be associated with another metal, such as an alloy form with a less active metal. Lithium may also be in contact with another metal structure, such as nickel or silver screen, which serves as the anode conductor.

The conductivity of lithium hexafluoroarsenate dissolved in methyl formate (MF) was tested at various concentrations. At room temperature (27°C ± 1°C) the following measurements were observed.

Molar Concentration LiAsF$_6$/MF	Specific Conductance, mmho/cm
1 M	29.7
2 M	40.2
3 M	33.5
Saturation (3 M to $>$ 4 M)	29.5

Example 1: The anode is constructed by pressing a lithium metal sheet 0.015 inches thick into an expanded silver mesh. The sheet is then cut into 2 x 1.5 inch rectangles and wire leads are attached. The cathode is made from a mixture of 24.3 parts of MnO$_2$, 2.44 parts of carbon and 0.24 part polystyrene. The mixture is formed into a paste by adding xylene and applied to an expanded silver support, whereupon xylene is removed by vacuum drying. The cathode is shaped into 2 x 1.5 inch rectangles and wire leads are attached.

A cell is constructed having an anode and a cathode and a glass filter mat, serving as a separator, positioned between the plates. The composite structure is enclosed in a polyethylene case. An electrolyte solution is prepared by dissolving lithium hexafluoroarsenate in methyl formate in a ratio of one mol of lithium hexafluoroarsenate per liter of methyl formate. The cell is activated by injecting a sufficient quantity of the electrolyte solution into the cell structure to place the electrodes into an electrolytic contact with each other.

An electric load is then connected to the cell and current withdrawn at 0.42 milliamp per square centimeter at 35°C. Under these conditions the cell operated for 159 hours to a final voltage of 2.8 volts. It operated for 195 hours to a final voltage of 2.0 volts.

Example 2: A cell is constructed in the manner of Example 1, except that the cathode composition includes a quantity of LiAsF$_6$. The cathode comprises 10 parts MnO$_2$, 5.3 parts LiAsF$_6$, 2.44 parts carbon and 0.25 part polystyrene. In the laboratory, the cell operated to a final voltage 2.5 volts at an average voltage of 2.8 volts for 109 hours and to a final voltage of 2.0 volts for 142 hours.

Example 3: A cell is constructed as in Example 1, except that silver chloride is substituted for the MnO$_2$ as the depolarizing material. The cathode composition includes 24.3 parts AgCl, 2.44 parts carbon and 0.25 part polystyrene. Upon activation, the cell built in this manner operated 200 hours to a final voltage of 2.5 volts at an average voltage of 2.6 volts.

Example 4: A cell is constructed as in Example 1, except that NiF$_2$ is used as the depolarizing agent. A cathode pad is made, in the manner explained in Example 1, from a mixture of 50 parts NiF$_2$, 10 parts carbon and one part polystyrene. The cell is connected to a 50 ohm load and activated by addition of the electrolyte. A cell according to this example was built in the laboratory, giving the following results. 0.026 ampere-hours were withdrawn at a rate of 1.29 mA/cm^2 and an average voltage of 2.58 volts. The load was then changed to 100 ohms and the cell was operated until the potential dropped below 2.0 volts. Under the 100 ohm load, a current rate of 0.60 mA/cm^2 was achieved at an average voltage of 2.40 volts. The cell life to a 2.0 final voltage was 10.1 hours, yielding a total output of 0.268 ampere-hours.

Covalent Inorganic Oxyhalide Solvent and Lithium Solute

A process described by *J.J. Auborn; U.S. Patents 4,012,564; March 15, 1977; and 3,926,669; December 16, 1975; both assigned to GTE Laboratories Incorporated* is directed to electrochemical cells having an oxidizable active anode material, a solid metallic cathode current collector, and an electrolytic solution between and in contact with the anode and the cathode current collector, the electrolytic solution comprising a liquid covalent inorganic oxyhalide or thiohalide solvent and a solute dissolved therein, the inorganic solvent being the sole oxidant material and sole solvent material in the cell.

The cathode comprises a solid, nonconsumable, electrically conducting, inert current collector upon the surface of which the inorganic oxyhalide solvent is electrochemically reduced, whereby the inorganic solvent is electrochemically reduced, whereby the inorganic solvent, in conjunction with the oxidizable anode, serves as a source of electrical energy during operation of the cell. It is believed that the inorganic oxyhalide or thiohalide solvent is electrochemically reduced on the surface of the cathode current collector to yield a halogen ion which reacts with a metallic ion from the anode to form a soluble metal halide, such as, for example, lithium chloride.

The overall effect is to electrochemically reduce the solvent by removal of a portion of its halogen content in conjunction with the oxidation of the anode metal and thereby obtain electrical energy therefrom. This energy can be attained however, in the absence of other cathode depolarizers or oxidant materials, such as sulfur dioxide, which are not needed in the cells of this process since the inorganic oxyhalide or thiohalide solvent also serves as the oxidant material. In addition, it is believed that the inorganic oxyhalide or thiohalide solvent passivates the anode material, whereby the need to provide an additive or a further material to passivate the anode is obviated.

Examples 1 through 18: In the following examples, the cells have a lithium anode, and an electrolytic solution comprising a saturated solution of lithium tetrachloraluminate in phosphorus oxychloride. The cathode components of each cell, the open circuit potential at 25°C, and the current density at 50% polarization at 25°C obtainable are given in Table 1 below.

TABLE 1

EXAMPLE	CATHODE	OPEN CIRCUIT POTENTIAL, VOLTS	CURRENT DENSITY AT 50% POLARIZATION (mA/cm^2)
I	Nickel	2.80	0.200
II	Gold	2.92	0.357
III	Tungsten	2.50	0.190
IV	Lead	2.05	0.170
V	Palladium	2.70	0.155
VI	Molybdenum	2.75	0.105
VII	Germanium	2.75	0.300
VIII	Silicon	2.25	0.022

(continued)

TABLE 1: (continued)

EXAMPLE	CATHODE	OPEN CIRCUIT POTENTIAL, VOLTS	CURRENT DENSITY AT 50% POLARIZATION (mA/cm^2)
IX	Cobalt	2.75	0.250
X	Silver	2.70	0.155
XI	Mercury	2.90	0.155
XII	Iron	2.70	0.185
XIII	304 Stainless Steel	2.80	0.170
XIV	Niobium	2.60	0.125
XV	Manganese	2.70	0.270
XVI	Tantalum	2.30	0.130
XVII	Titanium	2.50	0.053
XVIII	Platinum	2.90	0.130

Examples 19 through 33: In the following examples, the cells have a lithium anode and an electrolytic solution comprising a 1.8 M solution of lithium tetrachloroaluminate in thionyl chloride. The cathode component of each cell, the open circuit potential at 25°C, and the current density at 50% polarization at 25°C obtainable are given in Table 2 below.

TABLE 2

EXAMPLE	CATHODE	OPEN CIRCUIT POTENTIAL, VOLTS	CURRENT DENSITY AT 50% POLARIZATION (mA/cm^2)
XIX	Nickel	3.66	1.142
XX	Gold	3.72	>8.000
XXI	Tungsten	3.67	1.067
XXII	Palladium	3.74	1.225
XXIII	Molybdenum	3.69	1.573
XXIV	Germanium	3.44	0.939
XXV	Silicon	2.92	0.215
XXVI	Cobalt	3.56	0.644
XXVII	Silver	2.91	1.835
XXVIII	304 Stainless Steel	3.56	0.221
XXIX	Niobium	3.53	0.370
XXX	Manganese	3.58	2.082
XXXI	Tantalum	3.44	0.497
XXXII	Titanium	3.06	1.201
XXXIII	Platinum	3.66	1.017

Solutes having lithium cations and large anions which are stable to oxidation and reduction are particularly desirable. The preferred lithium solute compounds are: lithium tetrachloroaluminate, lithium tetrachloroborate, lithium tetrafluoroborate, lithium hexafluorophosphate, lithium hexafluoroarsenate, lithium hexafluoroantimonate, lithium hexachloroantimonate, lithium hexachlorostannate, lithium hexachlorozirconate, lithium hexachlorotitanate and lithium chlorosulfate. Other preferred compounds are Lewis acids, particularly aluminum chloride ($AlCl_3$), boron trichloride (BCl_3), boron fluoride (BF_3), tin chloride ($SnCl_4$), antimony chloride ($SbCl_5$), antimony fluoride (SbF_5), titanium chloride ($TiCl_4$), aluminum bromide ($AlBr_3$), phosphorus fluoride (PF_5), phosphorus chloride (PCl_5), arsenic fluoride (AsF_5), arsenic chloride ($AsCl_5$), zinc chloride ($ZnCl_2$) and zirconium chloride ($ZrCl_4$), in conjunction with a metal halide such as lithium chloride. In addition, Lewis bases having the general formula $A_m B_n$ where A is an element selected from the group consisting of lithium, sodium, potassium, rubidium, cesium, magnesium, calcium, strontium, barium and the rare earths and B is an element selected from fluorine, chlorine, bromine, iodine and oxygen are also useful. Included in this latter category are cesium chloride, rubidium chloride, and barium chloride.

Tetrahydrofuran Mixed Solvents

A. Dey and B.P. Sullivan; U.S. Patent 3,947,289; March 30, 1976; assigned to P.R. Mallory & Co., Inc. describe an organic electrolyte system suitable for use at low and ambient temperatures with active metals. The electrolyte binary or ternary solvent mixtures have tetrahydrofuran or N-nitrosodimethylamine or mixtures thereof as one component and propylene carbonate; gamma-butyrolactone; 1,3-dioxolane; dimethyoxyethane; bis-2-ethoxyethyl ether; or bis(2-methoxyethoxy)ethyl ether as the other component with active metal salts as the solute.

The preferred binary compositions are propylene carbonate (PC) and tetrahydrofuran (THF), gamma-butyrolactone (BL) and tetrahydrofuran (THF), tetrahydrofuran (THF) and dioxolane (DL), tetrahydrofuran (THF) and dimethoxyethane (DME). The preferred ternary compositions are tetrahydrofuran (THF) and dioxolane (DL) and bis(2-methoxyethoxy)ethyl ether (MEEE), tetrahydrofuran (THF) and dioxolane (DL) and propylene carbonate (PC), tetrahydrofuran (THF) and gamma-butyrolactone (BL) and propylene carbonate (PC), tetrahydrofuran (THF) and gamma-butyrolactone (BL) and bis(2-methoxyethoxy)ethyl ether (MEEE), tetrahydrofuran (THF) and N-nitrosodimethylamine (NDA) and propylene carbonate (PC).

Example 1: The electrical conductivities of the various organic electrolytes consisting of 1 M $LiClO_4$ solution in the various solvents and the solvent mixtures were measured using an AC bridge (at 1,000 cycles per second). The results are given in the table below.

Solvents and Mixed Solvents (vol percent)	Specific Conductivities in ohm^{-1}cm^{-1}		
	at Room Temp.	−15°C	−30°C
Tetrahydrofuran (THF)	3.2×10^{-3}	10^{-6}	—
Propylene Carbonate (PC)	4.6×10^{-3}	1.5×10^{-3}	4.4×10^{-4}
gamma-Butyrolactone (BL)	9.7×10^{-3}	4.6×10^{-3}	3.0×10^{-3}
N-Nitrosodimethylamine (NDA)	—	—	6.4×10^{-3}
Bis-2-ethoxyethyl ether (EEE)	1.7×10^{-3}	6.8×10^{-4}	—
Dimethyl Sulfoxide (DMSO)	1.1×10^{-2}	—	solidified
75% THF + 25% PC	8.7×10^{-3}	5.6×10^{-3}	—
50% THF + 50% PC	9.6×10^{-3}	4.6×10^{-3}	—
25% THF + 75% PC	7.7×10^{-3}	2.9×10^{-3}	—
75% THF + 25% BL	9.0×10^{-3}	6.0×10^{-3}	—
50% THF + 50% BL	1.1×10^{-2}	6.6×10^{-3}	—
25% THF + 75% BL	1.1×10^{-2}	5.9×10^{-3}	—
75% THF + 25% EEE	2.7×10^{-3}	1.7×10^{-3}	—
50% THF + 50% EEE	2.3×10^{-3}	1.7×10^{-3}	—
25% THF + 75% EEE	1.8×10^{-3}	9.5×10^{-4}	—

(continued)

Solvents and Mixed Solvents (vol percent)	Specific Conductivities in ohm^{-1}cm^{-1} at Room Temp.	$-15°C$	$-30°C$
50% THF + 50% DMSO	1.3×10^{-2}	—	4.5×10^{-3}
75% NDA + 25% PC	—	—	4.0×10^{-3}
50% NDA + 50% PC	—	—	1.8×10^{-3}
25% NDA + 75% PC	—	—	1.1×10^{-3}
75% NDA + 25% THF	—	—	6.4×10^{-3}
50% THF + 25% 1,3 Dioxolane (DL) +25% Bis-(2-methoxy)-ethoxy-ethyl ether (MEEE)	5.9×10^{-3}	4.1×10^{-3}	3.2×10^{-3}
50% THF + 25% DL + 25% PC	1.0×10^{-2}	6.6×10^{-3}	4.8×10^{-3}
50% THF + 25% BL + 25% MEEE	9.2×10^{-3}	4.6×10^{-3}	3.2×10^{-3}
50% THF + 25% BL + 25% PC	1.1×10^{-2}	6.3×10^{-3}	4.2×10^{-3}
50% PC + 25% NDA + 25% THF	—	—	2.9×10^{-3}
50% NDA + 25% PC + 25% THF	—	—	4.9×10^{-3}

Example 2: Li/V$_2$O$_5$ cells were constructed in a parallel plate configuration. The lithium anode was prepared by pressing the lithium ribbon on the stainless steel expanded metal. The V$_2$O$_5$ cathode was constructed by pressure molding a rubberized mixture of 70% V$_2$O$_5$ + 30% graphite with 5% colloidal Teflon (binder) on the nickel expanded metal. The cathode was then cured at 300°C for one-half hour. The cathode dimensions were 0.91 x 0.38 x 0.05 inch. The filter paper separator was used. The cells were placed in the polyethylene capped glass bottles and were discharged in the electrolytes consisting of 1 M LiClO$_4$ solutions in the various solvents and the mixed solvents. The results are given below.

Solvents and Mixed Solvents (vol.%)	Discharge Efficiency up to 1.5 Volts of V$_2$O$_5$ Cathodes (based on 4 equivalents/mole) at (C.D=1 ma/cm^2)		
	Room Temp.	$-15°C$	$-30°C$
THF	55%	0	0
PC	13%	—	—
BL	36%	—	—
DL	21%	—	—
50% THF + 50% PC	54%	13%	7.5%
50% THF + 50% DL	48%	27%	26%
50% THF + 50% DME	59%	2%	0.1%
50% THF + 50% DO	59%	0	0
50% THF + 50% MEEE	55%	—	—
50% THF + 50% EEE	45%	0	0
50% THF + 25% DL + 25% MEEE	62%	—	—
50% THF + 25% DL + 25% PC	53%	—	—
50% THF + 25% BL + 25% MEEE	45%	—	—
50% THF + 25% BL + 25% PC	56%	—	—
50% THF + 10% DL + 40% MEEE	53%	—	—
50% THF + 15% DL + 35% MEEE	54%	—	—
40% THF + 10% DL + 50% MEEE	50%	—	—
40% THF + 25% DL + 35% MEEE	50%	—	—
30% THF + 20% DL + 50% MEEE	55%	—	—
25% THF + 25% DL + 50% MEEE	49%	—	—
10% THF + 25% DL + 65% MEEE	36%	—	—
10% THF + 40% DL + 50% MEEE	53%	—	—

The above results indicated that the performance of the Li/V_2O_5 cell at room temperature was considerably improved in the mixed solvents compared to that in the single solvents except in the case of THF. The low temperature performance of the cell in certain mixed solvents is found to be substantially superior to that in pure THF.

Clovoborate Salt

A process described by *C.R. Schlaikjer; U.S. Patent 4,020,240; April 26, 1977; assigned to P.R. Mallory & Co., Inc.* provides an electrolyte salt as a method for ion transport including a cation of a metal and a clovoborate anion. A feature of the process is that the electrolyte salt is sufficiently soluble and stable in a fluid oxyhalide or nonmetallic oxide or nonmetalic halide and mixtures thereof to function in an electrochemical cell. The electrolyte salt, when used in electrochemical cells, helps reduce capacity losses at low operating temperatures, helps reduce passivation of the metal anode, and thus helps reduce the voltage delay associated with start-up after storage at elevated temperatures.

More specifically, the process provides an electrochemical cell having as an electrolyte salt a compound having a cation of a metal and a clovoborate anion with the formula $(B_m X_n)^{-k}$, where m, n and k are integers, B is boron, and X is selected from the group consisting of H, F, Cl, Br, I and OH wherein at least some of the substituents are halides.

In thionyl type cells with lithium anodes, specific preferred salts known and characterized in the chemical literature include $Li_2B_{10}Cl_{10}$, $Li_2B_{10}Br_{10}$, $Li_2B_{10}I_{10}$, $Li_2B_{12}Cl_{12}$, $Li_2B_{12}Br_{12}$, $Li_2B_{12}I_{12}$, $Li_2B_6Br_6$, and $Li_2B_{12}Br_8F_4$. Less preferred salts include $Li_2B_9Cl_8H$, $Li_2B_9Br_6H_3$, $Li_2B_{11}Br_9H_2$, $Li_2B_{12}H_8F_4$, $Li_2B_{12}H_7F_5$, and $Li_2B_{12}H_6F_6$ and $Li_2B_{12}F_{11}OH$.

Example 1: The preparation of electrolytically conductive lithium halogenated clovoborate salts is exemplified by the preparation of $Li_2B_{10}Cl_{10}$. The anion $B_{10}Cl_{10}^=$ is prepared starting with decaborane, $B_{10}H_{14}$. About 5 grams of $B_{10}H_{14}$ (41 millimoles) are dissolved in about 30 ml of anhydrous benzene in a 200 ml round bottom flask. In a separate vessel about 14.5 ml of distilled triethylamine (104 mm) and about 35 ml of anhydrous benzene are mixed. The mixture is then added slowly to the flask containing the solution of decaborane and the total mixture is refluxed for about 16 to 24 hours.

After this time, a white solid separated whose composition is $(NEt_3H)_2B_{10}H_{10}$, i.e., the bistriethylammonium salt of the anion $B_{10}H_{10}^=$. The procedure to this point is described by M.F. Hawthorne and A.R. Pitochelli, *J. Am. Chem. Soc.* 81, 5519, 1959. The $B_{10}H_{10}^=$ anion is then chlorinated in aqueous solution and the $B_{10}H_{10}^=$ is transformed to $B_{10}Cl_{10}^=$ by the following procedure which differs from the one described by Muetteries et al (*Inorganic Chem.* 3, 159, 1964) in that higher concentrations in the order of 2.5×10^{-1} M and the sodium salt are used in place of the ammonium salt and lower concentrations in the reference in the order of 4.3×10^{-4} M.

About 6.45 grams (40 mmol) of $(NEt_3H)_2B_{10}H_{10}$ are dissolved in a solution of about 3.52 grams of NaOH (88 mmol, a 10% excess) in about 60 ml of water. The released triethylamine is removed first by using a separatory funnel, then by using the funnel and washing the solution with several 30 ml portions of benzene. The remaining solution is then adjusted to about pH 5 using acetic acid, and diluted to about 160 ml with water. Chlorine gas is then slowly introduced into the agitated solution. After about 12 to 15 hours, excess chlorine is removed by boiling from the resultant deep green solution. An aqueous solution of about 12.3 ml of triethylamine (88 mmol) in a slight excess of 6 M hydrochloric acid is then added to the reaction vessel. The slightly soluble salt $(NEt_3H)_2B_{10}Cl_{10}$ is precipitated and is then filtered and washed to remove sodium chloride, sodium acetate, hydrochloric acid and water.

The lithium salt is then prepared and purified by the following method: About 2.11 grams of lithium hydroxide (88 mmol) are dissolved in 50 ml of water. About 26.7 grams (40 mmol) of $(NEt_3H)_2B_{10}Cl_{10}$ are then introduced and dissolved. The released triethylamine is extracted with about three 50 ml portions of benzene. The aqueous solution is then acidified with concentrated hydrochloric acid while agitating. The remaining amine precipitates and is filtered to substantially remove it from the solution.

The solution is then made basic again with a slight excess of solid LiOH. About 4.2 grams (100 mmol) of lithium chloride are then dissolved in the basic solution and the solution is mixed with about 50 ml of tetrahydrofuran in a separatory funnel. The lighter phase then contains essentially all of the desired product, along with some water and lithium chloride. This phase is mixed with one gram of solid lithium chloride. The salt removes most of the water from the ethereal solution, forming a heavier aqueous phase which is removed.

The process is repeated until solid lithium chloride remains after mixing and standing. The lighter phase after separation from all solids and droplets of aqueous phase is then transferred to a 100 ml flask which includes a small fractionating column leaving an azeotropic mixture of THF and water with the dissolved desired product. Fractional distillation of the mixture results in the removal of the wet azeotrope of the THF and water at a temperature of 64°C, rendering the solution gradually drier.

Crystals of the product form as the liquid distilled from the flask is replaced with dry tetrahydrofuran. These are filtered and washed with dry tetrahydrofuran, then dried at about 160°C (available oven temperature) for about 48 hours. The resulting $Li_2B_{10}Cl_{10}$ crystals are checked for purity by infrared spectroscopy.

Example 2: The preparation of electrolyte including thionyl chloride soluble cathode material is as follows: A solution is prepared by dissolving about 3 g of $Li_2B_{10}Cl_{10}$ in about 12 ml distilled thionyl chloride, refluxing in a glass vessel with strips of freshly cut lithium for about 2 hours, then adding distilled thionyl chloride to a total volume of about 25 ml. The resulting mixture is deep violet in color and contains no precipitate.

The bistriethylammonium salts of substituted clovoborate anions are only slightly soluble in water and may be easily recovered by precipitation. A sample of this amine salt is prepared in water from the original $Li_2B_{10}Cl_{10}$ and another from a hydrolized aliquot of the solution $Li_2B_{10}Cl_{10}$ in thionyl chloride. The two samples produced substantially identical infrared spectra, demonstrating that no harmful substitution or alteration of the anion occurs during reflux in $SOCl_2$ with lithium metal present. After hydrolysis, the solution of $Li_2B_{10}Cl_{10}$ in water was colorless, suggesting that the color is produced by a complex formation between the anion and $SOCl_2$.

Example 3: Figure 3.6a is a cross-sectional view of the electrochemical test cell assembly used for testing the electrochemical generator of this process. The test cells having C and D cell size configuration are substantially alike with the C cells being described as follows and the D cells differing by having metal cases and glass seals instead of the Teflon casing and rubber seals used in the C size test cell.

The assembly as shown in Figure 3.6a consists of an aluminum cell holder **2** holding the cell elements consisting of Teflon cell body **3** and Teflon cover **4** enclosing and sealing via O-ring **6** of neoprene and helically wound electrode stack **5**. The electrode stack consists of cathode collector sheet **8** of nickel expanded metal mesh and lithium ribbon **9** separated from each other by separator mats **7**. The electrode stack is convolutely wound. The dimensions are selected so that the wound stack assembly has the approximate dimensions of a commercial C cell, i.e. 0.90 inch in diameter and 1.75 inches in height.

FIGURE 3.6: LITHIUM-THIONYL CHLORIDE SYSTEM

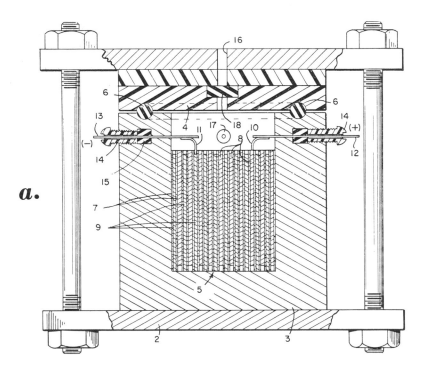

a.

(continued)

FIGURE 3.6: (continued)

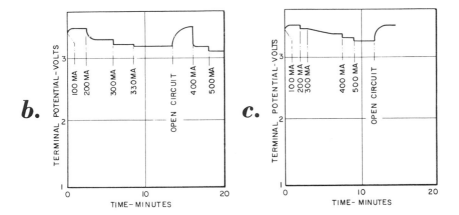

(a)   Test cell structure including a lithium anode, thionyl
      chloride electrolyte solvent/cathode depolarizer and
      clovoborate electrolyte salt system
(b)(c) Polarization data for the clovoborate electrolyte salt C
      size cells having different separators shortly after filling,
      on discharge at room temperature

Source: U.S. Patent 4,020,240

As the electrolyte and the anode are air-sensitive the cell and its elements are sealed from the external environment by the O-ring **6** and electrode septums **15** for cathode terminal **12** and anode terminal **13** respectively which reach the exterior of the cell through the insulating screws **14**. The terminals are connected via cathode lead **10** and anode lead **11** respectively to the cathode and anode elements **8** and **9**. The electrolyte is introduced into the cell via electrolyte fill port **16** through the self-sealing neoprene septum **18** and about 25 ml of the electrolyte are used to fill the cell.

The test cell is provided with a lithium reference electrode **17**, similar to the feed throughs **14**, to monitor electrode polarization. The anode **9** in the test cell contains about 6 ampere-hours of lithium. The open circuit potential of the Li/SOCl$_2$ cells of both the prior art and this process is about 3.62±0.05 volts.

When the electrolyte according to Example 2 is introduced into the above described cell having glass fiber separators between the anode **9** and the cathode collector **8** such cells show the polarization potentials under loads as in Figure 3.6b. When polypropylene separators are substituted for the glass fiber separators such cells exhibit the polarization potentials at the different loads as shown in Figure 3.6c.

Boron Trifluoride

A process described by *H. Lauck; U.S. Patent 3,915,743; October 28, 1975; assigned to Varta Batterie AG, Germany* relates to a battery with a negative lithium electrode, a positive sulfur electrode and an electrolyte of an organic solvent which contains boron trifluoride. In order to obtain high energy densities and high discharge voltages, it is generally necessary to make the free electrolyte space as small as possible and to maintain the distances between electrodes as short as possible.

In order to obtain high current yield, given the limited electrolyte volume available in the cell because of the desired high energy density, it has been found that the content of boron trifluoride concentration should be at least about 10%, but preferably at least 15 wt %. The relationship between the electrical output in a lithium sulfur cell and the ratio of sulfur to boron trifluoride is shown in the following table.

Weight Ratio of Sulfur:BF_3	Electrical Output (percent)
1:0.25	14.6
1:0.50	15.7
1:1.0	27.5
1:1.5	63.2
1:2.0	83.5

It has been shown that many solvents are attacked by boron trifluoride, causing changes which adversely affect the shelflife of the cell. The solvents of the process are not so attacked, even during extended cell storage, but remain stable in the presence of boron trifluoride. For instance, the dimethyl carbonate (DMC) and 1,2-dimethoxy ethane (DME) are stable in the presence of boron trifluoride. However, they form addition compounds with boron trifluoride whose stability is not so great as to prevent reaction of the boron trifluoride with the discharge product which is necessary to inhibit the formation of polysulfide during discharge of the sulfur electrode. This reaction is required to achieve nearly complete consumption of the sulfur during discharge.

The organic electrolyte is preferably composed of about 1 part by weight of the carbonate to 2 parts by weight of the ether, such as 1 part by weight of DMC and about 2 parts by weight of DME. However, the ratio of the carbonate to ether (DMC to DME) can be within the range of 1:0.5 to 1:3. It is preferred not to go beyond that range.

ANODE COMPONENTS

Silver Carbonate

A process described by *H. Lauck; U.S. Patent 4,016,338; April 5, 1977; assigned to Varta Batterie AG, Germany* relates to a galvanic element having a negative electrode of a strongly electropositive metal, a nonaqueous electrolyte and a positive electrode containing silver compounds. The process involves making the electrochemically reducible component of the positive electrode mass of silver carbonate. By using silver carbonate, a variety of significant advantages are obtained. Thus, a cell having a positive silver carbonate electrode and, e.g., a negative lithium electrode, has high discharge potential and high energy density relative to its volume. The silver carbonate electrode is capable of being heavily loaded and can be discharged at high current density. Because of the high discharge potential of the lithium/silver carbonate system, which is about double that of the conventional Leclanche cells, this primary element is suitable for assembly into batteries having high potentials which are interchangeable without difficulty with conventional batteries. For example, it is possible to build a 9-volt battery, for which six cells embodying the conventional Leclanche system had to be used, from only three cells. An important condition for good shelf-life of galvanic elements embodying the process is low solubility of the active, positive silver carbonate in the electrolyte.

Silver carbonate has low solubility in one molar solutions of lithium perchlorate in tetrahydrofuran, propylene carbonate, butyrolactone, dimethyl carbonate and dimethoxyethane. These electrolyte solutions also have adequate conductivity. Lithium perchlorate-containing mixtures of propylene carbonate or butyrolactone (about 40 to 80 percent by volume) with tetrahydrofuran, dimethyl carbonate or dimethoxyethane (about 20 to 60 percent by volume) have improved conductivity, but in some cases act as somewhat stronger solvents for silver carbonate. However, when such mixtures are used the yield increases. Particularly preferred are solutions of lithium perchlorate in a solvent mixture of propylene carbonate (about 70 to 85 percent by volume) and dimethyl carbonate (about 15 to 30 percent by volume).

FIGURE 3.7: DISCHARGE CURVE FOR SILVER CARBONATE-LITHIUM
CELL

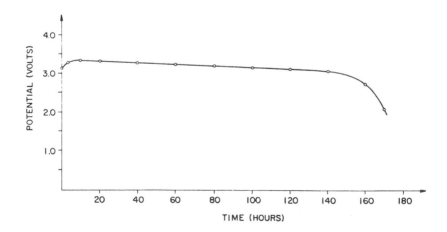

Source: U.S. Patent 4,016,338

To produce a positive electrode, silver carbonate is mixed with small quantities (about 1 to 6 wt %) of a conductive medium such as graphite or carbon black and if desired about 0.5 to 1.5 wt % of a binder such as polytetrafluoroethylene. This mass is compressed into an electrode tablet and inserted into a button cell container.

A mat-like material of about 1.5 mm thickness and made of glass or polypropyl-ene fibers may be used as the separator. The negative electrode is made by press-ing a lithium disc of suitable thickness into a piece of nickel expanded metal spot welded inside the lid of the cell container. The electrolyte is dripped onto the positive electrode and the separator. This electrolyte may be a one molar solution of lithium perchlorate in a mixture of about 80 percent by volume of propylene carbonate and about 20 percent by volume of dimethyl carbonate. The cell is then closed in the usual manner.

Cells of the process have open circuit potentials of 3.6 volts and average dis-charge potentials of 3.2 volts. A discharge curve for a cell loaded at 1 mA/cm^2 is shown in Figure 3.7. At the beginning of discharge the voltage is about 3.1 to 3.5 volts depending on the current load. During discharge about 95 wt % of the positive mass is used in the discharge process. The energy density of these cells exceeds 700 watt hours per liter.

High Purity Silver Chromate

A. Lecerf; U.S. Patent 4,032,624; June 28, 1977; assigned to Saft-Societe des Accumulateurs Fixes et de Traction, France describes a method for manu-facturing pure silver chromate in which a soluble chromate is made to react with a silver salt in a bichromate medium. The bichromate medium stabilizes the pH and makes it possible to operate in a concentrated medium. The pure product obtained may be used more particularly as a positive active substance in an elec-trochemical generator having a nonaqueous electrolyte, for it enables avoidance of losses in capacity during idle and storage periods of the generator.

Example 1: A solution of magnesium chromate and bichromate is prepared as follows: 3.35 mols of chromic oxide CrO_3 and 2.94 mols of magnesium oxide MgO are introduced into slightly less than one liter of water, which is agitated until the magnesium oxide is completely dissolved. Water is added to obtain a full liter of concentrated solution. This concentrated solution then contains 2.53 mols per liter of magnesium chromate $MgCrO_4$ and 0.41 mol per liter of magnesium bichromate $MgCr_2O_7$.

The use of an excess quantity of chromic oxide as compared with the stoichi-ometric amount permits rapid and complete dissolution of the magnesium oxide, and provides a concentrated solution, the magnesium chromate and bichromate contents of which can be calculated from the initial quantities. The presence of magnesium bichromate in the resultant concentrated solution has no effect on the remainder of the process. It would also be possible to use the stoichiometric quantities of chromic oxide and magnesium oxide in preparing the magnesium chromate and to filter the resultant solution to eliminate the undissolved mag-nesium oxide. This would provide a concentrated magnesium chromate solution virtually free of bichromate.

The concentrated magnesium chromate solution must be produced as just de-scribed as it is not available commercially. The magnesium chromate solution is used as follows for precipitating silver chromate. One hundred cc of a con-centrated solution of 0.26 M magnesium bichromate is brought to the boiling point. This concentrated solution has a pH of 4.5. There are progressively and simultaneously added in equivalent quantities to the solution, as by means of calibrated capillary tubes, 80 cc of the concentrated 2.53 M magnesium chromate solution derived as above described and containing 0.202 mol of chromate, and 112 cc of concentrated silver nitrate solution, containing 0.410 mol of silver. A precipitation of silver chromate occurs. The silver chromate precipitate thus obtained is washed several times in boiling distilled water and then dried for 24 hours at 150°C.

Example 2: Step 1 — Silver bichromate is prepared as follows: A solution of 0.250 mol of sodium bichromate and 0.125 mol of chromic oxide in 200 cc of water is brought to boiling point. Then 100 cc of a 5 M solution of silver nitrate is added to the boiling solution. As a result, a precipitate of silver bichromate is obtained which is filtered and centrifuged. The weight of the damp silver bichromate precipitate is 108.6 g.

Step 2 — Then silver chromate is precipitated as follows: The silver bichromate prepared as above described is placed in suspension in 800 cc of water, which is brought to boiling point. Then, 100 cc of the concentrated 2.53 M magnesium chromate solution derived as in Example 1 are added gradually thereto. The precipitate of silver chromate resulting is washed several times in boiling distilled water and dried for 24 hours at about 150°C as in Example 1. The dry weight of the silver chromate produced is 79.5 g, representing 0.240 mol of silver chromate.

The method used in Example 2 has an advantage over that of Example 1, in that in Example 1 it is necessary to carefully synchronize the addition of magnesium chromate and silver nitrate to the boiling 0.26 M magnesium bichromate solution to avoid an excess of magnesium chromate that would lead to an increase in the pH and resultant precipitation of contaminating silver oxide. Such synchronization in an industrial process requires complex measures, such as the use of dosing pumps, which are costly and subject to failure. No such requirement exists in the method of Example 2, which consists of but the two successive steps.

Vinyl Polymer Anode Coating

T. Kalnoki-Kis; U.S. Patent 3,993,501; November 23, 1976; assigned to Union Carbide Corporation describes a nonaqeuous cell comprising an active metal anode, such as a lithium, sodium, potassium or aluminum anode, a liquid cathode-electrolyte comprising a solute dissolved in a solvent which is an oxyhalide of an element of Group V or Group VI of the Periodic Table. The surface of the anode contacting the cathode-electrolyte is coated with a vinyl polymer film. The active reducible oxyhalide electrolyte solvent performs the dual function of acting as solvent for the electrolyte salt and as an active cathode depolarizer of the cell.

Although the active reducible liquid oxyhalides inhibit the direct reaction of active metal anode surfaces sufficiently to permit them to act as both the cathode material and as the electrolyte carrier for nonaqueous cells, they do cause formation of a surface film on the active metal anode during cell storage particularly at elevated temperatures, which consists of a rather heavy layer of crystalline material. This crystalline layer appears to cause passivation of the anode which results in voltage delay on initial discharge along with high cell impedance values in the range of 11 to 15 ohms for a standard C-size cell.

The extent of anode passivation can be measured by observing the time required for the closed circuit voltage of the stored cell to reach its intended voltage level after discharge has begun. If this delay exceeds 1 second, the anode passivation is considered excessive. What has been observed, e.g., in lithium-oxyhalide cell systems is that after a load is applied across the terminals of the cell, the cell voltage immediately drops below the intended discharge level, then increases at a rate depending on the extent of lithium passivation and the current density of discharge.

To substantially prevent this anode passivation which occurs in active metal-oxyhalide cathode-electrolyte cell systems during discharge, the surface of the active metal can be coated with a vinyl polymer film which will adhere to the metal, remain stable and not dissolve in the liquid cathode-electrolyte, and which will not effectively decrease the capacity of the cell during cell storage and discharge, and in some cases will even increase the cell capacity on discharge.

Example 1: Several specimens of pure lithium foil obtained commercially from the Foote Mineral Co. measuring 1 inch (2.54 cm) by 1.5 inches (3.8 cm) were coated with various concentrations of vinyl chloride-vinyl acetate polymer (obtained commercially from Union Carbide as VYHH consisting of 86% vinyl chloride and 14% vinyl acetate with an average molecular weight of about 40,000) in a liquid suspending medium (solvent) of thionyl chloride ($SOCl_2$). Each lithium sample was immersed in the coating solution for about 1 minute and when withdrawn the solvent was evaporated. The thickness of the polymer layer formed on each of the lithium samples was measured and is shown in the following table.

Concentration of VYHH (weight percent)*	Thickness of VYHH Layer on the Lithium Sample (centimeters)
1	0.00012
3	0.00015
6	0.00018

*Weight percent as based on weight of solvent

Example 2: Several lithium samples with and without a vinyl polymer film as prepared in Example 1 were immersed in a liquid cathode-electrolyte consisting of 1 M $LiAlCl_4$ in SO_2Cl_2 for various time periods and under various temperature conditions. The aged lithium samples were then placed under a scanning electron microscope and examined for crystalline formation. The results showed that the uncoated lithium foil had a surface crystalline layer, the crystals of which varied in size and shape depending on the time period and temperature conditions of the lithium exposure in the liquid cathode-electrolyte. Specifically, as the temperature increased, the amount of crystalline material formed increased as did the size of the crystals.

Contrary to the above, the vinyl polymer coated lithium samples after being aged in SO_2Cl_2 containing 1 M $LiAlCl_4$ for 7 days at temperature varying between 45° and 71°C showed no effective crystalline formation on the surface of the vinyl polymer film. This test demonstratively showed that coating the surface of a lithium surface with a vinyl polymer will prevent the formation of a heavy crystalline deposit which has been associated with lithium passivation in a liquid-cathode-electrolyte such as 1 M $LiAlCl_4$ in SO_2Cl_2.

CATHODES

Silver Permanganates and Periodates

A process described by *A.N. Dey and R.W. Holmes; U.S. Patent 3,904,432; September 9, 1975; assigned to P.R. Mallory & Co., Inc.* relates to high-energy-density electrochemical energy generators and more particularly to such generating cells and batteries having organic electrolytes and utilizing as cathodic material the permanganates and periodates of certain metals. The cathodic material or depolarizers used in the process are the heavy metal salts of the oxy acids of manganese and iodine.

The permanganates and periodates (MnO_4^{-5}) or (IO_6^{-5}) of copper, silver, iron, cobalt, nickel, mercury, thallium, lead, and bismuth and mixtures thereof are particularly suitable as the active cathodic or depolarizer materials for cells to be used in conjunction with active light metal anodes. Silver permanganate, $AgMnO_4$ and silver periodate, Ag_5IO_6 are preferred.

Suitable electrolytes may be made by dissolving organic or inorganic salts of light metals in the organic solvents. For example, 1 to 2 M solutions of lithium perchlorate or lithium aluminum chloride dissolved in tetrahydrofuran solvent constitute a suitable organic electrolyte. Other light metal salts are based upon the cations of the light metals such as lithium, sodium, beryllium, calcium, magnesium or aluminum with anions such as perchlorate, tetrachloroaluminate, tetrafluoroborate, chloride, hexafluorophosphate, hexafluoroarsenates, etc. dissolved in organic solvents. Among suitable organic solvents are tetrahydrofuran, propylene carbonate, dimethyl sulfide, dimethyl sulfoxide, N-nitrosodimethylamine, gamma-butyrolactone, dimethyl carbonate, methyl formate, butyl formate, acetonitrile and N,N-dimethylformamide.

Light metals suitable for the anode are selected from among lithium, sodium, potassium, beryllium, calcium, magnesium, and aluminum. In some cases, it is desirable to amalgamate the surface of the anode metal with mercury as the performance of certain light metal anodes for such cells can be considerably improved by such amalgamation. By surface alloying these metals with mercury by chemical displacement of the latter from solutions of a mercury salt, a notable improvement of the anodes is provided. The improvement is particularly striking with anodes formed of aluminum, magnesium, or their alloys.

Silver Chromate and Silver Phosphate

N. Margalit; U.S. Patent 3,981,748; September 21, 1976; assigned to ESB Inc. describes a nonaqueous, primary battery having an electrolyte solution consisting essentially of an organic solvent containing an electrolyte salt, a separator, a light metal anode and a blended cathode mix. The cathode mix comprises a mixture of silver chromate and a metallic phosphate having a discharge potential in the electrolyte solution higher than silver chromate. The battery preferably has a lithium anode and a nonaqueous electrolyte consisting essentially of methyl formate solvent containing lithium hexafluoroarsenate ($LiAsF_6$) which provides a single voltage plateau for the discharge of silver. Silver phosphate is the preferred metallic phosphate, having a discharge potential in the electrolyte solution higher than silver chromate, for the silver cation contributes to the capacity of the cell. The electrochemical system is particularly useful in small primary batteries commonly referred to as "button cells" due to its high volumetric energy density, retention of cell dimensions, and improvement of the rate of discharge characteristic of the silver chromate.

Tungsten Trioxide

M. Eisenberg; U.S. Patent 3,915,740; October 28, 1975; assigned to Electrochimica Corporation describes a cathode having an active material comprising a majority by weight of at least one compound of a Group VI-b metal of the Periodic Table of Elements selected from the group of compounds consisting of tungsten oxides, tungsten sulfides, sodium tungsten bronzes, polytungstates, polymolybdates and molybdenum sulfides. The cell also comprises an anode composed of a majority by weight of at least one light weight metal selected from the group consisting of lithium, sodium, calcium, magnesium and aluminum as an active material. The cell further includes an electrolyte comprising a solvent in liquid state at standard room temperature and a solute dissolved therein for electrolytic conduction between the cathode and anode.

In fabricating the tungsten and molybdenum compound cathodes described in the following examples the compounds were mixed in powdered form with graphite as a conductive diluent in a 75 to 25 wt % ratio. To this a 3 wt % addition of a binder was added such as polytetrafluoroethylene or Teflon. A sufficient amount of an organic solvent, such as isopropanol or methyl ethyl ketone was then added to the mix to form a thick paste. The mix then was pasted to an expanded nickel current collector, dried, and subsequently pressed between flat platens at pressure of from 5 to 10 tons psi. The cathode was then cured at a temperature of 300°C for one hour to enhance its mechanical integrity. In this method of cathode fabrication the weight ratio of the tungsten or molybdenum compounds to graphite can be varied from 20:1 to 1:1. The preferred ratio is 7.5:2.5.

For cell testing a current lead in each example was spot welded to a corner of a nickel current collector of a one by one inch cathode wrapped in a piece of separator, preferably nonwoven polypropylene. Two lithium anodes were prepared by pressing two rectangular pieces of a one inch square lithium metal of 0.02 inch thickness on expanded nickel metal. The electrode pack was then placed in an epoxy rectangular case with the current leads exiting the case top. The top was sealed by means of a fast curing epoxy. Following seal cure a small hole was drilled through the epoxy and the previously described electrolyte poured into the cell in an argon dry environment. Following seal cure the cell was discharged at a constant current of 5 mA.

Example 1: A cathode was prepared comprising 75% tungsten trioxide and 25% graphite. To this was added 2 wt % polyethylene powder Microthene. A small amount of xylene was added to enable thorough mixing and pasting. After the electrode was pressed and dried the rectangular cell was assembled as described above. The open circuit voltage measured 3.5 volts. When a current of 5 mA was applied the initial closed circuit voltage was 2.5 volts. The cell discharged for 27 hours to an end voltage of 2.0 volts. A duplicate cell was then discharged at a current of 10 mA. The initial closed circuit voltage for this cell was 2.7 volts and the cell operated for 11.2 hours to a cutoff voltage of 2.0 volts.

Example 2: A cathode mix was prepared using 70 wt % sodium tungstenate, 20% graphite and 10% acetylene black. 3.5% Microthene polyethylene binder dissolved in xylene was then added. After pasting, pressing and drying the electrodes, the cell was assembled as previously described. These cells showed open circuit voltages of 3.55 volts. When a load was applied to pass a current of 6 mA the initial closed circuit voltage was 2.75 volts. The cells operated for some 23 hours to a voltage cutoff point of 2.0 volts.

Sulfur

According to a process described by *H. Lauck; U.S. Patent 3,907,591; Sept. 23, 1975; assigned to Varta AG, Germany* a positive sulfur electrode is formed of amorphous, insoluble sulfur with an additive of conductive material. Commercial sulfur powder is heated for a short period to 150°, to 200°C and the resulting melt is then immediately poured into cooled water. The consequently solidified sulfur is then pulverized and dried in vacuum at ambient or slightly increased temperature. After completion of the drying operation, the sulfur powder is subjected to extraction by carbon disulfide until the weight of the insoluble sulfur no longer decreases.

The adhering carbon disulfide is removed in vacuum at ambient temperature, and the insoluble sulfur powder is mixed with a small amount of a stabilizer about 0.1 to 1 wt % referring to the total weight of the positive mass. Afterwards pulverized graphite is blended therewith, so that the ratio by weight of sulfur and graphite at maximum is 1:1. It is possible to substitute metal powder as conductive material for part of the graphite. After the admixture of a small amount of polyisobutylene (about 1 wt %) dissolved in ligroin and an addition of flakes of cellulose (about 1.5 wt %) the mass is intimately and thoroughly kneaded and then coated on a support or carrier, (0.1 g/cm²) such as expanded nickel metal or nickel metal mesh with conductors, for example. After pasting of the electrode, the material is again dried in vacuum. The dried electrode mass is thereafter pressed onto the support with a pressure of from 400 to 600 kg/cm².

To produce negative electrodes, light metal such as sodium, calcium, beryllium, magnesium, or aluminum is used; preferably, a layer of lithium having a thickness of about 0.5 mm is pressed onto a nickel mesh with conductors. For example, three of the above positive electrodes were placed into pockets formed of polypropylene fleece and united with two negative electrodes into a single package wherein the positive electrodes are separated from one another by the negative ones.

The electrode package was slipped into a rectangular casing formed of polystyrene. A 0.5 molar solution of boron trifluoride in propylene carbonate served as electrolyte. After having been filled with the electrolyte, the casing was airtightly sealed by a cover of polystyrene. The conductors were passed to the outside through two boreholes formed in the cover. The boreholes in the cover were then airtightly sealed with a resin.

Gel Composition

G. Feuillade, B. Chenaux and P. Perche; U.S. Patent 3,985,574; October 12, 1976; assigned to Compagnie Generale d'Electricite, France describe electrochemical elements comprising a stack of thin layers, successively a copper layer, a catholyte, a separator, a lithium layer; the catholyte and the separator being formed by a gel comprising formal or polyvinylic butyral, reticulated or otherwise and impregnated with a solvent such as propylene carbonate or N-methyl-pyrrolidone saturated with an ionically conductive salt such as ammonium perchlorate. A variant consists in forming the separator with a gel of reticulated polymer basically containing polyvinylidene fluoride and a mineral charge such as magnesia.

The catholyte is formed by a cupric sulfide powder CuS agglomerated by a gel consisting of a resin such as polyvinylic formal (PVF) and of a solvent such as propylene carbonate (PC) saturated with a conductive salt such as ammonium perchlorate (ClO_4Am). Such a gel is prepared by mixing the various ingredients mentioned above in the following proportion by weight:

$$\frac{5}{100} < \frac{PVF}{PC} < \frac{15}{100}, \qquad \frac{50}{50} < \frac{CuS}{PVF+PC+ClO_4Am} < \frac{90}{10}$$

Preferably, these values are 10/100 and 70/30 respectively. Such a gel, which has a weak consistency, is pasted onto a substrate consisting of a silver or tin-plated or silver-plated silver or copper foil on which a grid forming a honeycomb support is welded by thermocompression. The gel is therefore pasted on so as to pack the alveoli of the grid, then the excess gel overflowing from the grid is scraped off. It must be understood that the thickness of the grid and the dimensions of the alveoli are predetermined as a function of the required capacity of the element. It is also possible to produce the catholyte as set forth above, but replacing the PVF by polyvinylic butyral (PVB) and replacing the PC by N-methyl-pyrrolidone (NMP). Such a gel could possibly be nonreticulated or reticulated in the manner which will be described in the case of the separator.

It should be mentioned that the separator is formed by a polyvinylic acetal, more particularly PVF or PVB and a solvent saturated with a conductive salt such as ClO_4Am. Jointly with the PVF, the solvent generally used is PC. Jointly with the PVB, the solvent generally used is NMP. A gel which, according to the relative proportions of acetal and of solvent, has variable viscosities, is thus obtained. Such a gel could be either reticulated or nonreticulated and, in this latter case, the reticulation agent could be either directly incorporated with the ingredients forming the gel or incorporated with the gel after the forming thereof using the ingredients.

Example 1: PVF and PC are mixed in proportions by weight such that:

$$\frac{5}{100} < \frac{PVF}{PC} < \frac{15}{100}$$

and preferably equal to 10/100 the PC being saturated with ClO_4Am. A soft gel which is not very resistant, is run after a slight heating on to a porous support such as asbestos so as to impregnate it. The separator thus produced has excellent electrical conductivity and promotes good discharging of the element. Applications which are particularly advantageous are found for it in single cell batteries. It should be observed that it is possible to use, instead of ClO_4Am other salts, such as lithium perchlorate, potassium iodide or sodium fluoroborate. This is also true in the case of the catholyte.

Example 2: PVF and PC are mixed in proportions by weight such that:

$$\frac{20}{100} < \frac{PVF}{PC} < \frac{40}{100}$$

Preferably PVF/PC is 30/100. The PC is saturated with ClO_4Am. A gel having the consistency of India rubber is put into form either by hot casting on a support plate to form a thin film or by casting or pressing on an insulating woven fabric such as nylon. The separator obtained has a 70° to 120°C MP. A variant consists in substituting 50% of the PVF by bentone, an organophilic bentonite whose alkaline cations have been substituted by quaternary ammonium. A hard gel of excellent conductivity and resistance to element passivation during discharge is obtained.

GENERAL PROCESSES

Compact Battery

R.D. Alaburda; U.S. Patent 3,920,477; November 18, 1975; assigned to E.I. du Pont de Nemours and Company describes a thin, flat battery containing at least two galvanic cells, the cells being disposed in monoplanar configuration and the battery comprising at least one flat conductive plate which contacts the anode of one cell and the cathode of an adjacent cell.

In Figure 3.8a, second plate current collector **1** provides the negative pole of the battery, while another second plate current collector **2** provides the positive pole of the battery. **3** represents a space between **1** and **2** through which the first plate collector on the other side of the battery shows. In Figure 3.8b, **4** is the first plate current collector which is on the bottom side of the battery in Figure 3.8a, while nonconductive gasket **5** seals the edges of the conductive plate's sidewalls around the individual cells and **6** is the edge of the second plate current collector **1**. In Figure 3.8c, **4** again represents the first plate current collector at the bottom of the battery. Figure 3.8d shows the component piece of the battery of Figure 3.8a.

Element **11** is the first plate current collector with cathode active material **12** deposited directly thereon; **13** represents a nonconductive plastic gasket with windows **14** and **15** cut out therefrom; **16** is an anode of lithium on steel, lithium side up and fitting in cut out window **15**; **17a** and **17b** are nonconductive porous separators, **17a** placed over anode **16** and **17b** placed in cut out window **14**. The anode and separators are sized and shaped so as to fit snugly into cut out windows **14** and **15**.

In regular assembly of the batteries, when the two separators are in place, they are saturated with electrolyte solution, then assembly is continued. **18** is a second anode like **16** and is placed on separator **17b**, lithium side down; **20** is a nonconductive plastic gasket duplicating **13** and placed congruently over the already assembled parts of the battery, with cut out windows **19a** and **19b** framing porous separator **17a** and anode **18**; **21a** is a second plate current collector with cathode active material **22** coated thereon, and **21b** is an uncoated second plate current collector. **21a** is placed congruently on separator **17a**, with the side bearing the cathode-active material facing down within cut out window **19a**. **21b** is placed on top of anode **18** within cut out window **19b**.

The battery alternatively can be saturated with electrolyte by placing it in a container from which the atmosphere is then removed with a vacuum pump.

FIGURE 3.8: COMPACT BATTERY

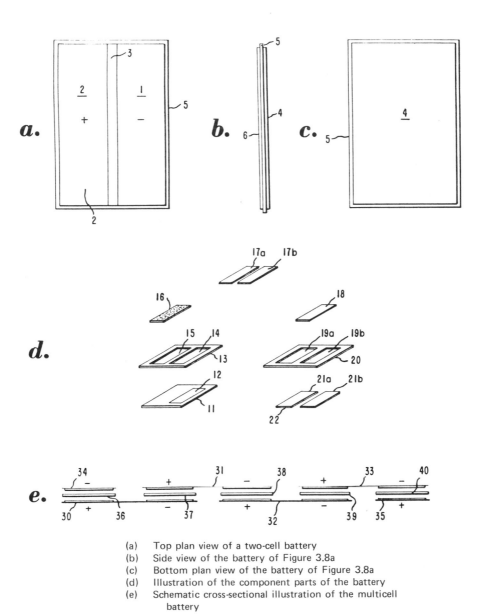

    (a)   Top plan view of a two-cell battery
    (b)   Side view of the battery of Figure 3.8a
    (c)   Bottom plan view of the battery of Figure 3.8a
    (d)   Illustration of the component parts of the battery
    (e)   Schematic cross-sectional illustration of the multicell
           battery

Source: U.S. Patent 3,920,477

Liquid electrolyte is then allowed to enter the container, suffusing the battery with electrolyte. The vacuum is released, and the battery removed for heat sealing. The seals are made by applying heat to the edges of the metal pieces, thereby sealing the gaskets to each other and to the edges of the three metal plates. Optionally the metal plates can be sealed to the gaskets in the beginning before further assembly. In this method only the two gaskets are sealed together after the electrolyte solution is added.

Figure 3.8e illustrates a five-cell battery of the process. In that figure, elements **30,31,32** and **33** represent conductive plates carrying anode material on one area of the inner side of the plate and cathode material on a second area of the inner side of the plate as indicated by the plus and minus signs. Second plate current collectors **34** and **35** are about half the size of first plate current collectors **30,31,32** and **33**. Plate **35** bears active cathode material and forms the cathode contact for the battery. Plate **34** bears active anode material and forms the anode contact for the battery. Separators **36,37,38,39** and **40** are between anodic and cathodic material. Sealing gaskets are not shown, but are applied in the same manner as for two-celled batteries.

The most preferred electrolyte composition for the cells of the process is: 58 parts 1,3-dioxolane, 25 parts 1,2-dimethoxyethane, 16.5 parts $LiClO_4$, and 0.5 part 3,5-dimethylisoxazole. Another particularly satisfactory electrolyte composition is lithium perchlorate in tetrahydrofuran.

Example: Two second plate current collectors are prepared from tin-coated low carbon steel foil plates 1.5 mils thick and 3.67 inches long by 1.67 inches wide. The two plates are cleaned to remove any oil or particulate matter. One of the two plates is coated over a rectangular area of 3.44 inches by 1.42 inches on one side with a deposit of blue tungsten oxide. A complete description of this material and the manner in which it can be adherently deposited on the steel plate are found in U.S. Patent 3,877,983 particularly in Example 2. The tungsten oxide coated plate forms the cathode of one cell of the battery of the process.

The cathode preparation was carried out as follows: a solution was prepared of 0.4 g of sodium carboxymethylcellulose in 10 ml of water, and to it were added 0.5 g of finely divided conductive carbon and 10.0 g of commercially available blue tungsten oxide of formula $WO_{2.7}$. The well-mixed, uniform suspension was poured into the 3.44 inch by 1.42 inch area of the one steel foil plate. A Gardner knife was drawn over it to obtain a smooth coating, and any excess removed from outside the 3.44 inch by 1.42 inch rectangular area. The coated plate was dried in an oven at 60°C for several hours.

One rectangular plate of the same material and thickness of those above is prepared as a first place current collector, this plate having an area of 3.40 inches by 3.67 inches. In the same manner as described above, cathodically active blue tungsten oxide is coated on a rectangular area 3.44 inches by 1.42 inches on one side of this plate. This plate thus contains a cathode for one call of a battery of the process. It will also contact the active lithium anode of the second cell, as shown in Figure 3.8d.

Two nonconductive, rectangular plastic gaskets or spacers of 7.5 mils thickness each 3.77 inches by 3.48 inches are cut out as shown in Figure 3.8d, the cut out holes being each 3.46 inches by 1.47 inches. The gasket material is a copolymer of ethylene and methacrylic acid, commercially available. The gaskets are

sealed to the metal plates by holding gasket and plate together in desired relationship and applying a metal block at 160° to 165°C to the metal for about 5 seconds under gentle pressure. Two rectangular anode plates of 2 mil carbon steel
foil each 3.44 inches by 1.42 inches are cleaned of foreign material. Each plate
is coated on one side with a 1 mil layer of lithium metal by drawing the plate
through a bath of the molten metal, and cooling the composite quickly. Lithium adhering to the second side of the plate is wiped off as the steel plate leaves
the molten metal. The two lithium coated steel plates form the anodes for the
two-cell battery.

Two rectangular nonconductive porous separators each 7 mils thick and 3.46
inches by 1.44 inches in area are prepared from ceramic fiber cloth. The separators are slightly larger than the anode plates so as to minimize any possibility
of cell short circuiting. An electrolyte solution was prepared to contain (by
weight) 58 parts of 1,3-dioxolane, 25 parts of 1,2-dimethoxyethane, 16.5 parts
of lithium perchlorate and 0.5 part of 3,5-dimethylisoxazole. The battery is assembled as follows:

 (1) The two second plate current collectors, one bearing tungsten
   oxide cathodic material, are heat sealed in symmetric fashion as shown in Figure 3.8a to one of the plastic gaskets,
   the cathode material facing the plastic.
 (2) The single large first current collector is symmetrically aligned
   with and heat sealed to the second plastic gasket, the side
   bearing the active cathodic material also facing the plastic.
   With the metal plate-plastic gasket from (2) forming a base,
   metal side down, the battery was built up as follows:
 (3) One anode, lithium side up, is placed in one cavity of the gasket, the uncoated side of the anode contacting the uncoated
   metal side of the first plate current collector.
 (4) One of the two porous separators is then placed over the
   anode as assembled in step (3) and the second separator is
   placed in the second, parallel cavity in the gasket.
 (5) About 0.9 g of electrolyte solution is then placed on and imbibed by each of the two porous separators.
 (6) The second anode is then congruently placed, lithium side
   down and facing the cathode material on the bottom plate
   and separated therefrom by the separator.
 (7) The two metal plate-plastic gasket composites prepared in
   step (1) are then placed symmetrically over the assembly,
   with the plate bearing the active cathode material facing
   the lithium coating on the anode plate from step (3).
 (8) The entire assembly is placed in a vacuum chamber, lightly
   clamped to prevent shifting of the various parts. The container is evacuated to remove air from the battery employing
   a vacuum pump. Then the battery is sealed, the gasket assembly of step (1) to gasket assembly of step (2) by applying heat to the two metal sides at the outline edge and to
   the slightly protruding gasket. The vacuum is released, the
   clamping mechanism removed, and the battery is ready for
   operation.

Batteries prepared in this manner furnish an open circuit voltage of 5.2 to 5.5

volts. They furnish about 4.2 to 4.4 volts when discharged under a 3.3 ohm load pulsewise, i.e., 1.1 seconds under load then a 3.0 second rest period before the next discharge period. Single cells of similar shape and size to the two-cell battery furnish energy at only 2.1 to 2.2 volts under the same discharge conditions.

Miniature Concentric Design

A.N. Dey; U.S. Patents 4,035,909; July 19, 1977; 3,945,846; March 23, 1976; and 4,028,138; June 7, 1977; all assigned to P.R. Mallory & Co., Inc. describes a primary electric cell employing a lithium/sulfur dioxide system, which can be made within the small dimensions defined by a diameter of 0.1 inch and a height or length of 0.75 inch. In accordance with the process, the cell is constructed with a container casing formed from thin hollow tubing, on whose inner surface a porous layer of carbon is formed as a cathode. A small elongated cylindrical anode of lithium metal is formed on a linear metallic collector, as a solid pin support for the lithium, for concentric coaxial disposition within the cathode.

The anode thus formed is enwrapped in a layer of thin insulating separator material, such as a sheet of porous paper or a sheet of microporous polypropylene, and this assembly, of the anode and the separator wrap, is then axially moved into the cylindrical central axial space within the surrounding cylindrical cathode layer. The lithium/sulfur dioxide electrolyte system of the cell consists of the lithium anode, the carbon cathode, and an electrolyte mixture of liquid sulfur dioxide, lithium bromide, acetonitrile and propylene carbonate.

During the preparation of the lithium anode for such disposition, the outer extending end of the anode collector pin is anchored in a rubber septum, which is essentially a rubber bead on the anode pin; and the rubber septum is sized so it will slidingly fit internally into the outer open end of a container can, as the entire anode assembly is axially inserted into operating position in the can. A hypodermic needle has been previously inserted in and through the rubber septum together with the anode collector pin, before sealing the assembly in the can, so that the hypodermic needle will be available as a conduit to conduct the electrolyte into the can after the rubber septum is sealed in place at the originally open end of the can, with the anode structure supported in proper position in the can.

The hypodermic needle is used to introduce the electrolyte under sufficient pressure to put the desired quantity of the electrolyte into the cell container which has been previously dry sealed. After that filling operation, the hypodermic needle is gradually withdrawn, so the rubber septum will reclose the opening behind the needle, as the needle is withdrawn, and will thus maintain the seal to prevent leakage of the electrolyte.

In another modification, the hypodermic needle serves both as the filling conduit and as the anode collector pin for supporting the body of the lithium anode material, and is permanently disposed and sealed in the rubber septum. The needle is utilized as a filling port to introduce the electrolyte to the amount wanted; and then the input end of the hypodermic needle, outside of the cell, is closed off and welded closed at its outer end to complete the seal for the cell.

Thus, the general features of this process involve the construction and the filling operation of a miniature cylindrical cell having a diameter of 0.1 inch and a length of 0.75 inch. The operating parameters of the cell will be understood from the energy output requirements which were set at 24 mWh, at 2 to 3 volts, for currents of 2 to 3 mA. The cell employs a lithium/sulfur dioxide (Li/SO$_2$) organic electrolyte system, consisting of a lithium anode, a carbon mix cathode, and an electrolyte mixture of liquid sulfur dioxide (SO$_2$), lithium bromide (LiBr), acetonitrile and propylene carbonate.

Two modifications of cell design are illustrated. The first modification **10** of the cell is shown in Figure 3.9a, and consists of a solid pin collector **12** for a central concentric axially disposed lithium anode **14**, a concentric outer carbon cathode **16** supported on the inner wall surface of an enclosing cylindrical aluminum can **18**, with the lithium anode **14** and the carbon cathode **16** separated by an insulating separator **18**. The electrolyte in liquid phase is introduced into the cell by injection through a hypodermic needle **20**, that extends through a rubber septum **22**, that is used to dry seal the can before the electrolyte is introduced. The needle is removed after the electrolyte is introduced into the cell.

FIGURE 3.9: MINIATURE CONCENTRIC BATTERY

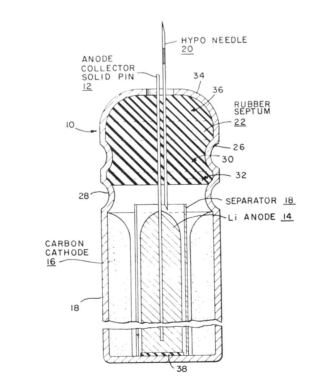

(continued)

FIGURE 3.9: (continued)

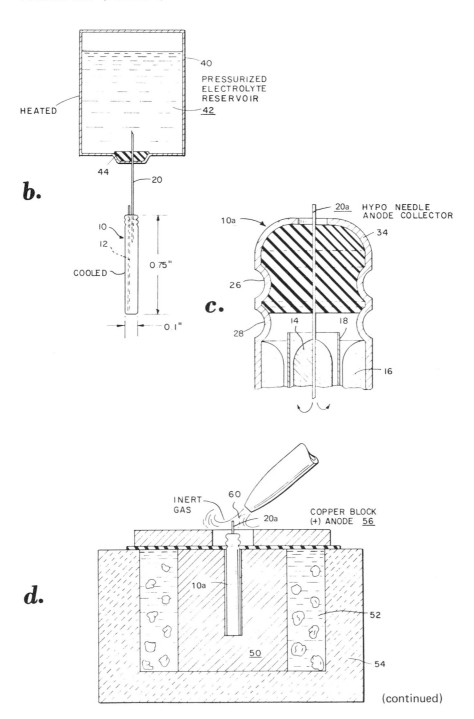

(continued)

FIGURE 3.9: (continued)

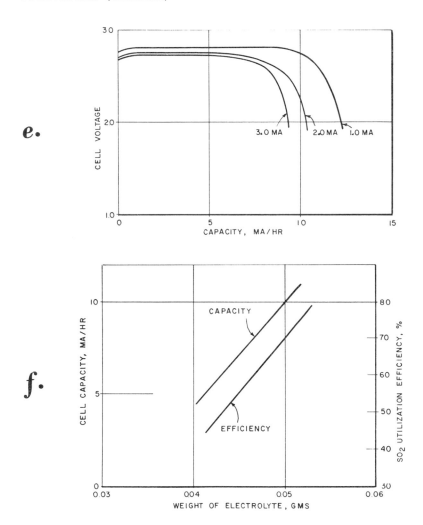

(a)   Vertical sectional view of one modification of the cell
(b)   Schematic diagram illustrating the electrolyte transfer
       system employed for filling the cell
(c)   Vertical cross-sectional view of a second modification
(d)   Schematic diagram of the apparatus and a welding sys-
       tem for completing the sealing of the cell
(e)   Graph showing several illustrative discharge curves of
       several of the cells
(f)   Graph showing the realized capacity and the efficiency
       of the cells as a function of electrolyte weight for
       a continuous discharge of two milliamperes

Source: U.S. Patent 4,035,909

Operational tests on the cells of the construction here shown indicate that the stoichiometric capacity is determined by the weight of the introduced electrolyte which contains the sulfur dioxide to serve as a depolarizer. The sulfur dioxide depolarizer is introduced into the cell which has been previously sealed closed. As shown in Figure 3.9a, the rubber septum 22 is compressed between two inwardly directed beads 26 and 28, on the can 18, which serve to radially tightly compress the rubber septum 22 in the region 30, and serve to crimp and clamp the rubber septum tightly around a circular peripheral region 32.

In addition, the upper initially open end of the aluminum container can 18 is tightly peened over radially inward at the region 34 to compress the upper portion 36 of the rubber septum 22 to impress a downward clamping pressure, by the peened-over section 34, onto the upper surface of the rubber septum 22. A Teflon disc 38 insulates the bottom of anode 14 from the metal can 18. The filling operation is performed outside the dry box. When the filling operation is accomplished, the hypodermic needle 20 is slowly and gradually withdrawn from the septum 22, in order that the rubber septum may reclose the opening which was temporarily filled by the hypodermic needle, and thus maintains the sealing action of the rubber septum.

The filling operation may be better explained upon considering the system shown in Figure 3.9b. As shown there, schematically, a closed container reservoir 40, which contains the liquified electrolyte 42, is provided with a rubber septum 44 through which the upper end of the hypodermic needle 20 of the cell of Figure 3.9a extends into the reservoir 40. Vapor pressure in the reservoir provides the pressure to force the electrolyte through the hypodermic needle into cell 10 to be filled. To simplify the filling operation, and to permit filling with maximum quantity, as desired, the electrolyte in reservoir 40 is heated and the cell 10 is kept cooled to a temperature from about -10° to -78°C.

The manner in which the cell is kept cooled during the filling operation is schematically indicated in the schematic diagram in Figure 3.9d. As there shown, the cell 10a body is closely thermally coupled in a metal block 50, of highly heat-conductive metal, such as copper, and the block 50, in turn, is immersed in a cold bath 52, of dry ice and acetone, to bring the temperature of the block and the cell 10a down to at least -10°C. The freezing bath 52 is contained in a suitable insulated tank 54. When the cell has been filled, the final sealing operation is then performed, by withdrawal of the hypodermic needle 20a, from the cell of the first modification, of Figure 3.9a.

However, with a relatively shallow rubber septum to serve as a seal, in that first modification, the vapor pressure that is generated within the cell, as some of the fluid electrolyte vaporizes and releases some of the sulfur dioxide, is sufficient to cause some of the sulfur dioxide to seep through the reclosed opening in the septum from which the hypodermic needle has been withdrawn. Because of that sealing problem, in that first modification, the sealing structure is modified, according to the second modification, as shown in Figure 3.9c, to permit the hypodermic needle to remain permanently embedded and sealed in the sealing septum.

As shown in Figure 3.9c, in cell 10a, the hypodermic needle here identified as 20a, is employed to serve, also ultimately, as the anode collector, to support the lithium anode material 14. The other elements of construction in the cell

10a are otherwise the same as in the cell **10** in Figure 3.9a, so that the change here has been to utilize the hypodermic needle in Figure 3.9c to serve also as the anode collector, after the hypodermic needle has served to introduce the electrolyte into the cell **10a**. In this structure, in Figure 3.9c, the fluid introduced into the combination hypodermic needle and anode collector, travels down through the needle as a tube and exits from the bottom of the tube into the operating space between the lithium anode and the cathode until the desired amount of electrolyte is introduced into the cell. After such filling operation, the combination needle and anode collector is separated from any filling source and closed above the septum seal, and sealed as shown in the operation in Figure 3.9d.

As shown in Figure 3.9d, the cell **10a**, after being filled, is sealed by closing off the upper end of the combination needle and anode collector **20a**. To illustrate a step in the process, the sealing operation is shown as performed by a plasma gas flame **60**. Tungsten Inert Gas welding is used for sealing the hypodermic needle, and welds the tube as close to the cell top as possible. A schematic diagram of the fixture is shown in Figure 3.9d. The cell is kept cold with dry ice during welding to avoid loss of SO_2 by local heating. The copper block used to make the electrical connection between the hypodermic needle and the ground terminal of the welder serves as a heat sink. The gas operated tungsten cathode is positioned just above the tip of the hypodermic needle which is to be sealed.

On triggering the welder, the plasma melts the hypodermic needle tip and forms a round bead of metal thereby sealing the tip. The same sealing operation may be performed, electrically, by a suitable contact pressure spot welding device, in which the two welding electrodes may first compress the top end of the hypodermic needle collector **20a**, and then apply a pulse of energy to effect the weld. During such sealing operation of the hypodermic needle to seal the cell **10a** closed, the cooling effect of the metal block **50** and the surrounding cold bath **52** serves to extract immediately any heat that may momentarily try to enter the cell as a result of such weld and sealing operation. The upper copper block **56** helps to cool the atmosphere around the top of the hypodermic needle as it is being sealed closed.

The double beading **26** and **28** provides a tight compression on the septum and a good seal. The height of the rubber septum employed here is 0.065 inch. The nature of the problems that are encountered in the manufacture and filling of such a miniaturized cell, will be realized from the dimensional parameters, previously stated as about 0.75 inch in length with a diameter built up on the needle as a collector of 0.036 inch, within a cathode bore of 0.046 inch to 0.52 inch.

The excellent results that may be achieved, nevertheless, from such a cell, in spite of its miniature dimensions, as long as the cell is properly combined and assembled, and adequately filled with electrolyte, may be realized from the graphs shown in Figure 3.9e, where the three curves show that the voltages of the tested cells remain excellently constant at their desired operating values, for a discharge capacity of 7 to 8 mAh, for drains from 1.0 up to 3.0 mA, with the voltage maintained at its normal value for even greater mAh operation where the drain is limited to 1 mA.

Figure 3.9f illustrates the relationship of the capacity and efficiency of the small cells as a function of the electrolyte weight at a medium discharge of 2 mA. It was possible to realize an energy output target of 25 mWh at 2 mA, with the miniature cells disclosed here.

Automatic Internal Safety Control

J.F. Zaleski; U.S. Patent 3,939,011; February 17, 1976; assigned to P.R. Mallory & Co. Inc. describes an electric cell with a lithium/sulfur dioxide electrolyte complex. The cell has automatic protective structural features responsive to excessive internal temperatures or pressures that would be due to internal fault conditions or to external short-circuit conditions. These protective features serve to selectively vent the cell or to break the internal cell circuit and to prevent explosive rupture of the cell and consequent possible damage to personnel in the adjacent surrounding environment by the corrosive ingredients of the cell.

Figure 3.10a shows a vertical sectional view taken along the central vertical plane that shows the construction and assembly arrangement of the final cell. As shown in Figure 3.10a, a cell **10** comprises an outer enclosing can **12**, generally of initially hollow cylindrical form, provided with a bottom closure shown as a metal cup **14**, closely fitting internally within the outer can **12** and peripherally welded thereto.

FIGURE 3.10: LITHIUM CELL WITH AUTOMATIC INTERNAL SAFETY CONTROLS

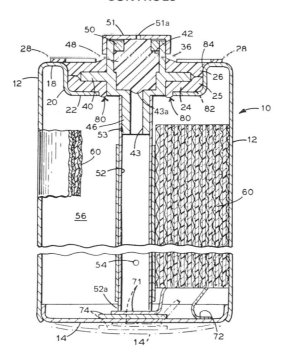

a.

(continued)

FIGURE 3.10 (continued)

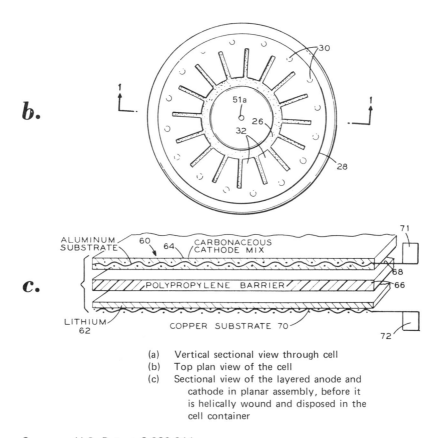

(a)   Vertical sectional view through cell
(b)   Top plan view of the cell
(c)   Sectional view of the layered anode and
       cathode in planar assembly, before it
       is helically wound and disposed in the
       cell container

Source: U.S. Patent 3,939,011

The upper end of the can 12, originally open, is formed and shaped to provide a reentrant portion consisting of three parts, an annular shoulder end rest 18, an axially inwardly extending circular limiting wall neck portion 20, and a transverse annular seat 22. The annular seat 22 serves to support a double layer sandwich 25, consisting of two layers 24 and 26 as annular rings of insulating material, preferably of an elastomer that will accept compression and retain its resilience in the chemical environment of the cell materials.

The double layer sandwich 25, consisting of the two layers 24 and 26 of insulating material, normally fits snugly into the circular space within the circular neck 20 and rests snugly and under pressure on the annular seat 22, due to the compression force impressed by a final assembly pressure ring 28, whose outer peripheral border is anchored by welding onto the annular reentrant shoulder end rest 18 of the upper end of the container 12, at arcuately spaced spots 30, as shown in Figure 3.10b. As shown in Figure 3.10b, the annular pressure ring 28 is appropriately slotted to provide a plurality of inwardly extending radial vent fingers 32 which are supported and disposed in prestressed condition in order to impress a downward pressure force on the top layer 26 of the insulating supporting sandwich 25. The insulating sandwich 25 serves to support a hollow cylindrical metallic supporting shell 36, that has an external annular peripheral flange 40 which seats and nests between the two sandwich layers 24 and 26 in order to be insulatingly supported by them. The metallic supporting shell 36 is axially hollow and has an upper bore 42 to define an upper inner chamber at its upper open end, that communicates, at its lower end, with an extending passage 43a of lower and smaller bore 43 within a coaxial extension 46 of the shell body 36.

The function of the supporting shell 36 is to support a plug seal 48, of suitable elastomeric material, to substantially fill and close the shell chamber 43. The plug seal 48 will later serve as a plug through which a hypodermic needle will be inserted for evacuating the container can and then conducting the sulfur dioxide electrolyte solution into the evacuated space.

In order to maintain the elastomeric seal plug 48 compressed for good pressure engagement with the inner wall surface of shell 36, a pressure and retainer ring 50 is employed, which is dimensioned to fit with a snug pressure fit into the upper open end of the supporting shell 36, to which the ring is then suitably bonded, as by brazing, with a material such as Wood's metal, which has a definite melting point at a predetermined temperature at which it is desired to permit the ring 50 to be released from sealing bond with the shell 36. The melting point of Wood's metal at 170°F is enough above the normal safe operating temperature of 160°F, to provide quick detection of a dangerous temperature condition.

The tube 52 is spot welded to the depending bracket 46, as indicated by the weld spot 53. The tube 52 has several functions. It is provided with one or more fill ports 54, which are small openings drilled through the wall of the tube 52, to provide communication with the main chamber 56 of the cell, that accommodates the working chemical components and the electrolyte in general. The tube 52 serves a primary function here as a stationary contact element of a circuit breaker within the cell. The lower circular end 52a of tube 52 serves as a stationary contact surface of such circuit breaker, at which the circuit of the cell is opened to disconnect the cell from an external circuit. The tube 52 also serves as a central axial spacer for the helical wraparound 60 which includes the working electrodes of the cell.

In Figure 3.10c, the working electrodes are shown in assembly at an intermediate stage of assembly, with a lithium sheet 62 as the working anode of the cell, and a sheet layer of carbonaceous mix 64 as cathode separated from the lithium by a perforated polypropylene barrier 66. The lithium sheet 62 is 0.010 inch thick, impressed in and supported on a copper substrate 70, which may be perforated or of mesh structure. The carbonaceous cathode depolarizer mix 64 is about 0.030 inch thick buttered on the aluminum substrate 68, also of perforated thin sheet or mesh structure. This layered assembly is provided with two tabs, electrically connected to the respective substrates. Tab 71 is connected to the cathode 64-68 and tab 72 is connected to the anode lithium and substrate 62 and 70.

During manufacturing assembly of the cell, the planar assembly 60 is wrapped into helical form for axial insertion into the cam 12 from the bottom, to be slipped onto and over the center tube **52**, into space **56**. The copper tab **72** has been welded to a convenient area on the bottom closure cup **14**. The other tab **71** is secured to an insulation disc **74**, which is secured to a central region of the bottom closure cup **14**, to take advantage of the maximum axial movement and displacement of that bottom closure cup when distended by excess pressure within the cell. In the initial dry assembling operation, the helical electrode assembly is thus physically connected to the bottom closure cup **14** as a subassembly, which is then inserted into the can **12** to place the electrode assembly in operating position in chamber **56**, and to place the closure cup in position to close the can bottom ready for welding and sealing. At the same time, the can bottom cup **14** is moved to proper position to place the depolarizer contact tab **71** of Figure 3.10c into contact with the bottom edge **52a** of tube **52**.

Normally, when the bottom end of the tube **52** engages the contact tab **71**, the circuit through the cell is from the top safety cap **50** down through the metallic shell support **36**, thence through the tube **52** down to and through the depolarizer contact tab **71** through the working cell elements including cathode depolarizer **64**, the electrolyte, and the anode **62** to the anode tab **72**, the bottom closure cup **14**, and the can **12** to which the closure cup is suitably sealed by welding.

The bottom of the tube **52** is kept in contact with the depolarizer contact tab **71** by the normal resilient pressure of the bottom cup **14** pressing upwardly through the insulator disc **74**, which has been appropriately secured to the inner surface of the can bottom cup **14**. During normal operations the pressure in the cell is not high enough to distend the can bottom cup **14**, and therefore the contact between the tube bottom **52a** and the depolarizer contact tab **71** is highly conductive and of minimum resistance, under the pressure of the bottom cup **14**.

Upon the occurrence of one of the undesirable conditions that results in an increased pressure within the can, that increase in pressure reacts on the flexible bottom cup **14** of the can and causes the cup to distend outward, as indicated by the dotted line **14'**, and the outward distended movement of the bottom cup **14** moves the insulator and the depolarizer contact tab away from, and out of contact with, the bottom edge **52a** of the tube **52**. The cell circuit is thus interrupted at the bottom edge of the tube **52**, to disconnect the operating elements of the cell from the tube **52**, and, consequently, from the top outer safety cap **50**, that serves also as the terminal connected to the depolarizer cathode material during normal operation of the cell.

Even though the circuit of the cell is thus internally opened, so that further operating cell current transfer is terminated, there may still be conditions within the cell that continue to generate heat that will raise the temperature within the cell and thus increase the pressure to an explosure value that could be dangerous to anyone within the proximity of the cell, if such pressure were permitted to continue to increase. There are, thus, two conditions that are indicative of a possible dangerous explosion, namely, the temperature developed within the cell and a possible resulting pressure.

If the pressure condition changes gradually, as might be caused by a normal loading of the cell, but without the occurrence of a necessarily hazardous situation, adequate protection may be provided to the cell by merely permitting and providing for slight venting. Upon the occurrence of conditions of that kind, the vapor pressure gas within the cell may press against the under surfaces of the bottom sandwich layer 24, as at the region indicated by the arrows 80, and then move out in thin gas layer or strip form along the under surface of that bottom sandwich layer 25, as for example, along a path indicated by the dotted line 82, up to the space 84 directly underneath the radially extending fingers 32. Those fingers 32 are supported as cantilever beams at their rear regions, adjacent the weld points 30, so that the forward ends of those cantilevered fingers 32 may rise slightly from the top surface of the upper sandwich layer 26, to permit a small bubble of the gas to exit from the cell and thereby relieve the pressure within the cell sufficiently to hold that pressure down below a dangerous pressure value within the cell.

Thus, under normal operating conditions, where the cell is properly doing the work for which it was intended, and the energy supplied by the cell is not creating a dangerously abnormal condition, simple occasional venting, of a small amount of the electrolyte vapor as a gas, may be sufficient to protect the cell from accumulating an excessive pressure internally, and the operation of the cell may continue without danger or hazard to anyone in the environment of the cell. If however, a condition develops within the cell, due to internal or external conditions, that causes a fast temperature rise, that would undoubtedly cause a subsequent pressure increase, the high thermal conductivity of the tube 52 and its supporting frame shell 36 are utilized to conduct the heat to the plug seal retaining ring 50, and to soften the bonding material of that ring sufficiently to release the ring from the shell 36.

Consequently, pressure on the elastomeric plug seal material 42 is relieved to permit the compressed plug seal to expand to full volume of its uncompressed condition, and, thereby, to relieve the pressure of that seal 42 on the inner wall surfaces of the shell 36, to permit the gases within the coil container to rush out quickly, through the openings or ports within the tube 52, and up through the central axial passage into the internal chamber of the cell 36, past the relieved loose plug seal 42, and out through the dimple opening 51a, formed in the top of the seal safety cap 51 after the spot weld closure during manufacture.

Thus, on the occurrence of dangerous temperature, an immediate rise in pressure, or even in case of a delayed rise in pressure, the cell is able to control itself by the three provisions noted, namely, first, by the switch opening as a circuit-breaker operation at the bottom of the tube 52, by the distension of the can bottom 14; or secondly, by the slow leakage along the path 82 up through and past the cantilever spring fingers 32; or, thirdly, by the fast release of the gas pressure out through the top plug seal and safety cap, where the temperature rise indicates an extremely dangerous condition that is probably likely to lead to an explosion.

Fluid Tight Metallic Sleeve

D. Coueille; U.S. Patent 3,971,673; July 27, 1976; assigned to Saft-Societe des Accumulateurs Fixes et de Traction, France describes an electric cell having a fluid tight casing comprising a metal can having a bottom and a lateral wall and at least one current output terminal insulated from that can. It is characterized more particularly, in that its casing comprises, moreover, a metal sleeve surrounding the lateral wall of the can, stopped up in a fluid tight manner at its first end by the can and at its second end by an undeformable electrically insulating seal through which the terminal crosses in a fluid tight manner. According to a preferred case, the seal is made of glass. As a result of the undeformability of that seal, the greatest thickness compatible with the dimensions of the cell is imparted to it. Figure 3.11 shows cells having as their positive active material, copper sulfide and as their negative active material, lithium. The cell in Figure 3.11a has a metal casing with a diameter of 3 mm, a length of 6 mm and two current output terminals 1 and 2 of opposite polarities constituted by wires situated in the axis of the casing.

The casing comprises a can 10 made of nickel-plated steel or of nickel, in the shape of a cup, welded to the positive output 1 and containing a copper sulfide positive electrode 11 topped by a separator 25 made of a porous plastic material impregnated with electrolyte which can be a solution of lithium perchlorate in propylene carbonate. The separator 25 is itself topped by a negative electrode 21 made of lithium in which one end of the negative terminal 2 penetrates; the anode 21 is surrounded laterally and at its upper part by an insulating cup 22 made of plastic material. The can 10 is surrounded by a metal sleeve 30 whose end 31 is turned down over a seal 12 made of plastic material applied to the bottom of the can 10 round the positive terminal 1.

FIGURE 3.11: LITHIUM-COPPER SULFIDE CELLS

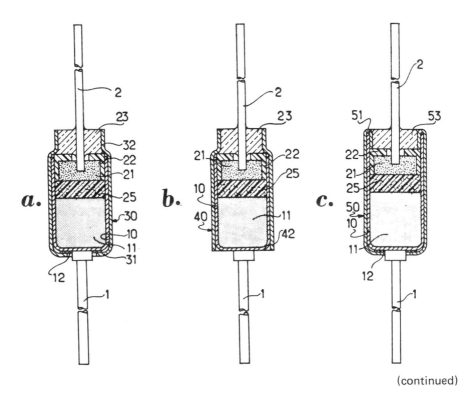

(continued)

FIGURE 3.11: (continued)

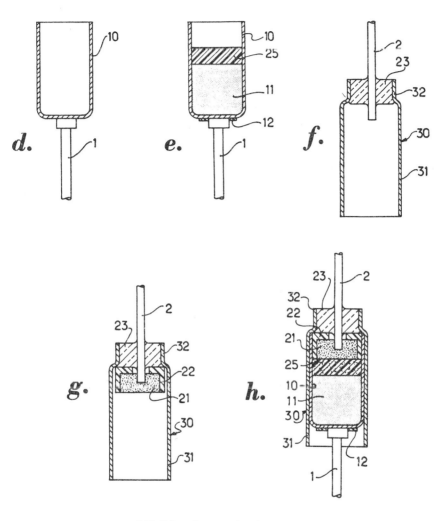

(a)(b)(c)   Cross-sectional views of three variants of
                 electric cells
(d)(e)(f)(g)(h) Phases of an assembly of a cell according
                 to Figure 3.11a

Source: U.S. Patent 3,971,673

The end **32** of the sleeve **30** is stopped up by a glass seal **23** in which the negative terminal **2** is embedded. The material of the sleeve **30** is chosen so that its
coefficient of expansion is substantially equal to that of glass; it can be a ferro-
nickel. On referring to Figures 3.11d through 3.11h, it will be seen that the assembly of the cell can be effected from two subassemblies.

To form the first subassembly, starting with a can **10** connected by welding to the positive output **1** (Figure 3.11d), the positive mass **11** is compressed in the can **10** and the separator **25** is arranged upon the latter; the positive terminal **1** is surrounded with the seal **12** which is applied to the bottom of the can **10** (Figure 3.11e). The second subassembly is formed by the metal sleeve **30** attached to the negative terminal **2** by embedding the latter in the seal **23** (Figure 3.11f). The forming of the seal through which the terminal **2** crosses is effected at a high temperature and could not be implemented on a completed cell during assembly, due to the risk of damaging its internal components. The insulation cup made of a plastic material **22** lined with lithium constituting the anode **21** (Figure 3.11g) is then inserted in the sleeve **30**.

The two preceding subassemblies are then fitted into each other (Figure 3.11h) in such a way that the upper edge of the can **10** abuts against a shoulder provided in the sleeve **30**. The end **31** of the sleeve **30** is then turned down over the seal **12** to close the cell in a fluid tight manner. It should be observed that sealing is effected between the can **10** which is positively polarized and the sleeve **30** which is insulated from the negative pole by the glass seal **23** which has great insulating power. The seal **12** therefore does not have any insulating function to fulfill.

The cell illustrated in Figure 3.11b is differentiated from that which has just been described in that the seal ring **12** is replaced by a soldering of the end **42** the sleeve **40** to the cup **10**; that soldering can be effected by means of an alloy having a low melting point, e.g., a tin and indium alloy. In the cell according to Figure 3.11c, it is the shape of the end **51** of the sleeve **50** at the level of its contact with the glass seal **53** which is different from that in Figure 3.11a. As will be seen, that end **51** is folded back so as to provide an annular space. The edge of the can **10** is then inserted in that annular space and the tank **10** bears against the fold at the time of the turning down of the other end of the sleeve **50** over the seal **12**.

Stainless Steel Casing

A process described by *G. Lehmann and A. Brych; U.S. Patent 3,925,101; December 9, 1975; assigned to Saft-Societe des Accumulateurs Fixes et de Traction, France* relates to alkaline metal (e.g., Li) and silver chromate primary cells. It provides for use as a material for the positive casing of such a primary cell, stainless steel containing at least 11% of chromium, or up to about 17% chromium, or about 18% chromium with about 9% nickel, all percentages being by weight. With such a positive casing a corrosion of the casing during cell storage or discharge is delayed considerably.

SODIUM AND LITHIUM
SOLID ELECTROLYTE SYSTEMS

SODIUM-HALOGEN

Battery Casing

A process described by *F.G. Will and H.J. Hess; U.S. Patent 3,959,020; May 25, 1976; assigned to General Electric Company* is directed to providing a battery casing and a primary sodium-halogen battery operable at temperatures of –48° to 100°C, which battery operates independent of orientation and allows larger currents to be drawn than from a battery of equivalent size employing a solid electrolyte disc. The battery casing includes an open-ended inner casing of a solid sodium ion-conductive material, an electronic conductor within the interior of the inner casing, an outer metallic casing including an opening therein, and a cover with a central opening therein surrounding the inner casing, fill tubes associated with the respective openings, and a reactant resistant glass sealing the outer casing to the inner casing.

In Figure 4.1a there is shown generally at **10** a battery casing which has an inner casing of a solid sodium ion-conductive material **11** with one open end **12**. An electronic conductor **13** is positioned within the interior of inner casing **11** and extending outwardly through open end **12** of casing **11**. An outer metallic casing **14** has an open end **15** an an opening **16**. While opening **16** is shown in the opposite closed end, such opening can be located at other points in casing **14**. A metallic fill tube **18** is affixed to closed end **17** and in communication with opening **16**.

An inwardly extending flange **19** is affixed to metallic casing **14** at its open end **15**. Metallic casing **14** surrounds inner casing **11** with the exterior surface of flange **19** flush with the surface of the open end **12** of inner casing **11**. Outer casing **14** includes a metallic cover **20** with a centrally disposed opening **21** therein. A metallic fill tube **22** is sealed to cover **20** and communicates with opening **21** therein. Cover **20** closes open end **15** of metallic casing **14**. A reactant resistant glass, sodium and halogen resistant glass, **23** seals outer casing **14** to inner casing **11** by sealing cover **20** to the exterior surface of flange **19**

FIGURE 4.1: BATTERY CASING

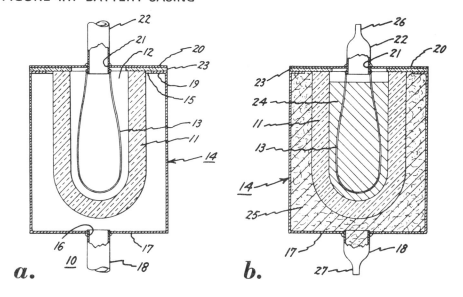

(a)   Sectional view of battery casing
(b)   Sectional view of battery

Source: U.S. Patent 3,959,020

and the upper surface of inner casing **11**. Electronic conductor **13** is affixed
to metallic fill tube **22**.

In Figure 4.1b there is shown a sealed primary sodium-halogen battery which
includes the above-described battery casing shown in Figure 4.1a. An anode
24 is positioned preferably in inner casing **11**. This anode, which is shown as
sodium metal, is selected from the class consisting of sodium, sodium as an
amalgam, or sodium in a nonaqueous electrolyte. A cathode **25** of a halogen
in conductive material is positioned preferably within outer casing **14** and in
contact with both casings **11** and **14**. After each casing is filled, the associated
fill tubes **18** and **22** are closed, for example, by welding at **26** and **27**, respectively. The resulting structure is a sealed sodium-halogen battery.   The following examples illustrate the process.

Example 1: A battery casing was assembled as above-described and shown in
Figure 4.1a by positioning an inner casing of an ionic conducting material of
sodium beta-alumina having an open end within an outer metallic casing with
an open end and an opening in the opposite closed end. The outer metallic
casing has an inwardly extending flange affixed to its open end. The inner and
outer casings were aligned so that the exterior surface of the flange was flush
with the surface of the open end of the other casing.

At the opposite closed end of the outer casing a metallic fill tube was affixed
and in communication with the opening in the closed end. A glass washer of

Corning Glass No. 7056, which is sodium and halogen resistant, was positioned on the upper surfaces of the open end of the inner casing and the exterior flange of the outer casing. A metallic cover of tantalum metal on a centrally disposed opening therein has attached thereto an extended metallic fill tube. An electronic conductor in the form of a 2 mil thick tantalum wire was formed into a loop and the free ends were welded to the inner edge of the tantalum fill tube. The cover was positioned on the opposite surface of the glass washer. This washer was then heated to a temperature of 1000°C in a furnace whereby the cover was sealed by the glass to the upper surfaces of the inner casing and the exterior surface of the flange.

Example 2: A sealed primary sodium-bromine battery is assembled by employing the battery casing described in Example 1. The inner casing is filled with sodium amalgam through the fill tube in the cover after which the fill tube was sealed by welding. The outer casing is filled with carbon felt material and bromine through the associated fill tube after which the fill tube is sealed by welding. The resulting structure is a sealed primary sodium-bromine battery. At room temperature this battery has an open circuit voltage of 3.7 volts.

In related work *H.J. Hess; U.S. Patent 3,918,991; November 11, 1975; assigned to General Electric Company* describes a sodium-halogen battery which comprises a casing with an anode positioned therein, the anode consisting of sodium in a nonaqueous organic electrolyte of pure 1,2-dimethoxyethane containing a dissolved sodium salt. The salt is selected from the class consisting of sodium perchlorate and sodium bromide. A solid sodium ion-conductive electrolyte is adjacent the anode, and a cathode is positioned adjacent the opposite side of the solid electrolyte.

Single Seal of Low Temperature Melting Glass

F.G. Will and R.R. Dubin, U.S. Patent 3,918,992; November 11, 1975; assigned to General Electric Company describe a battery casing and a sealed primary sodium-halogen battery wherein the battery casing includes a single seal of low temperature melting glass joining together the metallic anode cap, the inner casing of a solid sodium ion-conductive material, and the outer metallic casing. A sealed primary sodium-halogen battery has the above type of casing with a sodium anode in one casing and a cathode of a halogen in conductive material in the other casing.

In Figure 4.2a there is shown generally at 10 a battery casing which has an inner casing of a solid sodium ion-conductive material 11 with one open end 12. An outer metallic casing 13 has an upper portion 14 and a lower portion 15. Upper portion 14 has opposite open ends 16 and 17. An inwardly extending flange 18 is affixed to upper portion 14 at its first open end 16 and a flare 19 at its open end 17. Lower portion 15 has an open end 20, an opening 21 and a removable closed end 22. Upper and lower portions 14 and 15 are joined together by welding at 23 at their associated open ends 17 and 20.

Outer metallic casing 13 surrounds inner casing 11 with exterior surface 24 of flange 18 on a lower plane than surface 25 at open end 12 of inner casing 11. A cap 26 with a flange 26' is positioned adjacent to and spaced from flange 18. A single low temperature melting glass seal 27 seals together flange 18 of upper portion 14 of metallic casing 13 and flange 26' of cap 26 and seals flanges 18 and 26 to outer wall 28 of inner casing 11 adjacent its open end 12.

FIGURE 4.2: SEALED SODIUM-HALOGEN BATTERY

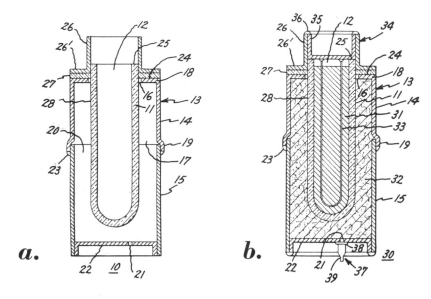

(a)   Sectional view of battery casing
(b)   Sectional view of battery

Source: U.S. Patent 3,918,992

In Figure 4.2b there is shown a sealed primary sodium-halogen battery **30** which
battery includes the above-described battery casing shown in Figure 4.2a. An
anode **31**, which is shown as sodium metal, is selected from the class consisting
of sodium, sodium as an amalgam, or sodium in a nonaqueous electrolyte. A
cathode **32** of a halogen in conductive material is positioned preferably within
outer casing **13** and in contact with outer wall **28** of inner casing **11** and with
the inner wall of casing **13**.

An electronic conductor **33** is positioned within inner casing **11** and extends
outwardly through open end **12** of inner casing **11**. A metallic closure consists
of cap **26** with flange **26'** and a cap insert **35**. This cap insert is positioned
within cap **26** and welded thereto at **36** to seal open end **12** of inner vessel **11**.
Electronic conductor **31** is shown welded to the interior surface of closure portion **35**. A fill tube **37** is shown affixed to closed end **22** by means of welding
flange **38** thereto. The opposite end of the fill tube is closed at **39**, for example,
by welding. The resulting structure is a sealed primary-halogen battery embodying this process. Examples of battery casings and sealed primary sodium-halogen
batteries made in accordance with the process are set forth below.

Example 1: A battery casing was assembled as above described and as shown
in Figure 4.2a by positioning an inner casing of a solid sodium ion-conductive
material with one end partially within the upper portion of an outer tantalum
metallic casing with opposite open ends and an inwardly extending flange affixed
to the upper portion at its first open end. A tantalum cap with a flange was

positioned so that the cap flange was adjacent the flange of the upper portion. A single low temperature melting glass seal was provided initially by positioning a glass washer of Kimble Glass No. N-51A, which is sodium and halogen resistant, between the flanges and in contact with the exterior surface of the inner casing. The associated flanges and glass washer were aligned so that the outer surface of the flange of the outer metallic casing was on a lower plane than the surface of the open end of the inner casing and spaced from the inner casing. The assembly was then heated in an argon atmosphere at a temperature of 1175°C whereby the glass washer provided a single glass seal, sealing together the adjacent flanges and sealing the flanges to the outer wall of the inner casing adjacent to, but spaced from, its open end.

The lower portion of the tantalum outer metallic casing had opposite open ends, a removable closed end, and an opening in the removable closed end. The upper and lower portions of the outer metallic casing were then joined together at their associated first open ends by welding. The removable closed end was positioned in the first open end of the lower portion.

Example 2: A sealed primary sodium-halogen battery was assembled as above described and shown in Figure 4.2b. The battery casing was assembled as described in Example 1 except that the lower portion of the outer casing was not welded to its upper portion. An insert of tantalum which forms part of the closure had attached thereto by welding on its interior surface an electronic conductor in the form of a 10-mil thick tantalum wire formed into a loop. 0.62 gram of mercury was placed within the inner casing to subsequently provide a sodium amalgam as the anode. The insert was then fitted within the cap whereby the electronic conductor extended within the inner casing, contacted the side wall and the mercury therein. The insert was then welded to the cap under high vacuum of 5×10^{-5} torr. The interior casing was then filled with sodium to provide a sodium amalgam anode within the interior casing as described in U.S. Patent 3,740,206.

The second open ends of the upper and lower portions of the outer metallic casing were welded together. Carbon felt in the form of a plurality of washers was slipped around the exterior surface of the interior casing. The removable closed end is then positioned in the first open end of the lower portion. The removable closed end was welded to the lower portion. A mixture of 89.0 wt % bromine, 10.0 wt % iodine, 0.5 wt % water, and 0.5 wt % sodium bromide was added through the fill hole in the closed end of the outer portion of the casing by means of a syringe. In this manner there was provided a cathode within and in contact with the outer casing and in contact with the exterior surface of the interior casing. The fill hole was then closed by welding.

Example 3: At room temperature the sealed primary sodium-halogen battery of Example 2 exhibited the polarization behavior which is shown below.

Current Density, ma/cm²	Voltage, volts
0	3.57
0.1	3.02
0.5	2.80
1.0	2.50
2.0	1.73
2.7	0.82
3.0	0

Glass Seals

R.F. Thornton; U.S. Patent 3,928,071; December 23, 1975; and U.S. Patent 3,959,011; May 25, 1976; both assigned to General Electric Company describes a hermetically sealed primary battery which includes two glass seals joining together an inner casing of a sodium beta-alumina ion-conductive material, an outer metallic casing, and a cap portion, a sodium type anode in one casing, and a cathode in the other casing.

In Figure 4.3 there is shown generally at **10** a hermetically sealed primary battery which has an inner casing of a solid sodium beta-alumina ion-conductive

FIGURE 4.3: HERMETICALLY SEALED BATTERY

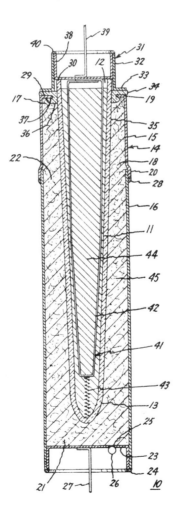

Source: U.S. Patent 3,928,071

material 11 with an open end 12 and a closed end 13. An outer metallic casing 14 has an upper portion 15 and a lower portion 16. Upper portion 15 has initially opposite open ends 17 and 18. An inwardly extending flange 19 is affixed to upper portion 15 at its first open end 17 and a flare 20 at its open second end 18. Lower portion 16 has initially a first open end 21, a second open end 22, a closure 23 for its first open end 21 welded at 24 to lower portion 16 and an initial opening 25 in closure 23 sealed at 26. An electrical lead 27 is shown welded to the exterior surface of closure 23. Upper and lower portions 15 and 16 are joined together as by welding at 28 at their associated second open ends 18 and 22.

Outer metallic casing 14 surrounds inner casing 11 with exterior surface 29 of flange 19 on a lower plane than surface 30 at open end 12 of inner casing 11 and spaced from inner casing 11. A metallic closure 31 consisting of a cap portion 32 with a flange 33 and a cap insert has its flange 33 positioned adjacent to and spaced from flange 19. A first glass seal 34 seals together flange 19 of upper portion 15 of metallic casing 14 and flange 33 of cap portion 32 of metallic closure 31 and seals flanges 19 and 33 to a portion of outer wall 35 of inner casing 11, which portion of outer wall 35 is nearly round and is adjacent open end 12. A second glass seal 36 seals together opposite surface 37 of flange 19 of upper portion 15 of outer metallic casing 14, and a portion of outer wall 35 of the inner casing 11, thereby providing strong seals and protection of the first seal by the second seal from corrosive action of chemicals in the outer casing.

Cap insert 38 with an electrical lead 39 welded to its exterior surface is shown positioned within cap portion 32 and welding thereto at 40, thereby closing open end 12 of inner casing 11. An electronic conductor 41 is affixed, as by welding, to interior surface of cap insert 38, thereby being positioned within interior surface of inner casing 11. Electronic conductor 41 is shown in the form of a closed wire hairpin 42 with a wire spiral 43 affixed to one end whereby conductor 39 is in contact with closed end 13 of inner casing 11. An anode 44 is positioned preferably in inner casing 11. Anode 44, which is shown as sodium metal, is selected from the class consisting of sodium, sodium as an amalgam or sodium in a nonaqueous electrolyte.

A cathode 45, which is shown as a halogen in conductive material, is positioned preferably within outer casing 14 and in contact with outer wall 35 of inner casing 11 and with the inner wall of outer casing 14. The resulting structure is a hermetically sealed primary battery embodying this process. The hermetically sealed primary battery is formed preferably by positioning an inner casing of a solid sodium beta-alumina ion-conductive material having an open end within the upper portion of an outer casing of a suitable chemically stable material such as tantalum. The upper portion has opposite open ends with an inwardly extending flange affixed at its first open end. The upper portion of the outer casing surrounds partially the inner casing with the exterior surface of its flange on a lower plane than the surface of the open end of the inner casing and the upper portion of the outer casing is spaced from the inner casing.

A metallic closure, such as of tantalum, consists of a cap portion with a flange and a cap insert which has the flange of the cap portion positioned adjacent the flange of the upper portion of the outer metallic casing. Other suitable metals can be employed for the outer casing and the metallic closure provided the metal bonded to the glass has approximately the same expansion coefficient and chemical stability toward the material in the outer casing. Two glass washers are positioned, respectively, between the adjacent flanges and in contact with the exterior wall of the inner vessel, and against the opposite surface of outer metallic casing flange, and a portion of the outer wall of the inner casing. Each glass washer is made of a suitable sodium and halogen resistant glass, such as Corning Glass No. 7052, General Electric Company Glass No. 1013, Sovirel Glass No. 747, or Kimble Glass No. N-51A. The glass washers and associated assembly are positioned in an inverted position held by a suitable jig fixture and heated to a temperature in the range of 1175° to 1250°C in an argon atmosphere whereby one glass washer seals together the adjacent flanges of the upper portion of the outer metallic casing and the cap portion of the metallic closure and seals the flanges to the outer wall of the inner vessel adjacent its open end.

The second glass washer seals together the opposite surface of the flange of the upper portion of the outer metallic casing and a portion of the outer wall of the inner casing. An amount of mercury is placed within the inner vessel, which amount is the amount required in the sodium amalgam to be used as the anode. The cap insert has an electronic conductor of a material such as tantalum or nickel affixed to its interior surface which is positioned within the interior surface of the inner casing. The preferred form of the conductor is a closed wire hairpin with a wire spiral affixed to its opposite end whereby the spiral contacts the interior surface of the closed end of the inner casing and contacts the mercury therein. The cap insert is then welded to the cap portion. In this manner the insert seals the open end of the inner casing. The inner casing is then filled with sodium to provide a sodium amalgam anode.

Sealed Lithium-Sodium Unit

G.C. Farrington and W.L. Roth; U.S. Patent 4,012,563; March 15, 1977; and U.S. Patent 4,027,076; May 31, 1977; both assigned to General Electric Co. describe a sealed lithium-sodium electrochemical cell with sodium beta-alumina ion-conductive electrolyte. The cell comprises a casing, an anode positioned within the casing, the anode selected from the class consisting of lithium-sodium, lithium-sodium as an amalgam, and lithium-sodium in a nonaqueous electrolyte, a cathode positioned within the casing and the cathode functioning with a lithium-sodium type anode and a solid sodium beta-alumina ion-conductive electrolyte. The solid sodium beta-alumina ion-conductive electrolyte is positioned within the casing between the anode and cathode and in contact with both the anode and cathode.

In Figure 4.4 there is shown generally at **10** a modified sealed lithium-sodium electrochemical cell with sodium beta-alumina ion-conductive electrolyte. An outer casing **11** comprising a lower casing portion **12** of glass and an upper casing portion **13** of polyethylene affixed tightly to the upper end of the lower casing portion **11** thereby provides a chamber **14** for a cathode **15** of a concentrated solution of bromine in a nonaqueous catholyte and a platinum electrode **16**. Electrode **16** extends to the exterior of cell **11** through the junction of the lower and upper casing portions **12** and **13**.

An inner casing **17** in the form of a tube of solid sodium beta-alumina ion-conductive electrolyte is positioned within lower casing portion **12** and immersed partially in cathode **15**. An opening **18** is provided in the top of upper

FIGURE 4.4: SEALED LITHIUM-SODIUM CELL

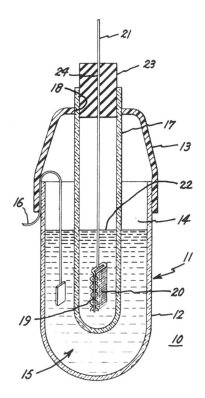

Source: U.S. Patent 4,012,563

casing portion **13** into which tube **17** fits tightly. An anode **19** of lithium-sodium in the form of a lithium-sodium ribbon is pressed onto a nickel mesh **20** which is folded together and attached to the end of a nickel electrical lead **21**. An anolyte **22** partially fills tube **17** and is in contact with lithium anode **19**. An electrically insulating closure **23** with a hole **24** therethrough is provided at the upper end of tube **17** to seal the initially open end of the tube. Lead **21** extends through the hole **24** in closure **23** to the exterior of cell **10**.

It has been found that one can form a sealed lithium-sodium electrochemical cell with sodium beta-alumina ion-conductive electrolyte by employing a casing having a cathode portion. These two portions are separated by a solid sodium beta-alumina ion-conductive electrolyte which will be further described below. Such a casing may be provided in various configurations.

One such cell employs an outer casing comprising a lower casing portion of glass and an upper casing portion of a plastic such as polyethylene affixed tightly to the upper open end of the lower casing portion, thereby providing a chamber for a cathode in a catholyte which functions with a lithium-type anode and a solid lithium beta-alumina ion-conductive electrolyte. An electrode extends from the cathode to the exterior of the cell through the junction of the lower and upper casing portions. An inner casing in the form of a tube of solid sodium beta-alumina ion-conductive electrolyte is positioned within the outer casing and immersed partially in the cathode. An opening is provided in the top of the upper casing portion into which the tube fits tightly. An anode of lithium-sodium in the form of a lithium-sodium ribbon pressed onto a nickel mesh is folded together and attached to the end of a nickel electrical lead. An anolyte partially fills the tube and is in contact with the lithium anode. An electrically insulating closure with a hole therethrough is provided at the upper end of the tube to seal the initially open end of the tube. The lead extends through the hole in the closure to the exterior of the cell.

Examples of lithium-sodium electrochemical cells with sodium beta-alumina ion-conductive electrolytes made in accordance with the process are set forth below.

Examples 1-14: 14 lithium electrochemical cells with sodium beta-alumina ion-conductive electrolytes were made, which are cells numbers 1-14. Each cell was formed of an outer casing having a lower casing portion of glass and an upper casing portion of polyethylene adapted to be affixed tightly to the upper end of the lower casing, thereby providing the chamber for a cathode in a catholyte with an electrolyte, which cathode functions with a lithium-sodium type anode in a solid sodium beta-alumina ion-conductive electrolyte. Each electrode was immersed in the cathode and extended to the exterior of the cell through the junction of the lower and upper casing portions.

The catholyte which was employed in cells 1-11 was a nonaqueous catholyte of 0.1 molar tetrabutylammonium fluoroborate in propylene carbonate saturated with $LiClO_4$. The catholyte for cell 12 was the cathode material thionyl chloride. in undiluted form. The catholyte for cells 13 and 14 was an aqueous solution of 0.1 molar nitric acid in water. The cathodes for cells 1-14 were, respectively, chlorine, bromine, iodine, sulfur, phosphorus, nickel, chloride, lead sulfide, silver oxide, cupric fluoride, lead iodide, sulfur dioxide, Fe(III), and oxygen. These cathodes and the manner that they were employed in their respective cells 1-14 are set forth below.

Cathodes	Manner Used
Chlorine and sulfur dioxide	Saturated solution in above-described nonaqueous catholyte used with platinum electrode.
Bromine and iodine	Concentrated solution in above-described nonaqueous catholyte used with platinum electrode.
Sulfur, phosphorus, nickel chloride, lead sulfide, silver oxide, cupric fluoride, and lead iodide	Each was pressed as anhydrous powder on an expanded nickel mesh cathodic electrode in above-described nonaqueous catholyte.
Thionyl chloride	Undiluted.
Fe(III)	Concentrated solution of $FeCl_3$ in above-described aqueous catholyte used with platinum electrode.
Oxygen	Saturated solution in above-described aqueous catholyte used with platinum electrode.

An inner casing in the form of a tube of solid sodium beta-alumina ion-conductive electrolyte was positioned within the outer casing of each of cells 1-14 and immersed partially in each of the respective cathodes by affixing tightly the upper casing portion to the lower casing portion. An opening was provided in the tube of each upper casing portion into which the tube fitted tightly. An anode of the same composition was employed for each of the cells 1-14.

The anode consisted of 99 wt % lithium and 1 wt % sodium, which were mixed together and pressed onto a nickel mesh which was folded together and attached to the end of a nickel electrical lead. Each anolyte filled partially the respective tube that was in contact with the respective lithium-sodium anode. An electrically insulating closure with a hole therethrough was provided at the upper end of each tube to seal the initially open end of each tube. Each lead extended through the respective hole in the closure to the exterior of each cell. Each of the resulting 14 devices was a sealed lithium-sodium electrochemical cell which was made in accordance with the process.

The table below sets forth for cells 1-14 the cell potential in volts as a function of current in milliamperes. Each cell is identified by number as well as the cathode employed therein. Each of the cells was operated at a temperature of $26°C$.

Current-mA	1 Cl_2	2 Br_2	3 I_2	4 S	5 P	6 $NiCl_2$	7 PbS
Open Circuit	3.90	4.08	3.72	2.87	2.90	2.89	3.17
0.020	3.78	4.01	3.65	2.46	2.57	2.40	3.10
0.040	3.67	3.92	3.55	2.17	2.37	2.05	3.00
0.060	3.58	3.86	3.47	1.97	2.23	1.76	2.92
0.080	3.51	3.80	3.42	1.83	2.10	1.53	2.84
0.100	3.45	3.76	3.37	1.72	2.00	1.44	2.76
0.200	3.26	3.64	3.22	—	1.72	—	2.61
0.400	3.01	3.46	3.02	—	1.28	—	2.32
0.600	2.84	3.32	2.88	—	0.93	—	2.11
0.800	2.69	3.22	2.74	—	0.59	—	1.94
1.0	2.56	3.14	2.56	—	0.31	—	1.80
2.0	1.90	2.70	—	—	—	—	—
4.0	1.25	2.21	—	—	—	—	—
6.0	0.80	1.91	—	—	—	—	—
8.0	0.39	1.62	—	—	—	—	—
10.0	0.00	1.39	—	—	—	—	—

Current-mA	8 AgO	9 CuF_2	10 PbI_2	11 SO_2	12 $SOCl_2$	13 Fe(III)	14 O_2
Open Circuit	3.39	3.22	2.76	3.20	3.90	3.12	3.02
0.020	3.31	3.17	2.24	2.87	3.51	2.83	2.88
0.040	3.21	3.12	1.99	2.58	3.28	2.78	2.74
0.060	3.13	3.07	1.87	2.41	3.11	2.75	2.63
0.080	3.07	3.02	1.82	2.29	2.95	2.46	2.53
0.100	2.99	2.98	1.76	2.21	2.81	2.36	2.45
0.200	2.87	2.84	1.54	—	2.02	2.13	2.31
0.400	2.61	2.61	1.18	—	1.17	1.97	2.00
0.600	2.44	2.44	0.96	—	0.49	1.77	1.83
0.800	2.30	2.29	0.77	—	—	1.48	1.69
1.0	2.18	2.17	0.60	—	—	1.32	1.57
2.0	1.69	1.59	—	—	—	0.71	1.06
4.0	1.00	0.60	—	—	—	0.22	0.53
6.0	0.66	—	—	—	—	—	0.20
8.0	0.22	—	—	—	—	—	—
10.0	—	—	—	—	—	—	—

LITHIUM

Sulfur Monochloride Cathode

W.L. Roth and G.C. Farrington; U.S. Patent 3,953,233; April 27, 1976; assigned to General Electric Company describe a sealed lithium-sulfur monochloride cell which employs a lithium anode, a sulfur monochloride cathode, and a solid lithium sodium aluminate electrolyte which is a lithium-ion conductor.

FIGURE 4.5: LITHIUM-SULFUR MONOCHLORIDE CELL

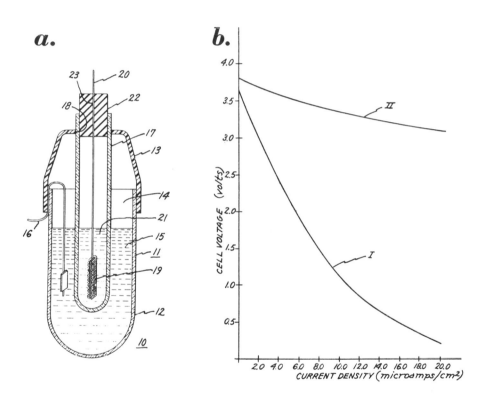

(a)   Sectional view of a lithium-sulfur monochloride cell
(b)   Set of polarization curves showing cell performances of the cell
        shown in Figure 4.5a

Source: U.S. Patent 3,953,233

In Figure 4.5a there is shown generally at **10** a sealed lithium-sulfur monochloride cell embodying this process. An outer casing **11** comprising a lower casing portion **12** of glass and an upper casing portion **13** of polyethylene affixed tightly to the upper open end of the lower casing portion **11**, thereby provides a chamber **14** for a cathode **15** of sulfur monochloride. An electrical lead **16** in the form of a platinum wire or graphite filament is immersed in cathode **15** and extends to the exterior of cell **10** through the junction of the lower and upper casing portions **11** and **12**. An inner casing **17** in the form of a tube of solid lithium-sodium aluminate electrolyte is positioned within casing **11** and immersed partially in cathode **15**. An opening **18** is provided in the top of upper casing portion **13** into which tube **17** fits tightly. An anode **19** of lithium metal in the form of a lithium ribbon pressed onto a nickel mesh which is folded together and attached to the end of a nickel electrical lead **20**. An anolyte **21** partially fills tube **17** and is in contact with lithium anode **19**. An electrically insulating closure **22** with a hole **23** therethrough is provided at the upper end of tube **17** to seal the initially open end of the tube. Lead **20** extends through hole **23** in closure **22** to the exterior of cell **10**.

In Figure 4.5b, performance of the cell shown in Figure 4.5a is provided by polarization curves. In this figure cell voltage in volts is plotted against current density in microamperes per square centimeter.

Thus, it has been found that one could form a sealed lithium-sulfur monochloride cell with a lithium ion-conductive electrolyte by employing a casing having a cathode portion and an anode portion. These two portions are separated by a solid lithium-sodium aluminate electrolyte in disc or tube form which will be further described below. Such a casing may be provided in various configurations such as, for example, shown in Figure 4.5a. The cell of Figure 4.5a employs an outer casing comprising a lower casing portion of glass and an upper casing portion of a plastic such as polyethylene affixed tightly to the upper open end of the lower casing portion, thereby providing a chamber for a cathode such as sulfur monochloride.

An electrical lead in the form of a platinum wire or graphite filament is immersed in the cathode and extends to the exterior of the cell through the junction of the lower and upper casing portions. An inner casing in the form of a tube of solid lithium-sodium aluminate electrolyte is positioned within the outer casing and immersed partially in the cathode. An opening is provided in the top of the upper casing portion into which the tube fits tightly. An anode of lithium metal in the form such as lithium ribbon pressed onto a nickel mesh is folded together and attached to the end of a nickel electrical lead. An anolyte partially fills the tube and is in contact with the lithium anode. An electrically insulating closure with a hole therethrough is provided at the upper end of the tube to seal the initially open end of the tube. The lead extends through the hole in the closure to the exterior of the cell.

Examples of sealed lithium-sulfur monochloride cells made in accordance with the process are set forth below.

Example 1: Two cells, No I and II, were assembled as generally described above and as shown in Figure 4.5a. The cells were constructed in the same manner with the exception that one lithium-sodium aluminate tube, No I, had an 84.7 percent lithium ion content while the other lithium-sodium aluminate tube, No II, had a 1.34 percent lithium ion content. The remaining alkali ion content of the tube was sodium ions.

The tube for cell No I was formed from a tube of sodium beta-alumina approximately 6.2 cm long, 1.1 cm o.d., and 0.15 cm wall thickness. The tube was baked out overnight at 1175°C prior to lithium ion exchange. The lithium ion exchange was made by immersion in lithium nitrate at 600°C for 13 hours. A resulting 3.12% weight decrease corresponded to 84.7% sodium substitution by lithium ions.

The tube for cell No II was formed from an identical sodium beta-alumina tube which was baked out in the same manner. The lithium ion exchange was made by immersion in 20 mol percent lithium nitrate and 80 mol percent sodium nitrate at 400°C for 72 hours. A resulting 0.049% weight decrease corresponded to 1.34% sodium substitution by lithium ions.

For each cell an outer casing was formed of a lower casing portion of glass and an upper casing portion of polyethylene affixed tightly to the upper open end of the lower casing portion, thereby providing a chamber for the cathode of sulfur monochloride in an electrolyte of propylene carbonate, 0.1 molar tetrabutylammonium tetrafluoroborate saturated with $LiClO_4$. An electrical lead in the form of a platinum wire was immersed in the cathode and extended to the exterior of the cell through the junction of the lower and upper casing portions. An inner casing in the form of a tube of solid lithium-sodium aluminate electrolyte was positioned within the outer casing and immersed partially in the cathode. The tube for cell No I contained 84.7% lithium ion content, while the tube for cell No II contained 1.34% lithium ion content.

An opening was provided in the top of each upper casing into which the respective tube fitted tightly. An anode of lithium metal in the form of a lithium metal ribbon pressed onto a nickel mesh was folded together and attached to the end of a nickel electrical lead. An anolyte of 0.1 molar tetrabutylammonium tetrafluoroborate in propylene carbonate saturated with $LiClO_4$ partially filled each tube and was in contact with the lithium anode. An electrically insulating closure with a hole therethrough was provided at the upper end of each tube to seal the initially open end of the tube. The lead extended through the hole in the closure to the exterior of the cell. These structures resulted in two sealed lithium-sulfur monochloride cells made in accordance with the process.

Example 2: The performance of the cells, No I and II of Example 1, is shown in the polarization curves in Figure 4.5b which were produced at a temperature of 36°C. The cell voltage in volts is plotted against current in microamperes per square centimeter for each cell. No attempts were made to minimize interfacial polarization at the lithium-sodium aluminate ion-conductive electrolyte interfaces.

W.L. Roth and G.C. Farrington; U.S. Patent 3,953,228; April 27, 1976; assigned to General Electric Company also describe a sealed lithium-reducible sulfur oxyhalide cell which comprises a casing, an anode positioned within the casing, the anode selected from the class consisting of lithium, lithium as an amalgam, and lithium in a nonaqueous electrolyte. The cathode positioned within the casing consists of a reducible sulfur oxyhalide and a reducible sulfur oxyhalide with an ionic conductivity enhancing material. The solid lithium-sodium aluminate electrolyte positioned within the casing between the anode and cathode and in contact within both the anode and cathode has an approximate composition of $LiNaO \cdot 9Al_2O_3$ of which 1.3 to 85% of the total alkali content is lithium.

In related work, *G.C. Farrington and W.L. Roth; U.S. Patent 3,953,231;*

April 27, 1976; assigned to General Electric Company describe a sealed lithium-solid sulfur cell for ambient temperature operation which comprises a casing, an anode positioned within the casing, the anode selected from the class consisting of lithium, lithium as an amalgam, and lithium in a nonaqueous electrolyte. The cathode positioned within the casing is of solid sulfur in a nonaqueous electrolyte with an ionic conductivity enhancing material. The solid lithium-sodium aluminate electrolyte positioned within the casing between the anode and cathode and in contact with both the anode and cathode has an approximate composition of $LiNaO \cdot 9Al_2O_3$ of which 1.3 to 85% of the total alkali ion content is lithium.

Chlorine Cathode

G.C. Farrington and W.L. Roth; U.S. Patent 4,004,946; January 25, 1977; assigned to General Electric Company describe a sealed lithium-chlorine cell which comprises a casing, an anode positioned within the casing, the anode selected from the class consisting of lithium, lithium as an amalgam, and lithium in a nonaqueous electrolyte, a cathode positioned within the casing, the cathode comprising chlorine with an ionic conductivity enhancing material. A solid lithium-sodium aluminate electrolyte is positioned within the casing between the anode and cathode and in contact with both the anode and cathode. The solid lithium-sodium aluminate electrolyte has an approximate composition of $LiNaO \cdot 9Al_2O_3$ of which 1.3 to 85% of the total alkali ion content is lithium.

Phosphorus Cathode

G.C. Farrington and W.L. Roth; U.S. Patent 3,953,230; April 27, 1976; assigned to General Electric Company describe a sealed lithium-phosphorus cell with positive separation of the anode and cathode by a solid lithium-sodium aluminate electrolyte which is lithium ion conductive.

Phosphorus is an inexpensive, light (31 g/mol) oxidant which exists in many allotropic modifications. The best known modifications are white (yellow) phosphorus, red (violet) phosphorus and black phosphorus. The white form melts at 44.1°C, boils at 280.5°C, dissolves in CS_2, benzene, and other organic solvents and reacts readily and violently with oxygen. Red phosphorus, which has a melting point of about 590°C, comprises a number of modifications which are formed by heating white phosphorus to temperatures above 250°C. Red phosphorus is more stable than white phosphorus and with reasonable care can be handled under normal conditions. Black phosphorus is formed by heating white phosphorus at 220°C and under a pressure of 1,200 kg/cm², or by heating to between 220° and 270°C with mercury on copper as a catalyst in the presence of a black phosphorus seed. The black phosphorus is an excellent conductor of electricity (0.71 ohm-cm at 0°C) and is stable in contact with the atmosphere.

The phosphorus in solid form which is obtained, for example, by pressing onto a screen such as a nickel screen, is immersed in a nonaqueous solvent, which solvent does not react with the phosphorus. While various nonaqueous solvents are suitable, propylene carbonate has been found to be a quite satisfactory solvent for the phosphorus. A satisfactory manner of suspending the phosphorus on a nickel screen in the solvent is by means of an electric current conductor in the form of a wire which is welded to the screen. The opposite end of the wire can then be extended through the battery casing to the exterior to provide a suitable electrical lead to the cathode. Various ionic conductivity enhancing materials which are suitable include chemically stable salts such as tetraalkyl-ammonium perchlorates, tetrafluoroborates and lithium perchlorate.

Example: One cell was assembled as generally described in U.S. Patent 3,953,232. For the cell, a lithium-sodium aluminate electrolyte disc was made by first preparing a cylinder of beta-alumina by firing $Na_2O + Al_2O_3$ plus 1% MgO at 1750°C. The density of the beta-alumina cylinder was 3.224 g/cm^3 corresponding to less than 1% void volume. A disc of 1 mm thickness was sliced from the cylinder and converted to a lithium-sodium aluminate electrolyte by immersion in molten $LiNO_3$ at 400°C for 24 hours. The exchange of the sodium ions for the lithium ions was accompanied by a 1.91% decrease in weight corresponding to approximately 50% sodium ion substitution by lithium ions, and the final density was 3.148 g/cm^3. X-ray diffraction showed that the electrolyte disc has a hexagonal crystal structure with lattice parameters a = 5.603±0.001 Å, and c = 22.648±0.003 Å.

A two-part Teflon polymer casing which included an anode portion and a cathode portion was employed to assemble the cell. Each portion had a chamber with an upper opening and a side opening. The side opening in the cathode portion was further recessed. A silicone washer was positioned in this opening. The above-prepared lithium-sodium aluminate electrolyte disc was positioned against the washer and within the recessed opening in the cathode portion. A silicone washer was positioned between the casing portions and the openings in the washer and in the casing portions were aligned. A pair of threaded fasteners were then employed to hold the casing portions together and tightened at one end by nuts.

The chamber of the anode portion for the cell was provided with an anode consisting of an electrolyte of propylene carbonate with dissolved lithium perchlorate and tetrabutylammonium tetrafluoroborate and a lithium foil anode inserted therein and held in position in the chamber and in contact with the electrolyte. A nonaqueous solvent of propylene carbonate containing lithium perchlorate and tetrabutylammonium tetrafluoroborate was placed in the cathode chamber of the cell.

Phosphorus pressed onto a nickel screen was immersed in the solvent in the cathode chamber resulting in a cathode. The cathode structure was supported by a current collector lead. The resulting device was a lithium-phosphorus cell made in accordance with the process which cell could be readily sealed. The open circuit voltage of this cell was 3.0 volts.

Nonstoichiometric Lithium Compound Cathode

W.L. Roth and G.C. Farrington; U.S. Patent 3,970,473; July 20, 1976; assigned to General Electric Company describe a solid state electrochemical cell which employs a lithium type anode, a nonstoichiometric lithium compound cathode, and a solid lithium-sodium aluminate electrolyte which is a lithium ion conductor.

In Figure 4.6a there is shown generally at **10** a solid state electrochemical cell. The cell has a plastic tube **11** and a solid plastic rod **12** inserted in the upper

FIGURE 4.6: SOLID STATE CELL

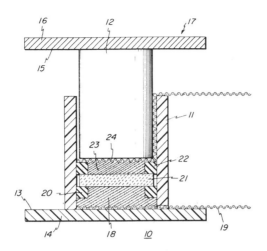

a.

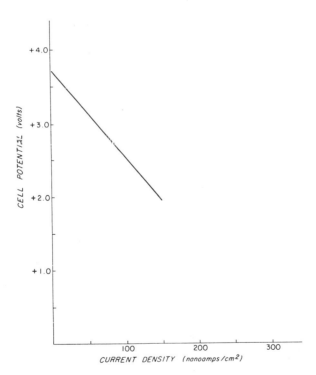

b.

(a)   Sectional view of a solid state electrochemical cell
(b)   Polarization curve showing cell performance of the cell shown
in Figure 4.6a

Source: U.S. Patent 3,970,473

end of the tube 11. The lower end of tube 11 is adjacent to the inner surface 13 of plastic plate 14. The upper end of rod 12 is in contact with the inner surface 15 of plate 16 of vise 17. At the lower end of the tube is positioned a cathode 18 of a nonstoichiometric lithium compound with a cation valence of x where $0 < x < 1$. Cathode 18 is in the form of $Li_{0.33}V_2O_5$ powder pressed into a soft nickel mesh screen lead 19 which is positioned on upper surface 13 of plastic plate 14 between the lower end of tube 11 and plate 14. Screen lead 19 extends externally of tube 11.

A silicone rubber washer 20 is pressed into the upper surface of cathode 18. A solid lithium-sodium aluminate electrolyte 21 in the form of a disc is positioned against the upper surface of washer 20 and in contact with the upper surface of cathode 18. A silicone washer 22 is positioned on the upper surface of electrolyte 21. A lithium anode 23 in the form of lithium metal powder is pressed against and in contact with the upper surface of electrolyte 21 and against and in contact with the inner periphery of washer 22. A soft nickel mesh screen lead 24 is pressed into the upper surface of anode 23 and against the upper surface of washer 22. Solid plastic rod 12 is positioned against screen lead 24 which extends upwardly in tube 11 between rod 12 and tube 11 and extends then externally of tube 11. Plate 16 of vise 17 holds cell 10 together.

In Figure 4.6b performance of the cell shown in Figure 4.6a is provided by a polarization curve at a temperature of 26°C. In this figure cell voltage in volts is plotted against current density in nanoamperes per square centimeter. It has been found that one could form a solid state electrochemical cell with a lithium ion-conductive electrolyte by positioning a plastic plate 14 on a surface, such as a workbench. Tube 11 has its lower portion positioned on the upper surface of plate 14.

Prior to positioning tube 11 on plate 14, a nickel screen lead 19 is placed under tube 11 and extends externally therefrom to provide an electrical lead. Lithium vanadate in the form of $Li_{0.33}V_2O_5$ powder is placed in tube 11 adjacent its lower portion and in contact with nickel screen 19. Silicone washer 20 is then positioned on the upper surface of the powder. Rod 12 is inserted in the upper end of tube 11 and pressed against the upper surface of the powder 18 and washer 20. Vise 17 is tightened to press powder 18 into nickel screen lead 19 and to press washer 20 into the upper surface of the pressed powder. In this manner the upper surface of pressed powder 18 is flush with the upper surface of washer 20.

Vise 17 was loosened and rod 12 removed. Electrolyte 21 in the form of a disc was positioned on the upper surface of a washer 20 and in contact with the upper surface of pressed powder 18. A similar washer 22 was then positioned on the upper surface of electrolyte 21. Lithium metal 23 in the form of powder was placed within tube 11 and generally within washer 22. A soft nickel mesh screen lead 24 is positioned against the upper surfaces of lithium metal 23 and washer 22 and extended upwardly in tube 11 against its inner surface and extended externally of tube 11. Rod 12 was then inserted in the upper end of tube 11 and pressed against the upper surfaces of powder 23, nickel screen lead 24, and washer 22. A portion of lead 24 is located between tube 11 and rod 12 and another portion of lead 24 extended externally of tube 11. Vise 17 was tightened to press nickel screen lead 24 into powder 23 and to press powder 23.

Example 1: The cell was assembled as generally described and shown in Figure 4.6a. A lithium-sodium aluminate electrolyte disc was made by first preparing a cylinder of beta-alumina by firing $Na_2O + Al_2O_3$ plus 1% MgO at 1750°C. The density of the beta-alumina cylinder was 3,224 g/cm^3 corresponding to less than 1% void volume. A disc of 1 mm in thickness was sliced from the cylinder and converted to a lithium-sodium aluminate electrolyte by immersion in molten LiNO$_3$ at 400°C for 24 hours. The exchange of the sodium ions for the lithium ions was accompanied by a 1.91% decrease in weight and the final density was 3.148 g/cm^3. X-ray diffraction showed that the electrolyte disc has a hexagonal crystal structure with lattice parameters a = 5.603±0.001 Å, and c = 22.648 ±0.003 Å.

Example 2: A plastic plate was positioned on a surface. A plastic tube at its lower portions was positioned on the upper surface of the plate. Prior to positioning the tube on the plate, a nickel screen lead was placed under the tube and extended externally therefrom to provide an electrical lead. Lithium vanadate in the form of $Li_{0.33}V_2O_5$ powder was placed in the tube adjacent its lower portion and in contact with the nickel screen. A silicone washer was then positioned on the upper surface of the powder.

A solid plastic rod was inserted in the upper end of the tube and pressed against the upper surface of the powder and the washer. The inner surface of the plate of a vise was tightened against the end of the solid plastic rod to press the powder into the nickel screen lead and to press the washer into the upper surface of the pressed powder. The resulting upper surface of the pressed powder was flush with the upper surface of the washer. The vise was loosened and the rod was removed from the tube.

The above-described electrolyte in disc form was positioned on the upper surface of the washer and in contact with the upper surface of the pressed powder. A similar washer was then positioned on the upper surface of the electrolyte disc. Lithium metal in the form of powder was placed within the tube and generally within the washer. A soft nickel mesh screen lead was positioned against the upper surfaces of the lithium metal powder and the upper washer and extended upwardly in the tube against its inner surface and extended externally of the tube. The rod was then inserted in the upper end of the tube and pressed against the upper surface of the lithium powder, the second nickel screen lead and the second washer.

A portion of the lead was located between the tube and the rod and another portion of the lead extended externally of the tube. Vise **17** was tightened to assert pressure against the end of the rod, thereby pressing the nickel screen lead into the lithium powder and pressing the powder. The resulting device was a solid state electrochemical cell made in accordance with the process.

Example 3: The performance of the cell of Example 1 is shown in the polarization curve in Figure 4.6b. The cell voltage in volts is plotted against current in nanoamperes per square centimeter at a temperature of 26°C.

Metal Chalcogenide Cathode

C.C. Liang and L.H. Barnette; U.S. Patent 3,959,012; May 25, 1976; and U.S. Patent 3,988,164; October 26, 1976; both assigned to P.R. Mallory & Co., Inc.

describe a cathode-active material for use with a solid electrolyte in solid state electrochemical cells. The anode-active material is preferably one of the electro-chemically-active anodic materials capable of displacing hydrogen from water. Such anode metals comprise the light metals and include among others alumi-num, magnesium, lithium, calcium, sodium and potassium.

Preferably the solid electrolyte material for this cell should comprise a compo-sition consisting essentially of lithium iodide and aluminum oxide, with or without lithium hydroxide. Suitable and preferred solid electrolyte systems for this cell are more completely described in U.S. Patent 3,713,897. These solid electro-lytes have a conductivity ranging between 5×10^{-6} and 1×10^{-5} ohm^{-1} cm^{-1} at room temperature.

The cathode-active materials comprise, as active materials, mixtures of a metal halide and at least one metal chalcogenide. The metal for the halide and chalcogenide is selected from metals of the group consisting of lead, silver, copper, mercury, nickel, chromium, iron, cobalt, arsenic, bismuth, antimony, molybdenum and tin. The chalcogenides are the sulfides, selenides and tellurides of the above metals. The preferred halide species are the bromides and iodides, with the iodides being preferred. The suitable metal chalcogenides include PbS, Ag_2S, HgS, Cu_2S, Sb_2S_3, As_2S_3, As_2Se_3, MoS_2, Bi_2S_3, $PbSe$, $PbTe$, Sb_2Te_3, and FeS_2.

Fluid Oxidizer Cathode

A process described by *E.T. Seo, H.P. Silverman and R.J. Day; U.S. Patent 4,020,246; April 26, 1977; assigned to TRW Inc.* relates to a cell which employs a solid beta-alumina electrolyte together with a solid alkali metal anode and a fluid oxidizer cathode. The cathodic oxidizer may be an oxidizing gas dissolved in an organic solvent, or it may be a liquid organic electrolyte solution of an inorganic salt or a metal no higher in the electromotive series than the alkali metal being used for the anode, or it may be a liquid organic electrolyte solu-tion of an organic oxidizer.

These cells exhibit voltages in the range of 2.5 to 3.5 volts, depending upon choice of reactants. Thus, the cells provide a relatively cheap source of elec-trical power which may be used for small electronic devices, such as electronic watches, heart pacemakers, C-MOS circuits, and other similar devices. The follow-ing examples illustrate the process.

Example 1: Approximately 0.3 mol of tetracyanoethylene (TCNE) dissolved in 2.8 g of 0.5 molal solution of sodium hexafluoroarsenate in propylene carbonate were placed into a clean, dry, cylindrical glass vessel. A cylindrical cup made of beta-alumina was filled with sodium and placed in an oven. The sodium-filled cup was heated above 300°C until the electrical resistance dropped substantially. The cup was removed from the oven and the liquid sodium poured out of the cup. The beta-alumina cup was placed in a furnace and fired at 800°C for at least one hour. The cup was then removed and cooled in a protective atmosphere of helium, nitrogen or argon.

Platinum gauze with a wire conductor soldered thereto was then immersed in the organic liquid in the glass vessel. A cylindrical cup of beta-alumina having an outside diameter slightly less than the inside diameter of the platinum gauze was placed in the center of the platinum gauze cylinder. A nickel wire conductor was immersed in the sodium and extended above the top of the cylindrical glass vessel. The cylindrical glass vessel was sealed with epoxy resin. The following tables provide a comparison of power and energy outputs of the cell of this example with prior art mercury and silver oxide cells.

TABLE 1

Type of Cell	Hg*	Ag$_2$O**	Na/TCNE***
Open circuit voltage, V	1.40	1.60	3.20
Nominal operating voltage, V	1.32	1.50	3.00
Total capacity, mAh	1.60	165	57.7
Output power, μW	30	30	30
Specific power, μW/g	15.1	11.7	28.8
Power density, μW/cm^3	59.5	59.5	70.6
Total energy, mWh	224	247	173
Specific energy, mWh/g	113	96.7	167
Energy density, mWh/cm^3	454	502	408
Operating life at 30 μW, hours	7,460	8,230	5,770

*No. 675E mercury cell
**No. 303 silver oxide cell
***Na/TCNE cell of comparable packaging at 220°C

TABLE 2

Na/TCNE Cell

Capacity, based on active material	380 coulombs; 106 mAh (±5%)
Open-circuit voltage:	
Initial	3.2 V
After half discharge	2.1 V
Load, after half discharge	100 kΩ
Current, after half discharge	3.0 μA
Voltage, after half discharge	0.30 V
Total discharge, 50 months	397 coulombs; 110 mAh

Table 2 shows a 50 month performance of the cell in this example. When placed in operation, the cell had an open-circuit voltage of 3.2 volts. At the end of the 50 month period, during which the load specified in Table 2 remained constant, the open-circuit voltage was 2.1 volts.

Example 2: A cell substantially identical to that of Example 1 was constructed except that the catholyte consisted of 0.2 mol of sodium hexafluoroarsenate in 10 grams of dimethylsulfoxide. Air was bubbled through this solution. In addition, the platinum gauze was replaced by a gauze of gold-mercury amalgam. The beta-alumina electrolyte has a surface area of 1.54 cm and a thickness of 0.14 cm.

The cell was operated in a temperature region of 90°C. Open-circuit voltage of the cell was 2.6 to 2.8 volts and the internal resistance was approximately 350 ohms. The short circuit discharge current was approximately 9 mA and the discharge current across a 350 ohm load was 4.5 mA at 1.3 volts.

Powdered, Hydrated Cupric Sulfate in Cathode

A process described by *Y. Ito; U.S. Patent 3,909,296; September 30, 1975; assigned to Daini Seikosha KK, Japan* relates to cells having solid electrolytes such as beta-alumina and including compounds having water of crystallization as the cathodic-active materials.

According to this process, the free water produced at the cathode is absorbed by a hygroscopic substance, positioned adjacent to the cathode structure. The chemical reaction of the cell is not prevented by this substance. Thus the co-efficient of utilization for the cathodic-active material is increased, and consequently the capacity of the cell is increased. The cathodic-active material is a mixture of powdered cupric sulfate pentahydrate ($CuSO_4 \cdot 5H_2O$) and graphite in a weight ratio of 70:30. Binders may be included. Other cathodic materials containing water of crystallization such as cupric sulfate trihydrate may also be used.

Hydrated Copper Salts in Anode

K. Takeda; U.S. Patent 3,977,899; August 31, 1976; assigned to Daini Seikosha KK, Japan describes a solid electrolyte cell having beta-alumina as a solid electrolyte. In this cell, positive-active material and negative-active material are opposed to each other with an intervening beta-alumina solid electrolyte layer. The utilization of the positive-active material is increased in order to prolong the life of the heightened cell by mixing a conductive metal powder into the positive-active material or by inserting an electrical conductor into the positive-active material which is inert to the active material. The conductor is conductively connected to the electrode terminal of the cell.

The positive-active material is preferably a hydrated copper salt and is prepared as follows. Cupric sulfate pentahydrate, cupric sulfate trihydrate, and graphite in powdered form are mixed in a weight ratio of 10:10:2. Then metal powder in $\frac{1}{10}$ the weight of the mixture is added as current collector and mixed further. This cathodic composition is formed into pellets by hydraulic pressure in a suitable mold.

In this case, copper or silver powders are used as the current collector; however the process is not limited to these powders. Other conductive forms of the material such as small pieces, pieces of metal nets, wire pieces of filaments, screens, turnings, chips or other subdivided elongated metal particles, or other metal powders electrically conductive and inert or nonreactive to the chemical reactions of the cell may be used.

Alkali Metal Solution in Organic Solvent

R. Galli and F. Olivani; U.S. Patent 3,912,536; October 14, 1975; assigned to Montecatini Edison SpA, Italy describe an alkali metal negative electrode for solid electrolyte electrochemical devices, consisting essentially of a solution of an alkali metal in one or more aprotic organic solvents. Complexing agents for the alkali metal may be present in the solution. The alkali metal solution may be saturated with respect to the alkali metal, which may be present as solid phase. The aprotic organic solvents may be selected from the class consisting of ethers, polyethers, N-hexalkylphosphotriamides and tertiary amines.

Example 1: A solution containing 2 g of naphthalene in 20 cc of tetrahydrofuran (THF) is introduced into a little flask provided with stirrer, reflux cooler, nitrogen inlet and outlet tubes and feed funnel. Under stirring and in a nitrogen stream, 0.5 g of lithium in small pieces is then added. In a short time the lithium dissolves and the solution turns dark; the lithium in excess is present as solid phase. The solution thus obtained is kept in a nitrogen atmosphere. The conductivity of the solution at 20°C is 1.5×10^{-3} ohm^{-1} cm^{-1}.

In an inert and anhydrous atmosphere of argon, the solution is contacted with a thin LiI sintered disc (surface about 0.8 cm^2; thickness 0.3 mm) which, in its turn, contacts a disc having the same dimensions and consisting of an intimate mixture of iodine (80% by weight) and graphite (20% by weight). The cell thus obtained generated a 2.62 V electromotive force which is checked for one week and is found to be constant. The theoretical value of the electromotive force of a lithium/iodine cell, calculated from the value of the LiI formation free energy, is 2.79 V. Therefore the value actually obtained is in good agreement with the calculated value and proves that the lithium saturated solution actually behaves as a lithium electrode.

Example 2: By using the same lithium solution as in Example 1, some cells are assembled utilizing various solid electrolytes and various cathodic materials for the positive electrode. More precisely, lithium iodide and lithium bromide are employed as solid electrolytes.

The following materials are used as cathodic materials for the positive electrode: $AgO-Ag_7NO_{11}$ (a mixture of AgO and Ag_7NO_{11} obtained by anodic oxidation of a silver nitrate aqueous solution on graphite anodes); CuS; S + graphite (a mixture containing 78% by weight of sulfur and 22% by weight of graphite); MnO_2 + graphite (a mixture containing 79% by weight of manganese dioxide and 21% by weight of graphite); bromine-impregnated graphite. The following are the values of the electromotive force (EMF) measured when the various cells are thus assembled:

Solid Electrolyte	Positive Electrode	EMF
LiI	$AgO-Ag_7NO_{11}$	2.50 V
LiI	CuS	1.80 V
LiI	S + graphite	2.75 V
LiI	MnO_2 + graphite	2.05 V
LiBr	Br_2 + graphite	3.01 V

Example 3: According to the method of Example 1, a lithium-saturated solution is prepared by dissolving 0.5 g of lithium in a solution obtained by mixing 25 cc of THF with 7 cc of quinoline. The alkali metal solution thus obtained is employed to prepare a cell of the type $(Li)_S/LiI/I_2$, where $(Li)_S$ means the lithium-saturated solution constituting the negative electrode, while the iodine electrode consists of an intimate mixture of I_2 (80% by weight) and graphite (20% by weight). The value of the thus-generated electromotive force (EMF) is 1.45 V.

Example 4: According to the same method of Example 1, sodium-saturated solutions in different solvent systems are prepared. Such solutions are used to prepare cells of the type $(Na)_S/ES/I_2$ where $(Na)_S$ means the sodium-saturated solution which is the negative electrode, the solid electrolyte consists of a sintered body of beta-alumina (obtained by hot-pressing a powder of commercial beta-alumina refractories), and the iodine positive electrode consists of an intimate mixture of I_2 (80% by weight) and graphite (20% by weight).

The following table shows the conductivity values (measured at 20°C) of the various solutions as well as those of the electromotive force (EMF) of the various cells.

Negative Electrode	Solution Conductivity, ohm^{-1} cm^{-1}	EMF, volts
1.7 g Na + 2 g naphthalene + 25 cc THF	8.5×10^{-4}	3.1
3 g Na + 0.8 g phenanthrene + 20 cc THF	1.8×10^{-3}	3.0
0.5 g Na + 2.8 g naphthalene + 20 cc HMPT*	1.5×10^{-3}	2.7
3 g Na + 1 g anthracene + 20 cc THF	6.4×10^{-4}	2.6
1.5 g Na + 1 g stilbene + 20 cc THF	1.5×10^{-4}	2.29

*N-hexamethylphosphorotriamide

Lithium Bromide Electrolyte

W. Greatbatch, R.T. Mead, R.L. McLean, F. Rudolph and N.W. Frenz; U.S. Patent 3,994,747; November 30, 1976; assigned to Eleanor & Wilson Greatbatch Foundation describe a solid electrolyte primary cell comprising a lithium anode, a bromine cathode and a lithium bromide electrolyte. A solid lithium element operatively contacts the cathode material, and one form of cathode material is a charge transfer complex of an organic donor component material and bromine. The organic donor component material can be poly-2-vinylpyridine. Another cathode material is liquid bromine.

The surface of the lithium anode element which operatively contacts the cathode material can be provided with a coating of an organic electron donor component material. When the lithium anode operatively contacts the bromine cathode, a solid lithium bromide electrolyte begins to form at the interface and an electrical potential difference exists between conductors operatively connected to the anode and cathode.

Referring to Figure 4.7, the lithium-bromine cell is generally designated **10** and includes a housing or casing element having a generally cup-shaped base portion **12** and a peripheral rim or flange portion **14**. The base portion can be of rectangular or circular configuration, and the casing is of a material which is nonreactive with bromine. One form of material found to perform satisfactorily is a fluoropolymer material commercially available under the name Halar. The cell includes an anode in the form of a solid lithium element **20** and a current collector element **22** contacting a surface of lithium element **20**.

An anode lead **24** connected such as by welding at one end to current collector **22** extends out through an aperture in the housing base portion **12** making external electrical connection to a load circuit. In forming the anode for the cell, current collector **22** is moved into position adjacent the inner surface of the base portion **12** and lead **24** is inserted through the opening and used to draw or pull current collector **22** tightly against the surface of the housing. If desired, an element or button **26** of anode material, i.e. lithium, can be placed

FIGURE 4.7: LITHIUM-BROMINE CELL

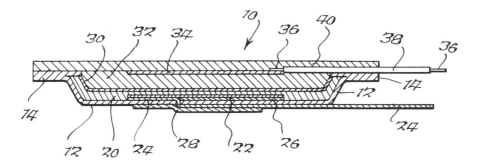

Source: U.S. Patent 3,994,747

between collector **22** and the surface of casing **12** as shown in Figure 4.7. The current collector **22** can comprise No. 12 zirconium mesh having a thickness of about 0.004 inch and lead **24** can be a relatively thin strip of zirconium. Then lithium element **20**, initially in plate or sheet form, is placed in casing portion **12** adjacent collector **22**. The entire assembly then is positioned in a suitable holding fixture and then force is applied to the exposed surface of lithium element **22** in a manner forcing or extruding it along the inner surface of casing portion **12** and along the inner surface of portion **14** so that it conforms to the inner surface of the casing with a resulting shape as shown in the drawing.

A seal or patch **28** of suitable material, for example a fluoropolymer material commercially available under the trademark Tefzel, can be placed over the outer surface of the housing around the aperture through which lead **24** extends and sealed in place by a suitable cement such as the cyanoacrylate cement commercially available under the designation Permabond 100. In addition, the exposed surface of lithium element **20** preferably is provided with a coating **30** of an organic electron donor component material.

The cell further comprises a bromine cathode including a region of cathode material **32** within the assembly and operatively contacting lithium element **20** and a cathode current collector **34** operatively contacting the cathode material **32**. According to a preferred mode of the process, the cathode material **32** comprises a charge transfer complex of an organic donor component and bromine. A preferred organic donor component is polyvinylpyridine polymer, and in particular poly-2-vinylpyridine polymer. Cathode material **32** preferably comprises a mixture of bromine and poly-2-vinylpyridine in a weight ratio of 6:1 bromine to polymer. The mixture is allowed to stand until it develops a rubbery consistency and is of a generally brown-red coloration.

A quantity of the cathode material then is placed in the assembly in contact with the coated lithium element **20** and in an amount filling the open interior region. A cathode current collector and lead combination is positioned in the assembly and in contact with the cathode material. Cathode current collector **34**, which can comprise No. 12 mesh platinum metal, is secured at the periphery such as by welding to one end of a cathode lead **36**, which can be a thin strip of platinum iridium alloy, which is enclosed by a sheet of insulating material **38**, for example the aforementioned Halar material, which lead **36** extends out from the periphery of the casing for making external electrical connection. Then a casing closure element **40** in the form of a sheet of suitable material is placed over the end of the assembly in contact with the peripheral rim or flange **14** and the components are then heat sealed together.

The marginal or peripheral portion of sheet **40** and the rim or flange **14** therefore must be of a material which is heat-sealable, such as Halar. Heat sealing is performed by placing the assembly in a suitable fixture and applying a heated platen to the peripheral end or flange portion at a temperature of about 495±5°F and at a force of about 60±10 pounds. This has been found suitable to provide an adequate seal. While heat is being applied to the periphery of the assembly, the remainder of the cell assembly can be subjected to low temperature refrigeration or gas to prevent expansion and leakage of the cathode material **32**.

The lithium-bromine cell operated in the following manner. As soon as the bromine-containing cathode material **32** placed in the assembly operatively contacts lithium element **20**, a solid lithium-bromine electrolyte begins to form at the interface, and an electrical potential difference will exist between the anode and cathode electrical leads **24** and **36**, respectively, when the current collectors are in operative position. The mechanism by which the foregoing is accomplished is believed to include migration of lithium ions through the electrolyte whereby lithium is the ionic species in the cell.

The following table presents electrical data obtained from a lithium-bromine half cell according to the process as a function of cell life in days. For example, the data entered in the first row of the table were obtained one day after the half cell was placed in operation. The impedance quantities indicate impedance measured at 1,000 hertz, and impedance measurements were made with a 100 kilohm resistance connected in parallel with the cell under test.

Cell Life, days	Open Circuit Voltage, volts	Cell Impedance, ohms
1	3.456	79
5	3.457	120
7	3.457	128
14	3.459	163
19	3.458	190
26	3.457	217
33	3.458	250
41	3.452	318
51	3.451	349

The cathode material **32** comprising a charge transfer complex of an organic donor component and bromine is prepared in the following manner. A preferred organic donor component material is poly-2-vinylpyridine. The mixture is prepared in a pressure-tight container having a pressure-tight closure. The polymer material is placed in the container and then the liquid bromine is added, the preferred ratio by weight of bromine to polymer being 6:1. The container is closed so as to be pressure-tight and is allowed to stand for about one-half day at room temperature. The result is a rubberlike, semisolid plastic mass with no liquid bromine result. The material is removed from the container, this generally requiring some tool or implement, and is placed into the cell assembly in a manner as described previously. It has been found that mixing the bromine and polymer in a somewhat greater weight ratio of bromine to polymer, for example 7.5:1, is not satisfactory. With such a weight ratio the liquid bromine was observed not to combine readily with the polymer but to remain in liquid form. Upon standing, when the mixture was solidified, it was observed to be a very sticky plastic mass which would adhere strongly to glass containers and would release copious amounts of bromine vapor and then upon heating would release liquid bromine. When the mixture is prepared with a bromine to polymer weight ratio considerably less than 6:1, it was observed that not all of the polymer would react with the liquid bromine.

The material of coating **30** on lithium element **20** is an organic electron donor material of the group of organic compounds known as charge transfer complex donors. The material of the coating can be the organic donor used in preparing the charge transfer complex of the cathode material **32**, but other materials can be employed. A preferred material for the coating is polyvinylpyridine and it is applied to the exposed surface of lithium element **20** in the following manner.

A solution of poly-2-vinylpyridine polymer in anhydrous benzene or other suitable solvent is prepared. The poly-2-vinylpyridine is readily commercially available. The solution is prepared from 2-vinylpyridine present in the range from about 10 to 20% by weight with a strength of about 14% by weight of 2-vinylpyridine being preferred. While 2-vinylpyridine, 4-vinylpyridine and 3-ethyl-2-vinylpyridine can be used, 2-vinylpyridine is preferred because of its more fluid characteristics in solution. When the solution is prepared at a strength below about 10% the resulting coating can be undesirably thin, and when the solution is prepared at a strength greater than about 20% the material becomes difficult to apply.

The solution is applied to the exposed surface of each lithium plate in a suitable manner, for example simply by application with a brush. The presence of the anhydrous benzene serves to exclude moisture thereby preventing any adverse reaction with the lithium plate. The coated anode then is exposed to a desiccant in a manner sufficient to remove the benzene from the coating. In particular, the coated anode is placed in a chamber with barium oxide solid material for a time sufficient to remove the benzene, which can be in the neighborhood of 24 hours.

Lithium Haloboracite Electrolyte

T.A. Bither, Jr. and W.K. Jeitschko; U.S. Patent 3,980,499; September 14, 1976; assigned to E.I. du Pont de Nemours and Company have found that lithium haloboracites of the formula $Li_4B_7O_{12}X$, where X is Cl, Br, I or mixtures thereof, are useful as solid electrolytes. The process provides a device for transporting lithium ions which comprises (a) means for supplying lithium ions; (b) means for removing lithium ions, the means for supplying and removing lithium ions being separated by (c) a solid electrolyte consisting essentially of a lithium haloboracite $Li_4B_7O_{12}X$, where X is Cl, Br, I or a mixture thereof; and (d) electrode means to complete an electrical circuit between the supply and the removal means.

The lithium haloboracites are crystalline solids in which unusual ionic conductivity has been found to occur. It has been found that the lithium ions in these haloboracites, unlike the divalent metal ions in ordinary haloboracites, are quite mobile and permit substantial ionic conductivity, especially at slightly elevated temperatures. Moreover, the haloboracites of the formula $Li_4B_7O_{12}Cl_{1-n}Br_n$, where n is 0.2 to 0.05, exhibit unusually high Li^+ mobility at unexpectedly low temperatures.

The use of lithium haloboracites as solid electrolytes makes possible, for example, a primary galvanic cell to provide a lightweight source of stable voltage for low current drain at ambient temperatures. In other applications the lithium haloboracite solid electrolytes may be used at higher temperatures with correspondingly higher conductivity in a lightweight power source employing, for example, liquid lithium anodes. In still other applications the lithium haloboracites may be used as a diaphragm in the electrolytic preparation of lithium metal. The following examples illustrate the process.

Example 1: A mixture of 0.551 g LiCl (13 mmol), 0.932 g LiOH·H_2O (22 mmol), and 3.627 g B_2O_3 (52 mmol) was sealed into a one-half inch i.d. gold tube about 5 inches in length. This tube was maintained for 4 hours at a temperature of 700°C under an external argon pressure of approximately 3 kbar, was then slowly cooled at a rate of approximately 25°C/hr. The resultant solids were extracted with water to remove soluble impurities and clear, colorless, polyhedral crystals of variable size up to several mm across were isolated. X-ray diffraction powder data obtained with a Hagg-Guinier camera upon these crystals indicated the pattern of the boracite-type compound $Li_4B_7O_{12}Cl$ which could be indexed on the basis of a cubic cell of unit dimension a = 12.144 A.

Optical examination of these crystals under polarized light showed that they were anisotropic at room temperature but became isotropic upon gentle warming. By differential scanning calorimetry, two endotherms peaking around 37° and 75°C were observed on upheat. Upon cooldown, the upper transition was rapidly reversible, but the lower transition was sluggish. The change from optical anisotropy to isotropy accompanied the lower of the two transitions. By differential thermal analysis, the onset of decomposition of $Li_4B_7O_{12}Cl$ was observed at a temperature around 870°C.

Example 2: A mixture of 0.721 g LiCl (17 mmol), 1.458 g $LiBO_2$·$2H_2O$ (17 mmol), 1.785 g B_2O_3 (25 mmol), and 1.054 g H_3BO_3 (17 mmol) was sealed into a one-half inch i.d. gold tube about 5 inches in length and reacted as in Example 1. The resultant solids were extracted with water to remove soluble impurities and a mixture of crystals, some clear and some having a crazed appearance, and having growth habits varying from pseudo-cubes to polyhedra to platelets up to several mm across, was isolated. X-ray diffraction powder data indicated this mixture of crystal habits to have the pattern of the boracite-type compound, $Li_4B_7O_{12}Cl$, described in Example 1.

Analysis on these crystals confirmed the composition $Li_4B_7O_{12}$. Calculated: Li, 8.39%; B, 22.87%; O, 58.03%; Cl, 10.71%. Found: Li, 8.05%; B, 23.40%; O, 58.62%; Cl, 10.57%. The crystals were optically anisotropic, but became isotropic after gently warming on a hot plate. After standing without heat for 2 hours the crystals slowly began to exhibit anisotropic behavior.

Examples 3 through 12: Lithium haloboracites having various proportions of Cl/Br were prepared in the same manner as described for Example 2 by mixing in each case 1.458 g $LiBO_2 \cdot 2H_2O$, 1.785 g B_2O_3, and 1.054 g H_3BO_3 with the amount of LiCl and LiBr shown in the table below. Following extraction of the product with water, mixtures of crystals of varying growth habits and degrees of clarity were obtained. Clear crystals having an irregular shard-like habit generally proved to be $Li_2B_4O_7$, whereas those of cubic-like or polyhedral habit were observed by x-ray diffraction powder data to have patterns isotypic with the boracites $Li_4B_7O_{12}Cl$ and $Li_4B_7O_{12}Br$ and from their interplanar spacings to have unit cells intermediate in size between those of the two end members, indicating formation of the mixed chloro-bromo lithium boracites $Li_4B_7O_{12}(Cl, Br)$. Differential scanning calorimetry demonstrated the presence of a transition in these boracites. The temperature at which the maximum endotherm appears on upheat is indicated in the final column of the following table. These materials were all anisotropic below the indicated transition temperature.

$Li_4B_7O_{12}Cl_{1-n}Br_n$ Boracites

| | . . . . . . Reactants . . . . . . | | | . . . . .Product. . . . . | | |
Ex. No.	LiCl (g)	LiBr (g)	Atom Ratio Cl/Br	Cubic Cell Edge a	Fraction Br n*	Transition Temperature (°C)
3	0.649	0.148	90/10	12.150	0.13	42
4	0.541	0.369	75/25	–	–	46
5	0.361	0.739	50/50	12.159	0.32	46
6	0.180	1.107	25/75	–	–	47
7	0.144	1.181	20/80	12.163**	0.41	58
8	0.108	1.255	15/85	12.166**	0.48	91
9	0.072	1.329	10/90	12.166**	0.48	102
10	0.072	1.329	10/90	12.168**	0.52	116
11	0.036	1.403	5/95	12.176**	0.69	160
12	0.018	1.440	2.5/97.5	–	–	197

*From Vegard's law.
**Values from cube root of rhombohedral cell volume.

Example 13: The ionic transport of lithium through a lithium haloboracite was shown by the use of an electrochemical concentration cell. This cell was composed of a saturated lithium amalgam in one leg, and a 0.1 saturated lithium amalgam (1 part saturated amalgam, 9 parts mercury) in the other leg. The two amalgams were separated by a crystal platelet of $Li_4B_7O_{12}Cl_{0.68}Br_{0.32}$ of Example 5. This cell produced an open circuit voltage of 68 mV measured by a high impedance voltmeter between electrodes in the two legs and showed a direct current conductance of about 10^{-7} ohm^{-1}.

When the solid electrolyte crystal was replaced by a liquid ionic conductor comprising $LiClO_4$ dissolved in an organic solvent, the open circuit voltage was scarcely different (70 mV) indicating ionic conduction in the lithium haloboracite. Opposing surfaces of the crystal were then sputtered with gold and the direct current conductance now measured less than 10^{-10} ohm^{-1} indicating that the electronic component of the dc conductivity was no more than a thousandth of the ionic component. Thus the lithium ionic transport number is greater than 0.99 and the lithium haloboracites are useful as battery components.

Lithium-Sodium Aluminate Electrolyte

W.L. Roth and G.C. Farrington; U.S. Patent 3,953,232; April 27, 1976; assigned to *General Electric Company* describe a sealed lithium-reducible metal salt cell for ambient temperature operation which comprises a casing, an anode positioned within the casing, the anode selected from the class consisting of lithium, lithium as an amalgam, and lithium in a nonaqueous electrolyte. The cathode positioned within the casing comprises a reducible metal salt in a nonaqueous electrolyte with an ionic conductivity enhancing material. The solid lithium-sodium aluminate electrolyte positioned within the casing between the anode and cathode and in contact with both the anode and cathode has an approximate composition of $LiNaO \cdot 9Al_2O_3$ of which 1.3 to 85% of the total alkali ion content is lithium. Various suitable salts include the transition metal fluorides, chlorides, iodides, sulfides and oxides.

Conductive Paint Surface Treatment of Electrolyte

T. Harada and S. Funayama; U.S. Patent 3,963,522; June 15, 1976; assigned to Daini Seikosha KK, Japan describe a method of treatment relating to the surface of an electrolyte for a solid electrolyte cell of the Na^+ ion conduction type. In the process, conductive paint containing unoxidized metal powder such as silver, gold or mixtures is coated on the surface of the solid electrolyte and sintered in the range of 700° to 900°C so that faradaic impedance at the electrode-electrolyte interface is lowered. After sintering, the solid electrolyte is treated with fused sodium nitrate salt to prevent characteristic changes during extended storage.

Porous Boron Nitride Separator

J.E. Battles and F.C. Mrazek; U.S. Patent 3,915,742; October 28, 1975; assigned to the United States Energy Research & Development Administration describe a high-temperature electrochemical cell which includes an anode containing an alkali metal, a cathode containing a chalcogen, and an electrically insulative separator wetted with electrolyte between the electrodes. The separator is a porous layer of boron nitride processed to be substantially free of B_2O_3. This purification prevents the formation of an electrically conductive layer resulting from reactions involving boron oxide. The following examples illustrate the process.

Example 1: A sample of about 6 cm^2 of boron nitride cloth woven from 30-mil boron nitride yarn was found to contain about 2.2 wt % B_2O_3 before treatment. The cloth was exposed to dry flowing nitrogen gas at 1650°C for 2.5 hours. After this treatment, the cloth was found to have a B_2O_3 concentration of 0.4 wt %. Other tests conducted on similar cloth for 4 hours at 1750°C showed a B_2O_3 reduction from 2.2 to 0.3-0.5 wt %. Cloth having 0.4 wt % B_2O_3 was submerged in molten lithium for 600 hours at 400°C. On inspection following the test, the cloth was found to be of good integrity and nonconductive.

Boron nitride cloth purified in the above manner and having a B_2O_3 concentration of <0.5 wt % was assembled as an interelectrode separator within an electrochemical cell. The separator fabric was about 1.5 mm thick and of sufficient area to enclose a 12 cm diameter cathode. A molten LiCl-KCl salt was included as electrolyte for wetting the separator and providing ionic conduction between the electrodes. The cell included two anodes of solid lithium-aluminum alloy disposed on either side of a cathode formed of FeS_2 with an expanded mesh of molybdenum serving as a current collector. The cell operated for over 1,390 hours and 41 charge and recharge cycles. During this interval, no electrical shorting between the anode and cathode was noted that could be attributed to the breakdown of the interelectrode separator.

Example 2: An experimental cell including a single anode of porous stainless steel matrix impregnated with molten lithium-25 wt % copper solution was tested with a slightly modified interelectrode separator. The anode was covered with a 40 mesh molybdenum screen of 5-mil wire which was plasma spray-coated with about 5 mils of calcium zirconate to produce an overall thickness of about ½ mm. The cathode was enclosed in a pouch of 1½ mm thick boron nitride cloth purified in essentially the same manner as the separator of Example 1. The cathode was originally spaced about 9 mm from the anode to allow for expansion.

After operating the cell for over 520 hours and 105 cycles, it was found that the cathode had expanded sufficiently to bring the boron nitride cloth into contact with the plasma-coated screen. However, no evidence of electrical shorting attributable to lithium wicking through the boron nitride cloth was observed.

SEAWATER, RESERVE
AND THERMAL BATTERIES

SEAWATER BATTERIES

Bipolar Electrode

H.N. Honer; U.S. Patent 3,966,497; June 29, 1976; assigned to ESB Incorporated describes reserve batteries which operate by immersion in seawater. Seawater batteries of the type relating to the process are single discharge devices often used in considerable quantity and thus are expected to be as low cost in design as is consistent with reliability in all its aspects. The batteries are usually prismatic in shape with rectangular plates and barriers, the plane of the plates and barriers being vertical.

In Figure 5.1a, **10** represents in perspective a completed battery. A sidewall **12a**, a top **14** and an endwall **16** are shown. A second sidewall **12b**, bottom **15** and second endwall **18** (see Figures 5.1d and 5.1e) complete the external features of the battery. In one form and that of Figure 5.1a, the endwalls are made of stiff plastic sheet and the sidewalls, the top and bottom are one or more ribbons of adhesively coated insulating tape wrapped around the endwalls and thus containing the battery. In endwall **16**, a top port **20** is shown near top **14** and sidewall **12a**. A second pot **22** is also shown near the bottom of the battery. Ports **22** and **20** provide for the ingress and egress respectively of electrolyte, e.g., seawater, to the electrochemically active materials of the battery.

Figures 5.1b and 5.1c illustrate a typical bipolar electrode subassembly of the process. In the figures, **24a** represents a barrier made of a sheet of flexible plastic nonconductive material. For seawater electrolyte batteries, any waterproof dielectric sheet material may be used for the barriers, such as polyethylene, polypropylene, polystyrene, etc. However, polyethylene terephthalate has been found to have better handling properties compared to other materials tested. In order to save space in the battery, or in other words to make the battery as small as possible, the barriers should be thin. However, they must be stiff enough to be self-supporting and this puts a limit on the thinness of the barrier. It has been found that polyethylene terephthalate has the property of being stiff and other-

wise more suitable than other available plastic sheet materials in the thickness
range of 0.05 to 0.25 mm.

FIGURE 5.1: SEAWATER BATTERY

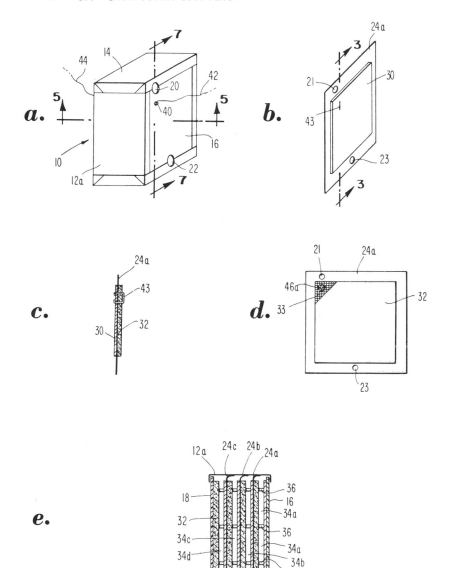

(continued)

FIGURE 5.1: (continued)

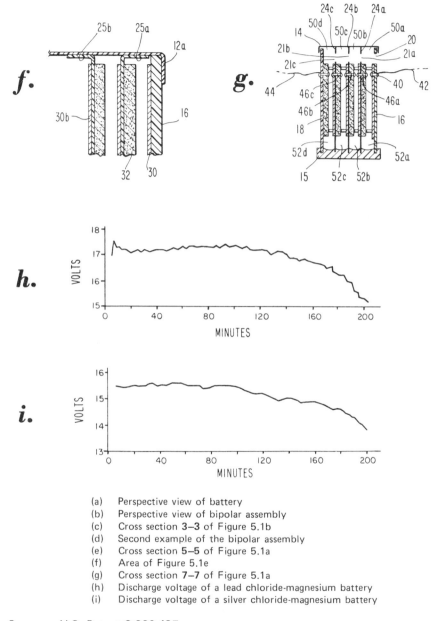

(a)   Perspective view of battery
(b)   Perspective view of bipolar assembly
(c)   Cross section 3–3 of Figure 5.1b
(d)   Second example of the bipolar assembly
(e)   Cross section 5–5 of Figure 5.1a
(f)   Area of Figure 5.1e
(g)   Cross section 7–7 of Figure 5.1a
(h)   Discharge voltage of a lead chloride-magnesium battery
(i)   Discharge voltage of a silver chloride-magnesium battery

Source: U.S. Patent 3,966,497

30 represents an anode located next to one face of the barrier and **32** presents a cathode next to the opposite or second face of the barrier. Normally, the anodes or negative electrodes of seawater-type batteries are metallic and are chosen from magnesium, zinc and aluminum, the active material of the cathodes or positive electrodes are sheets of chemically active material including the halides of metals, such as lead chloride, silver chloride, copper chloride, etc., with a conductive metal screen or other conductive grid embedded therein. A fastening device **43**, in this case a staple, penetrates the anode, the barrier and the cathode. It serves both to provide an intercell electrical connection from anode to cathode and to fasten both electrodes to the barrier, thus forming the bipolar assembly.

Other forms of intercell connections passing through anode, barrier and cathode include rivets and metallic clips. These also provide the double function noted above. Alternately, a metallic or conductive plastic ribbon may be folded over an edge of the barrier to which the electrodes are contacted by pressing the battery together at the time of final assembly. These do not provide the support of the mechanical connector.

It should be noted that, in other forms of bipolar electrodes where a metallic barrier is used, the electrical conductivity of the connector between anode and cathode will be many times greater than that of the fastening devices used in the process. However, since the devices to which the process relates are used only at comparatively low rates of discharge, the fastening devices as discussed have proven to be satisfactory. Ports for transmission of electrolyte, such as those shown at **21** and **23**, complete the bipolar subassembly.

It has been found that, when batteries employing fasteners such as rivets or staples for intercell connectors passing through anode, barrier and cathode are discharged, particularly for long duration discharges, the cathode materials are electrochemically reduced and occupy less volume. This reduces the tightness of the contact and may cause a reduction in the battery performance. Therefore, it has been found advantageous in batteries built for long duration discharges (8 hours or more) to have a portion of the cathode grid structure free of the cathode active material, and to make the electrical contact to the bare grid. This assembly is shown in Figure 5.1d in which a bare corner **33** of the cathode **32** is shown having a rivet **46a** passing therethrough.

Figure 5.1e is a horizontal cross section of the battery of Figure 5.1a along the line **5—5**. This battery comprises four cells. The interior of the battery is divided into four cell compartments by the barriers **24a**, **24b** and **24c**, each barrier being located between each adjacent pair of cells of the battery, and each with its electrodes attached being a bipolar electrode subassembly. It is to be noted that endwall **16** supports an anode **30** on its inner face and that endwall **18** similarly supports a cathode **32** thus completing the first and fourth cell of the battery. Electrolyte spaces **34a**, **34b**, **34c** and **34d** are provided between the anode and cathode of each cell.

The electrolyte space **34** is preserved by plate spacing means. In the design of Figure 5.1e, the means comprises a series of plastic buttons **36** adhered to the face of one of the electrodes, preferably the anode. Alternatively, sheeted materials such as nonwoven fabrics may be used for plate separation. In Figure 5.1g, a connector **40**, in this instance a rivet, electrically connected to the anode adjacent to endwall **16** passes through endwall **16** and is in turn connected to terminal wire **42**, a similar arrangement provides a connection from the cathode attached to endwall **18** to terminal wire **44**. In seawater batteries of this type, there is a natural circulation of electrolyte from bottom to top of each cell. The circulation is due to a combination of temperature differentials, gas bubble formation and increase of density due to formation of partially soluble end products. The circulation of electrolyte is necessary to wash out the end products and provide fresh electrolyte for the duration of the discharge. This is known in the art and numerous patents have been issued relating to means of porting of cells and batteries.

In order to obtain circulation in a battery of the type here discussed, it is necessary that the battery operate in a generally vertical position. This can be identified by being a position in which a first electrolyte port is above the second port. The orientation of the battery as shown in Figure 5.1a is the normal operating position as shown by the location of top port 20 and bottom port 22. The sidewalls 12a and 12b (Figure 5.1e) and endwalls 16 and 18 (Figure 5.1e) are in vertical planes and the top 14 and bottom 15 (Figure 5.1g) are in horizontal planes.

The width of the barriers is somewhat greater than the width of the endwalls. At the time of battery assembly, two opposing edges of each barrier are folded over to form flanges. It is desirable that the flanges do not interfere with the ports such as 21, 23 (Figure 5.1b) and, therefore, in the construction shown, the flanges are formed on the vertical rather than the horizontal edges of the barriers. The flanges 24 extend the entire height of the barriers and on each edge thereof. The flanges of one cell approach the flanges of an adjacent cell so that the flanges form an inner part of the vertical end walls of each cell except for one or the other of the end cells. One or more adhesively coated tapes 12a and 12b, for sealing the sidewalls of the battery, are wrapped around endwalls 16 and 18 and contact the outer faces of each of the several flanges of the barriers.

The faces of the flanges provide an adherence surface for the adhesive tape and serve to locate and hold the several bipolar assemblies with respect to the battery assembly as well as sealing off each cell compartment from the cells adjacent thereto. The flanges 24 and the tape or tapes 12a and 12b alone form the endwalls of the individual cells. The tapes 12a and 12b also overlap the endwalls 16 and 18 of the battery and are sealed thereto by the adhesive coating. The actual sidewalls of the completed battery are a composite of the flange material 24a to 24(n-1) and tapes 12a and 12b, except for the side walls of the first cell or alternatively the nth cell depending upon the orientation of the flanges. The sidewalls of this one cell are formed by the adhesive tape or tapes alone. The construction is shown in Figure 5.1f, an enlargement of the top corner of Figure 5.1e.

In Figure 5.1f, 25a represents a flange formed on the vertical edge of barrier 24a, and 25b represents a flange formed on the vertical edge of barrier 24b. Figure 5.1g depicts in section the battery of Figure 5.1a along the line 7—7. The anode 30 and cathode 32 are somewhat shorter than the full height of the battery so as to leave top passages 50a, 50b, 50c and 50d and bottom passages 52a, 52b, 52c and 52d within the structure. The top passage 50a runs from sidewall 12a to sidewall 12b and is defined by the first endwall 16, the first barrier 24a and the top wall 14. Further, it is continuously open to the electrolyte space 34 of the first cell. The bottom passage 52a likewise runs from sidewall 12a to sidewall 12b and is defined by the bottom wall 15, the first endwall 16 and the first barrier 24a. The first bottom passage is continuously open to the electrolyte space 34 of the first cell. The top passages 50b, 50c, 50d and bottom passages

52b, 52c and 52d defined in turn by the barriers and the second endwall are associated with each of the succeeding cells of the battery. The several passages are connected together and to the outside by ports 20 and 21a, 21b and 21c.

Rivets 46a, 46b and 46c, Figure 5.1g, serve the same purpose of intercell connector and fastener as the staple 43 of Figure 5.1c. Alternate constructions are shown in Figure 5.1g for the top wall and bottom wall of the battery. In the first construction, as shown at 14, a strip of adhesive tape is placed over the top or bottom of the battery and pressed down to contact the top or bottom edges of the barriers, the sidewalls and the endwalls of the battery. Alternately, as shown at 15, the battery is dipped into a liquid plastic or hot-melt material which is then permitted to set so as to form a solidified material. One type of suitable plastic is epoxy resin. A typical hot melt may be a mixture of waxes, natural rosins, etc., as known in the art. The second of these constructions is more rugged than the first.

In the manufacture of the cathodes described above, it is found that the support grid may be embedded into a sheet of the chosen cathode mix in such a manner that the grid is exposed on one surface of the cathode and distant from the opposite face. When such a cathode is used in a battery of the construction noted above, it has been found that the position of the grid, with respect to the electrolyte and the barrier flanges, is an important factor in achieving rapid activation. It has been found that the battery will activate more rapidly upon immersion if the face of the cathode, having the grid exposed on its surface, is in direct contact with the electrolyte and away from the barrier.

The flanges of the barriers may be folded either toward the anode side of the barrier or the cathode side. Where the cathode is thick and rugged, there is little choice as to the direction of the fold. However, when the cathode is thin and fragile, it is desirable to fold the flange toward the anode. By this practice, there is less chance of breaking the edges of the cathode which might result in poor performance and possible short circuiting of one or more cells. In Figure 5.1f, flange 25a is shown folded toward the cathode 32. Flange 25b is shown folded toward anode 30b.

Seawater batteries as built, using the constructions noted above, have been completely successful in test and in actual service discharges. Of particular interest has been the ability of the battery to successfully withstand extreme environmental exposure including shock and vibration testing.

Figure 5.1h shows the voltage during discharge of a 16-cell lead chloride-magnesium battery made in accordance with the process. From the 3 minute point to the end of the discharge, the battery was immersed in a 3% saline solution at 30°C and the load resistance was 87 ohms. Figure 5.1i shows similar information resulting from the discharge of a 13-cell silver chloride-magnesium battery under similar conditions.

In related work, H.N. Honer; U.S. Patent 3,953,238; April 27, 1976; assigned to ESB Incorporated describes a similar multicell seawater battery which is provided with a top venting means. The venting means comprises a series of passages, each located above and communicating with a particular cell. The several passages are connected by ports alternately at one end and at the other end of the cells to form a zig-zag conduit at the top of the battery. A similar passage may be located at the bottom of the battery. The improved porting reduces electrical losses to ground and also reduces the background noise of the battery.

Cast Electrode

A process described by *T.J. Gray and J. Wojtowicz; U.S. Patent 4,016,339; April 5, 1977; assigned to Minister of National Defence, Canada* relates generally to seawater batteries, utilizing electrochemically active material such as lead chloride, cuprous chloride or a lead chloride/cuprous chloride mixture as the cathode electrode material, and more specifically relates to a method of forming cathode electrodes from such materials.

In the process, there is provided a battery electrode structure having a substantially flat configuration comprising a cast mass of electrochemically active material, the mass having contained therein and exposed opposite major surfaces thereof an open-mesh or perforated electrically conductive structure adapted for connection to a battery terminal. The cast mass further contains an open-mesh electrically conductive support member in contact with the exposed open-mesh electrically conductive structure.

Preferably, the electrochemically active material is lead chloride, cuprous chloride or a mixture of both. Since the structure is formed as a cast mass, the structure is of superior mechanical strength to the compressed powder electrodes of the prior art.

Referring to Figure 5.2a, a demountable steel mold construction **10** comprises side walls **11**, a bottom wall **12** and end walls (not shown). Adjacent each wall **11**, inside the mold, is placed a copper wire screen **13**. The screens **13** are separated and supported by an intermediate copper screen **14**, formed into corrugations. Portions **13a** of the screens **13** freely extend above the top of the mold **10**, and the screens form the current collection elements for the subsequently cast electrode. Typically, the screen size is 14 mesh.

For casting the electrode, lead chloride **15** is melted in a graphite crucible and brought to a temperature of about 520°C. The melt **15** is then poured into the mold **10** and allowed to cool. After cooling and solidification into the plate-like mass **15a**, the mold is dismantled and the cast electrode **16** removed (see Figures 5.2b and 5.2c).

The electrode is trimmed and cleaned, and a copper wire **17** is then placed between the freely extending screen portions **13a** which are formed down around the wire **17** and soldered thereto as shown at **18**. Finally, the surfaces of the plate are lightly sanded to enlarge the area of the screen wires exposed to the electrolyte.

Thus, the cast electrode comprises a highly conductive perforated support structure (screens **13**) to which the connector wire **17** is directly connected, the support structure being surrounded and in intimate contact with the active mass (the lead chloride mass **15a**). Since the support structure is exposed at the surfaces of the electrode, there is provided a well-developed three-phase boundary (electrolyte-active mass-conductor) when the electrode is placed in an electrolyte, thus giving a highly efficient, low-loss system.

FIGURE 5.2: CAST ELECTRODE FOR SEAWATER BATTERY

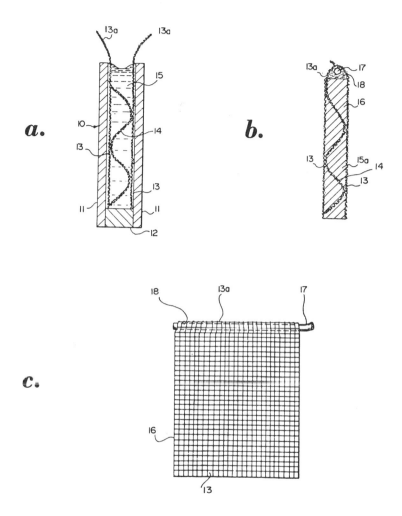

(a) Cross-sectional view of a mold construction from which
a battery electrode structure may be formed
(b) Cross-sectional view of a battery electrode structure which
has been cast from the mold of Figure 5.2a
(c) Side elevation of the battery electrode structure of Figure
5.2b

Utilizing the foregoing method, thick lead chloride plates (⅛ to 1" thickness)
were made by casting from the melt into molds in which current collectors (copper gauze) were previously inserted. The finished electrodes were smooth and
could be produced in various geometrical sizes depending on the proposed utilization. They also exhibited good mechanical strength by virtue of their cast
structure. Optimization studies were done so that lead chloride plates of pre-

determined thickness to meet power requirements and geometric factors of a particular battery application could be fabricated. It has been found that for long life operation (i.e., in excess of several hours duration) Al/PbCl$_2$ batteries should consist of the single cell configuration coupled with an efficient d/c-to-d/c converter. This arrangement avoids intercell shorting problems but provides higher voltages than could be supported by a single one-cell battery.

Cells, containing four PbCl$_2$ cathodes of 4" x 8" size and five aluminum alloy anodes, have been tested in flowing seawater under intermittent load conditions. The battery operated for 50 days with a specific energy yield of about 40.7 watt hours per pound. An interesting aspect of these tests is that in the initial stages of battery life fast attainment of high power delivery could be enhanced by shorting the cell through a small resistance load. During later stages, the rise to higher power levels took place without shorting; apparently the self-activation process is related to the surface state of the anode. Results of these tests are briefly summarized in the following table. The flow rate of the seawater was 7.5 liters per minute.

Daily Operating Level	EMF (volts)
22.5 hours at 0.375 watt	0.66 to 0.8
1.0 hour at 3.25 watts	0.47 to 0.64
0.5 hour at 3.75 watts	0.42 to 0.63

Lead Chloride Cathodes

A process described by *H.N. Honer, F.P. Malaspina and W.J. Martini; U.S. Patent 3,943,004; March 9, 1976; assigned to ESB Incorporated* relates to primary batteries having cathodes containing lead chloride and which are activated by immersion in seawater.

The seawater battery comprises a metallic anode and a cathode having an active material of lead chloride containing a small portion of lead oxide. The lead oxide may be litharge (PbO), litharge containing lead where the lead is about 20% to 30% of the lead-lead oxide mixture, red lead (Pb$_3$O$_4$) or lead peroxide (PbO$_2$). The ratio of lead oxide to lead chloride may vary between about 1 part by weight lead oxide and 100 parts by weight lead chloride and about 10 parts lead oxide to 100 parts lead chloride. A preferred range is from 2 to 3 parts by weight lead oxide to 100 parts lead chloride. Other ingredients in the cathode may include carbon powder up to about 15% as a conductivity agent and a binder, such as polyfluoroethylene (PFE), up to about 5%.

In a preferred method of making the cathode mix for the battery, the several ingredients needed for the cathode are blended and then extruded in sheet form as with an extruder or with calender rolls. The sheet so obtained, is cut to size and a metal mesh grid is cold pressed therein to provide a completed cathode.

It will be seen that the cathode of the process differs from the prior art in the addition of lead oxide to the lead chloride cathode mix. The lead oxide addition has been found to materially reduce the activation time of lead chloride cathodes upon immersion, particularly when the solution (seawater) in which it is immersed is at a low temperature (i.e., near 0°C). The three common oxides of lead, litharge, red lead and lead peroxide, all act to reduce the activation time of lead chloride. Red lead appears to have the least effect, weight for weight, of the three oxide types. It has been found that lead-lead oxide (i.e., litharge containing lead), a material used in great quantity in the battery industry, also provides the beneficial effect, and it is to be included in the term lead oxide as used in this discussion.

The amount of the lead oxide addition is preferably about 1 to 10 parts per 100 parts by weight of the lead chloride. This differs substantially from the prior art use of lead powder in lead chloride cathodes where the optimum ratio of lead to lead chloride is shown to be 30:70 or 30% of the total mix weight. The lead powder addition of the prior art is largely inert so that the electrode containing it has a large quantity of a heavy, nonactive filler. In this process, the lead oxide is present in smaller quantities, and it can be discharged whereby it adds to the total electrical output of the cathode.

The cathode is characterized by having a rapid rise in voltage upon immersion in seawater, even after long storage periods, adverse temperature and humidity conditions, and under a heavy electrical load at the time of activation.

Lead Chloride Grid

According to a process described by *L.J. Burant; U.S. Patent 4,021,597; May 3, 1977; assigned to Globe-Union Inc.* a lead chloride cathode, suitable for use in seawater batteries, comprises a supporting electrically conductive grid to which lead chloride is applied by dipping the grid in a molten lead chloride bath. The grid is preferably made of mesh or expanded metal and, after allowing the lead chloride to solidify, a conductive matrix is formed in the cathode by inserting it in a salt solution and partially discharging it to form electrically conductive lead pathways within the cathode adjacent the points at which the grid material is closest to the surface and the grid itself. The preferred method comprises dipping the mesh or grid material in molten lead chloride, removing the grid and the adhering lead chloride, allowing the lead chloride to solidify, and partially discharging the cathode to form conductive lead portions.

Steel Wool Fiber Cathodes

A.E. Ketler, Jr.; U.S. Patent 3,907,596; September 23, 1975; assigned to Ocean Energy, Inc. describes a seawater battery comprising thin sections of metal wool held under high contact pressure by a conducting metal, with the sections electrically connected to each other to form a cathode, and a dissimilar metal anode between two or more of the cathode sections.

Referring to Figure 5.3, the cell includes thin, high contact pressure cathodes comprising steel wool fibers 2 held in place between electrically conducting support grids 3. Other metal fibers may be substituted for the preferred steel, and the grid is preferably of steel. Bolts, rivets, or any other means of maintaining the high contact pressure between the grid 3 and the metal fiber 2 are used to increase the electrical conductivity, which creates an especial advantage over the prior art basket cathodes. Preferred are pig-ring fasteners 5 for clamping the fiber 2 in place. The cathode grid 7 has substantially flat surfaces parallel to the anode 1 surfaces hereinafter described so as to provide low internal electrical resistance in the device. One very suitable cathode arrangement is to provide four cathode grids disposed at right angles with respect to the adjacent grids so as to form a box-like arrangement with open top and bottoms, as shown in Figure 5.3e.

The use of high density of steel fibers provides low electrical loss between contacting fibers and is an advantage over the loosely packed baskets in prior art devices.

A generally flat anode **1** is mounted between the above-described cathode grids **7**. The anode is preferably of substantially square-like cross section when the cathode is of the box-like design as illustrated in Figure 5.3e, as previously described. The preferred anode metal is magnesium, due to its availability and workability, but use of this metal is not critical. The anode metal is selected so as to be dissimilar from the cathode metal and so as to create an electric potential in seawater electrolyte. A special feature is the use of helically-wound insulation masking **4** in Figure 5.3d of the anode for extended life operation, in the preferred example. The insulation **4** may be masking tape, electrical insulation tape, or any other insulating material which is easily applied and not destroyed or affected by seawater.

FIGURE 5.3: SEAWATER BATTERY

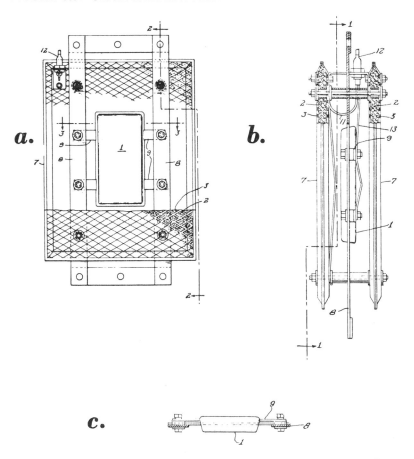

(continued)

FIGURE 5.3: (Continued)

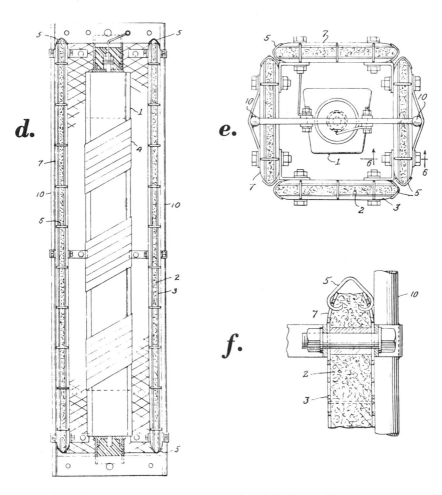

(a)   Front elevational view with a portion of the front grid broken away, the view being taken along the line **1–1** in Figure 5.3b
(b)   End view partly in section taken as indicated by the line **2–2** in Figure 5.3a
(c)   Plan section taken on the line **3–3** of Figure 5.3a
(d)   Front elevational view of another form of the process with the front grid omitted for the sake of clarity
(e)   Plan view of Figure 5.3d drawn on a somewhat enlarged scale
(f)   Fragmentary section taken on the line **6–6** of Figure 5.3e and drawn on a still larger scale

Source: U.S. Patent 3,907,596

In the preferred case, 80% of the anode is masked. This barber-pole winding **4** serves to slow down the corrosion of the anode. The anode **1** is, of course, mounted so as to be electrically insulated from the cathode **7**, for example, by means of nonconducting strap **8** and lug **9** arrangement as shown in Figure 5.3a, or supporting rod **10** arrangement as shown in Figure 5.3e.

Electrical connections to the anode and cathode are shown in Figure 5.3b wherein socket **12** is the terminal for cathode connection **11** and anode connection **13** secured respectively to the anode and cathode by any conventional means. The two conductors may be connected to any suitable load, such as a converter which lowers or raises the voltage or changes it to alternating current, if desired, so as to operate lights, radio beacons, communication equipment, or other devices as desired.

A cell having a generally rectangular magnesium anode approximately four inches on a side by thirty-seven inches long surrounded by four thin high contact pressure steel wool grid cathodes performed at a high energy level when immersed in seawater. This cell, when immersed in seawater, will produce useful long term output voltages in the range of 0.2 to 0.7 volts for periods up to an estimated 1.5 to 2 years. Cells, with total energy capacity levels 233 watt hours per pound of magnesium consumed, may be constructed and give a reasonably steady power output for from six months to two years. Higher power levels are obtainable by increasing the size of the anode and cathodes. Since the cell is open to circulation of seawater, it needs no pressure case or enclosure of any kind.

In both galvanic batteries shown, metal anode means **1** is provided which, in Figure 5.3a to Figure 5.3c, has a pair of vertically elongated, substantially planar, opposed, wide face portions. Each of the pairs of cathodes **7** has a pair of inner and outer, vertically elongated, substantially planar, wide face, grid-like or expanded metal portions **3** with metal wool **2** securely interposed therebetween under high contact pressure. As shown in both examples, the anode means **1** is in a centrally disposed and substantially uniform spaced-apart relationship with respect to the electrodes **7**, and has its wide face portions in an opposed and spaced-apart relation with wide face portions of the cathodes **7**. The cathodes **7**, as shown particularly in Figures 5.3b and 5.3f, are electrically connected together, either by bolt and sleeve means or by angle members, in a spaced relation about centrally positioned anode means **1** and, in the embodiment of Figure 5.3e, define a box-like enclosure.

As also shown particularly in Figures 5.3b and 5.3e, the anode means **1** is securely positioned between and within the spacing between the cathodes **7** by crossextending arms means, such as shown in Figure 5.3f, or by crossextending lug means **9** secured to strap means **8**, such as shown in Figures 5.3a to 5.3c. Importantly, the construction is such that both the upper and lower vertical end portions of the vertically elongated battery elements are substantially fully open to the spacing between the anode means and the cathodes (Figures 5.3b and 5.3f). This enables a flushing through circulation or movement of the saline solution and minimizes clogging difficulties which have been encountered in connection with previous constructions.

Perforated Anode

According to a process described by *D.C.P. Birt, R. Holland and L.J. Pearce; U.S. Patent 4,020,247; April 26, 1977; assigned to Secretary of State for*

Defense, United Kingdom, the cell of a water-activated primary battery includes a cathode and an anode separated and insulated from one another by a porous membrane of paper, cloth or other like material, which is permeable to the flow of liquids and gases. The anode is perforated to permit entry of electrolyte to the cathode during initial priming of the cell as well as subsequent ingress of fresh electrolyte through the perforated anode during discharge of the cell, and to facilitate the escape of gas from within the cell during its operation.

In one arrangement, the anode material is aluminum, magnesium or zinc, and the cathode material is manufactured from an organic oxy-halogen material mixed with carbon black. One such organic oxy-halogen is trichlorotriazinetrione. An alternative organic oxy-halogen material is 1,3-dibromo-5,5-dimethylhydantoin.

Referring to Figure 5.4a, the battery comprises a series of cells, each having a cathode 10 mounted on a backing plate 11, a perforated anode plate 12 disposed in a plane substantially parallel to the plane of backing plate 11 and, sandwiched between, a porous membrane 13 permeable to the flow of liquids and gases.

Electrical connection between the perforated anode 12 of one cell and the cathode backing plate 11 of an adjacent cell is effected by an array of integral conductive dimples or depressions 14 formed in the perforated anode sheet 12.

A water-activated battery, made in accordance with the process, comprised 7 cells, each having as the anode plate 12 a circular sheet of aluminum alloy 0.015 inches thick and 9.25 inches in diameter, the sheet being perforated and dimpled with an approximate density of 9 holes and 9 dimples per square inch respectively. The holes were 0.01 inches in diameter and the dimples were of a size to provide a clear 0.02 inch gap between the anode plate 12 and the parallel cathode backing plate 11.

The cathode 10 material was made from a mixture of 50 to 80% by weight of the common bleaching agent trichlorotriazinetrione, also known as trichloroisocyanuric acid, with carbon black together with a binding agent such as cellulose or viscose fibers. These materials were mixed in the form of a slurry with a dispersing agent such as a volatile hydrocarbon or trichlorethylene which was subsequently evaporated off. Alternatively, the materials in powder form could have been dry mixed and then compressed to the required shapes. The cathode was 9.00 inches in diameter and 0.08 inch thick. The cathode backing plate 11 must be made of a conducting material inert to chlorine in an acidic environment.

Titanium can be used but the presence of a passivate oxide film on the side facing the cathode can give an unacceptably high electrical resistance. This problem may be overcome by scratch brushing the surface of the titanium or by partially coating the titanium with graphite by rubbing it with a graphite rod. The preferred material however, and that used, was zinc sheet 9.25 inches in diameter and 0.013 inch thick painted on the side facing the cathode with a conducting carbon rich paint. This is a lower cost material and gives good electrical contact to both the cathode 10 and the dimpled perforated anode plate 12. The porous membrane 13 was a 9.25-inch diameter sheet of paper although cloth or other similar material is equally suitable. The assembled cells were compacted together under a pressure of 50 pounds per square inch to consolidate the electrode material and to ensure good electrical contact between the cells. The compacted assembly was then sealed by cementing the cells together around their peripheries with an epoxy resin.

FIGURE 5.4: WATER ACTIVATED BATTERY

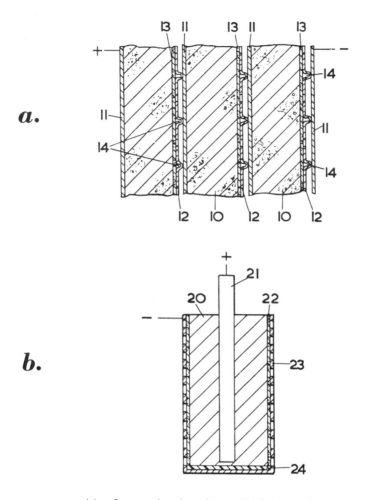

(a)   Cross section through a multicell primary battery
(b)   Section through a single cylindrical water activated cell

In use and for high rate/high power discharges at current densities of over
0.05 A/cm², the battery was activated by initially filling it with a solution of
aluminum chloride of about 1 molar concentration. At lower current densities,
this initial priming is unnecessary and the battery will operate when filled with
a salt solution, such as seawater.

The aluminum chloride formed at the surface of the anode plate **12** diffuses
through the porous membrane **13** into the porous cathode **10** and the perforations in the anode plate **12** allow gas to escape into the space behind the adjacent cathode backing plate **11** where, if necessary, cooling water may be circu-
lated. The perforations also allow entry of electrolyte to the cathode during the

initial priming as well as subsequent ingress of fresh electrolyte during the discharge of the cell. The typical output of a battery based on the above components was 500 watts for 23 minutes.

A further water activated battery comprised rectangular cathode plaques 2.4 inches by 2.4 inches by 0.05 inch thickness made by the dry compaction of mixtures of varying proportions of 1,3-dibromo-5,5-dimethylhydantoin and carbon black with about 2½% of chopped viscose fiber. For improved storage capability, these cathode plaques were used in preference to the trichlorotriazinetrione and carbon black cathodes described above. The ratio of cathode reactant to carbon black was varied from 2:1 to 8:1 with equally satisfactory results. Aluminum, magnesium or zinc can be used as the anode material, preference being given to those alloys of these metals which are known to be electrochemically active in seawater.

With seawater flowing past the back face of the anodes, electrochemical utilization of the reactant was typically in the range of 80 to 90%. Satisfactory discharge curves with constant voltage plateaus have been obtained at 20°C at current densities up to 0.01 A/cm^2. By raising the temperature to 70° to 80°C, the current density could be increased to over 0.10 A/cm^2.

One specific advantage in the use of 1,3-dibromo-5,5-dimethylhydantoin as the cathode material over trichlorotriazinetrione is that the cells can be temporarily drained without evolution of excessively noxious vapors.

Referring to Figure 5.4b, a cylindrical water activated cell comprises a cylindrical cathode 20 (of either of the cathode mixtures referred to with reference to Figure 5.4a) formed around a carbon rod 21 which acts as the cathode current collector. The cathode 20 is surrounded by a porous membrane 22 of paper or cloth which is in turn encased in a perforated can 23 of electrochemically active aluminum alloy which acts as the anode. An insulating plug 24 of plastic material may be fitted at the base of the cathode 20 to simplify construction of the cell. This cell is activated simply by immersing it in a salt solution such as seawater.

Lithium-Aluminum Cell

A process described by *D. Bass; U.S. Patent 3,928,075; December 23, 1975; assigned to Canada Wire and Cable Limited, Canada* relates to a consumable fuel element for battery applications, and more particularly to a water activated fuel element using an alkali metal anode and a corrosive cathode.

The fuel element comprises an alkali metal consumable anode, a consumable cathode in intimate contact with at least a portion of the surface of such anode and which corrodes in aqueous solutions of alkali metal hydroxide of a pH higher than 9 which is formed by the electrochemical reaction of water with the alkali metal of the anode, and an insulating layer separating the anode from the cathode along the contact area of anode and cathode and which is capable of being dissolved in aqueous solutions of alkali metal hydroxide. The insulating layer is of a minimum thickness to insure intimate contact but no electrical short-circuit between the anode and cathode. The thicknesses of the anode, cathode and insulating layer are selected to assure constant mean rate of consumption of all parts of the fuel element. One end of the fuel element is provided with two electric contacts, one to the anode and the other to the cathode for withdrawing electric energy from the fuel element and sealed. The opposite end of the fuel element is open so as to permit water to contact the alkali metal anode and so create and maintain the working environment of the fuel element as long as the element is in contact with water. During storage, a temporary seal may be placed on the open end of the fuel element to keep moisture from contacting the anode but the seal must be removed prior to immersion into water, e.g., seawater.

In a preferred example of the process, lithium is used as the anode material and aluminum or aluminum alloys are used as cathode materials. The thickness of the aluminum material may be in the range of 5 to 15 microns for an anode cross-sectional area ranging between 0.02 and 0.2 cm². When aluminum or aluminum alloys are used as cathode material, the insulating layer may be an oxide layer obtained by exposing aluminum to air. The thickness of the air-formed oxide layer on aluminum materials varies between 50 and 100 A. The thickness of the oxide layer may be increased by anodizing treatment up to one micron or more.

The anode may be in the shape of a cylindrical rod surrounded by a cathode shell of corrosive material such as aluminum or aluminum alloys. The anode may be in the shape of a thin strip surrounded with a cathode shell of corrosive material. The cathode may alternately be in the shape of a matrix of corrosive shell material forming a plurality of channels which are filled with consumable alkali metal anode material. The anode may, in another possible configuration, take the shape of a hollow tube, the inside of which is cladded with a cathode shell and the outside covered with an insoluble insulating layer which may be metallized on the side facing the anode.

Alternately, the anode may take the shape of a plate, one side of which is cladded with a cathode sheet and the other side covered with an insoluble insulating sheet which may be metallized on the side facing the anode. Still another form of the fuel element may consist of a plurality of anode strips connected in parallel by a bridging member and separated by inert spacing strips. The composite plate is covered on both sides with a thin layer of cathode material. One of the covering layers may be a cathode sheet and the other an insoluble insulating sheet which may be metallized on the side facing the anode strips. Plural fuel elements may be electrically connected in series or in parallel or both to make out a battery.

E.L. Littauer, R.P. Hollandsworth and K.C. Tsai; U.S. Patent 4,007,057; Feb. 8, 1977; assigned to Lockheed Missiles & Space Company, Inc. describe a reactive metal-water electrochemical cell with an alkaline electrolyte containing soluble inorganic ions which are reduced preferentially to water at the cathode. More particularly, the additives substantially reduce or eliminate reduction of water and evolution of hydrogen at the cathode resulting in significant improvement in current efficiency, gravimetric energy density and cell potential at a given current.

Additionally, suppression of hydrogen evolution makes possible a sealed system with venting of gases from the cell no longer necessary. The inorganic ions of the process are nitrite, hypochlorite, chlorate, bromate, dinitrogen trioxide and sulfite ions which are formed when soluble salts of nitrite, sulfite, hypochlorate, bromate, chlorate, soluble gases such as dinitrogen trioxide and sulfur dioxide and other soluble inorganic compounds are dissolved in the alkaline electrolyte.

Hydrogen Peroxide as Lithium Anode Modifier

W.R. Momyer; U.S. Patent 4,001,043; January 4, 1977; assigned to Lockheed Missiles & Space Company, Inc. has found that the efficiency of reactive metal-water cells is significantly improved by the use of alkaline electrolytes containing inorganic soluble ions which enhance efficiency of the reactive metal anode. More particularly, it has been found that peroxide ions reduce the anode's sensitivity to changes in hydroxyl ion concentration of the electrolyte, reduce the anode's sensitivity to changes in electrolyte flow rate, and improve the operating temperature characteristic of the cell. The peroxide ions are formed from soluble peroxide compounds such as hydrogen peroxide, sodium peroxide, sodium super oxide, lithium peroxide, potassium peroxide, potassium super oxide and the like.

The use of hydrogen peroxide as a cathode reactant in alkaline solutions to improve cell voltage is well-known in prior art nonconsuming anode-fuel electrochemical cells. Background data on the use of hydrogen peroxide as a cathode reactant is given in *Electrochemical Processes in Fuel Cells* by Manfred Breiter, Springer Verlag, New York, New York, 1969. The use of hydrogen peroxide as a cathode reactant to improve cell voltage in a reactive metal-water electrochemical cell would not appear feasible, however. Hydrogen peroxide is a strong oxidizing agent and reactive metal anodes, such as lithium, are strong reducing agents. The expected reaction upon mixing the two is the direct chemical combination to produce lithium hydroxide and the release of a large amount of heat such that the lithium anode becomes molten. The net result is that no useful energy would be derived.

It has been found, however, that the hydrous oxide film on the reactive anode surface prevents the violent hydrogen peroxide-anode reaction and accordingly permits enhancement in cell voltage. In so permitting the peroxide to be utilized in conventional fashion as a cathode reactant, the improvement realized in voltage enhancement in a reactive metal-water cell is illustrated by the following reactions 1 through 6, where reactions 1 through 3 are illustrative of the basic lithium-water reaction of U.S. Patent 3,791,871 and reactions 4 through 6 exemplify the lithium peroxide reactions of this process. In reactions 5 and 6, HO_2^- represents the peroxide ion which forms when hydrogen peroxide is dissolved in the alkaline electrolyte.

Anode	$2 Li \longrightarrow 2 Li^+ + 2e$	$E° = 3.05V$ (1)
Cathode	$2 H_2O + 2e \longrightarrow 2 OH^- + H_2$	$E° = 0.83V$ (2)
Cell	$2 Li + 2 H_2O \longrightarrow 2 Li^+ + 2 OH^- + H_2$	$E° = 2.22V$ (3)
Anode	$2 Li \longrightarrow 2 Li^+ + 2e$	$E° = 3.05V$ (4)
Cathode	$HO_2^- + H_2O + 2e \longrightarrow 3 OH^-$	$E° = 0.88V$ (5)
Cell	$2 Li + HO_2^- + H_2O \longrightarrow 2 Li^+ + 3 OH^-$	$E° = 3.93V$ (6)

The net result of reaction 6 is the consumption of 2 mols of lithium per mol of peroxide ion with the generation of lithium hydroxide. Whereas it appears that water is also consumed in the process, this is not the case since water is generated by a neutralization reaction:

$$H_2O_2 + OH^- \rightarrow H_2O + HO_2^-$$

The additional voltage and energy potentially available with peroxide is readily apparent from a comparison of reactions 6 and 3. It is generally accepted, however, that voltage much greater than 2 V cannot be achieved in aqueous electrolytes because decomposition of water into hydrogen and oxygen gases will occur. In the process, voltages somewhat greater than 3 V have been achieved, and it has been possible to harness the large amount of available energy between lithium and peroxide to produce useful electrical work. Elimination of hydrogen gas as a reaction product also has significant advantages with safety considerations, gas separation problems from the electrolyte, and in cell/battery design.

Although voltages of 3 V or greater have been achieved with lithium in nonaqueous (organic) electrolytes, only rather insignificant amounts of power (product of cell potential in volts and current in amperes yields power in watts) can be generated. This process yields both high voltage (and therefore high energy density) at high power densities (1.0 W/cm^2 achieved versus 0.01 to 0.03 W/cm^2 in nonaqueous electrolytes). Consumption of lithium is also less than in the lithium-water cell because high power densities can be achieved from voltage rather than current.

It has been further determined that hydrogen peroxide acts as an anode moderator in reactive metal-water cells and, by enhancing efficiency of the reactive anode, reduces the anode's sensitivity to changes in hydroxyl ion concentration of the electrolyte and changes in electrolyte flow rate and improves the operating temperature characteristics of the cell. In conventional nonconsuming anode-electrochemical cells, hydrogen peroxide does not improve anode efficiency and, in some circumstances, has been found to degrade anode performance. Conventional fuel cells normally operate on hydrogen as the anode reactant and oxygen or air as the cathode reactant.

The anode and cathode compartments are separated by a membrane to prevent the direct chemical combination of the reactants. When peroxide is used as the source of oxygen for the cathode, performance is generally not as good as with oxygen gas and anode performance is degraded because it is easier for the soluble peroxide to diffuse through the membrane and react with hydrogen gas.

Lithium-Aluminum Alloy

H.B. Urbach, D.E. Icenhower, M.C. Cervi and R.J. Bowen; U.S. Patent 3,980,498; September 14, 1976; assigned to U.S. Secretary of the Navy describe the use of an electrochemical cell having a lithium-aluminum alloy anode, a suitable cathode, and an aqueous electrolyte such as available seawater. The cell is a liquid-tight package having an electrolyte inlet and outlet for circulating the electrolyte. The electrochemical cells are combined to form a stack and electrically interconnected to provide higher voltage and current output. An electrolyte circulation system, including a pump, reservoir, temperature controller, heat exchanger and waste filters is connected to the stack.

Lithium Chloride in Electrolyte

According to a process described by *J.R. Moser; U.S. Patent 3,928,076; Dec. 23, 1975; assigned to U.S. Secretary of the Army*, the efficiency of a cell containing a magnesium anode and an aqueous magnesium perchlorate electrolyte is increased by incorporation of lithium chloride in the electrolyte.

The mechanism by which the lithium chloride inhibits the self-discharge reaction at the anode is not exactly known. It is well known that chloride ions are readily adsorbed on practically all surfaces. In the process, it is believed that the self-discharge reaction is inhibited by physical adsorption of chloride ions on the surface of the magnesium containing anode, thereby forming a semipermeable shield against excess electrolyte reaching the anode surface while the cell is in the wet condition. When a load is placed across the cell terminals, the chloride ions are expelled from the anode surface so that the majority of the active anode surface area is available to carry and sustain the electrochemical reaction.

Since a considerably lower amount of oxide film is formed due to the lower self-discharge rate, undesired resistance increases due to oxide film are minimized. When the load is removed from the cell, the chloride ions will again adsorb on the active anode surface. Only after long periods of time does this effect of the chloride ions become negligible due to the overwhelming effect of the chemical and electrochemical reaction products.

Copper Oxalate Depolarizer

R.F. Koontz; U.S. Patent 4,007,316; February 8, 1977; assigned to The Magnavox Company describes a deferred action seawater battery having a magnesium alloy anode and a cathode depolarizer. The depolarizer can comprise one of the following materials: copper oxalate, copper formate, or copper citrate and copper tartrate. In addition, the depolarizer contains a form of carbon, a binder, and a metal grid used as a current collector and a base for the cathode.

An example of a battery made in accordance with the process comprises a magnesium or magnesium alloy anode, a cathode having a depolarizer, such as copper oxalate, sulfur, and carbon, such as acetylene black or graphite formed on a conductive metal grid. The anode of the battery could be in the form of a flat sheet or any other convenient configuration and could consist of alkali and alkaline earth metals. A commercially available magnesium alloy suitable for the anode carries the designation AZ61 and has the approximate composition of 6.5 percent aluminum, 0.7 percent zinc, 0.2 percent manganese with the remainder magnesium. A copper oxalate depolarizer has approximately 70 to 80 percent copper oxalate, 15 percent sulfur and 5 to 15 percent carbon.

In most cases, cathode depolarizers are produced from powders and, in many situations, it is desirable to increase the electrical conductivity of the powder. One may add various portions of nonreactive conductive materials to obtain the desired electrical conductivity. Carbon is a preferred material for this purpose because of its low cost and ready availability. Any of the various forms of carbon, such as acetylene black, graphite or petroleum coke, can be used. A binder is also required to hold the powder together. Sulfur is a preferred binder since it has been found to be more efficient than an epoxy resin. The use of sulfur in battery cathodes is old in the art of batteries. An electrical conductor, such

as a metal grid which may be in the form of a screen, expanded metal, or perforated sheet stock, is used to form the cathode. The powder is pressed on and
into the metal grid. The metal grid not only performs as an electron collector
but also lends strength and rigidity to the pressed powder cathode.

In addition to the anode and cathode, a spacer must be provided to separate the
electrodes from one another. Yet the spacer must be in such a form as to allow
free access of electrolyte between the electrodes and to allow corrosion products
resulting from the electrochemical reaction to exit from the cell. This spacer
must be nonconductive and can be in the form of a small disc, rods or mesh.
Chemical reaction between a magnesium anode and the electrolyte produces hydrogen gas and magnesium hydroxide. The gaseous products should be allowed
to escape from between the electrodes. The escape creates a pumping action
which helps pull the solid corrosion products (magnesium hydroxide) out of the
space and causes new electrolyte to enter.

Ordinary tap water can be used as an electrolyte although seawater is preferred.
Maximum power level will be reached faster with salt water than with distilled
water. Salt increases the conductivity of the electrolyte by reducing the resistance of the electrolyte.

The battery assembly is completed by the attachment of lead wires to the electrodes and enclosing everything within a suitable encasement. The lead wires
must extend from the encasement and the encasement must have openings so
that the electrolyte can be allowed to enter between the electrodes.

Figure 5.5a shows polarization curves for batteries having depolarizers of the
prior art and a battery having an improved depolarizer of copper oxalate. The
voltage out of the battery is shown in volts along the ordinate while the current
density in amperes per square inch is shown along the abscissa. Curve **10** is for
a battery having a depolarizer comprised of heavy metal derivatives of aliphatic
dicarboxylic acids such as copper oxalate. The battery was submerged into an
electrolyte maintained at a temperature of zero degrees centigrade with the electrolyte having a salinity of 1.5 percent by weight. The cathode area was eight
square inches with a separation from the anode of 0.052 inches.

Curve **11** is the polarization curve for a battery having a silver chloride depolar
izer. Curve **12** is for a battery having a cuprous iodide depolarizer, while curve
13 is for a battery having a lead chloride depolarizer. It will be noted that curve
10, which uses the improved depolarizer, maintains a higher output power than
the prior art batteries although all the batteries have similar physical dimensions.

Figure 5.5b shows performance curves for a battery using a copper oxalate depolarizer. The output voltage in volts is shown along the ordinate while the
time in minutes is shown along the abscissa. Curve **16** represents the output
when the battery was submerged into a 35 degree centigrade electrolyte having
3.6 percent by weight salinity. Curve **17** is for the battery at zero degrees centigrade with an electrolyte of 1.5 percent salinity by weight. For this example,
the cathode area was eight square inches with a separation of 0.052 inches
from the anode. The depolarizer had approximately 80 percent copper oxalate, 15 percent sulfur, and 5 percent carbon. The load on the battery was
6 ohms.

FIGURE 5.5: SEAWATER BATTERY PERFORMANCE CURVES

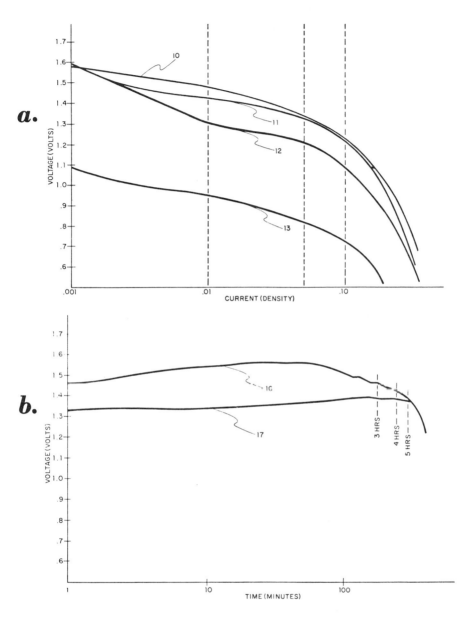

(a) Polarization curves comparing a copper oxalate depolarizer with various prior art depolarizers
(b) Performance curves of a copper oxalate depolarizer

Source: U.S. Patent 4,007,316

Electrode Gap Control

S.A. Black and S.S. Sergev; U.S. Patent 3,920,476; November 18, 1975; assigned to U.S. Secretary of the Navy describe a simple and economical means for controlling the electrode gap in an electrochemical cell that depends on the gap for control of power output. As the anode is consumed, a conical spacer attached to the cathode is allowed to move into a conical hole in the anode, thus controlling the electrode gap.

The spacing control rate can be preset to adjust for other cell parameters which might influence cell output over a period of time. The rate of spacing control is dependent only upon the basic reaction proceeding at any rate and upon the geometry of the spacers which is independent of the basic reaction rate. This technique offers a compact and lightweight system which provides a means of minimizing cell size and weight such as is needed in the case of using the magnesium/iron seawater cell to produce heat at a constant rate for divers.

Flushing Pump

I. La Garde; U.S. Patent 3,959,023; May 25, 1976; assigned to Sanders Associates, Inc. describes a seawater battery which includes a housing formed to define an electrolyte inlet and an electrolyte outlet and a pump hydraulically connected for pumping seawater into the inlet connection. The system includes a light electrical load together with means for selectively energizing the two loads from the battery, and means for actuating the pump concurrently with the energization of the heavy load, whereby, upon such actuation, seawater enters the battery through the inlet and is expelled through the outlet.

It has been found that a system in accordance with the process improves the output voltage of a seawater battery and, in addition, allows a larger amount of power to be extracted therefrom. The intermittent operation of the pump allows the inclusion of a pump of adequate capacity without drawing power unnecessarily during periods of light load. Additionally, the battery can be designed for the best compromise between periods of light load and periods of heavy load.

Injection System

E.E. Huhta-Kowisto; U.S. Patent 3,941,616; March 2, 1976; assigned to Puolustusministerio, Finland describes a seawater battery which has a battery cell system with electrode plates, preferably magnesium and silver chloride plates, and, using as electrolyte, a solution of chemicals introduced from the outside into its cell system. According to a particular characteristic of the process, the pipe or duct by which water is supplied into the cells contains an ejector mechanism drawing from a particular tank a solution of chemicals and mixing this solution with the feed water. The chemicals thus introduced into the battery serve one or several of the following purposes: to increase the conductivity of the electrolyte; to counteract excessively vigorous dissolution of the electrode materials; and to improve the efficiency of the electrolyte.

The equipment forms, in combination, a complete battery unit comprising control equipment and control means, such as valves, and/or a heat exchanger for exchanging heat between the spent electrolyte emerging from the cell system and the entering supply water, these various means being arranged so as to counteract undesirable changes of the battery's voltage output.

Homopolar Motor-Battery

A process described by *G.H. Gill; U.S. Patent 4,024,422; May 17, 1977* relates to a homopolar motor, and more particularly to a practical homopolar motor that is mated with a low voltage seawater activated battery to act as the power source of a scuba assist device.

The homopolar (acyclic, or Faraday) motor is fundamentally a high conductivity disk rotatable in a magnetic field. When current is applied between the shaft and rim of the disk, the current reacts with the magnetic field to produce a torque in the disk. The homopolar motor is also a low voltage, high current device, since the disk is essentially a one turn coil only, and disks cannot be connected in series for higher voltages without considerable difficulty.

In the original version by Faraday, current was conducted into the rim of the disk at one restricted area where the disk dipped into a pool of mercury. A horseshoe magnet applied a magnetic flux across the current zone, between the pool and the shaft of the disk. The remainder of the disk did not contribute to the motor torque since it contained neither current nor magnetic flux.

Modern magnetic materials, such as ceramics, are capable of producing flux over the entire area of the disk, but then the current must also be introduced around the entire periphery of the disk. This peripheral current introduction is a limitation of past designs of homopolar motors; only a centrifugally maintained mercury ring or multiple brushes can be used to introduce the current. The centrifugal ring is impractical for units in mobile service or where there is frequent stopping and starting. Brushes must be applied to the highest velocity portion of the disk, so the friction, wear, vibration and arcing are very difficult to control.

This process provides a homopolar motor-battery combination which eliminates the current introduction problem by making the rim of the disk the anode of a battery, whereby, current is generated in the disk itself by the battery electrolyte. No additional contacts are required. The full potential of the low impedance characterisitic of both elements, the homopolar motor and the seawater battery, are realized when the combination is immersed in electrolyte, since there is an individual battery cell for each disk, with all connected in parallel.

Figure 5.6a shows a plurality of disks **10** mounted on a conductive shaft **12**. The disks **10** are made of a strongly electrochemically positive material, such as gold, nickel, copper, silver or the like (which may be only a plating since the disks are not corroded by the operation). A plurality of permanent disk magnets **14, 14a**, which are axially magnetized, preferably of ceramic material, although other materials, such as Alnico, may be used, are arranged alternately with the disks **10**. A yoke **16** of soft magnetic material is connected to the end magnets **14** to complete the magnetic circuit. The shaft **12** and magnets **14, 14a** are supported within the yoke **16** by any suitable means.

A shroud member **18** of a highly electrochemically negative material, such as calcium, zinc, magnesium or an alloy or amalgam thereof, encloses the rims of the disks, formed to closely conform to the configuration of the disks. The shroud member **18** is supported within the yoke **16** by any suitable means. The magnets **14a** may be attached to the shaft **12** and rotate therewith, or may be attached to the shroud **18** or yoke **16** with clearance allowed so that shaft **12** and disk **10** can rotate freely within the clearance without affecting the operation of the motor.

FIGURE 5.6: HOMOPOLAR MOTOR-BATTERY

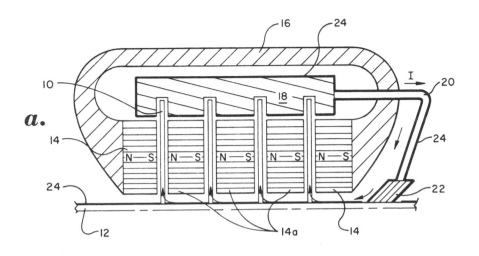

a.

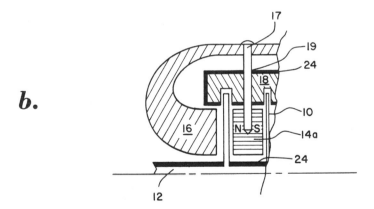

b.

(a) Cross-sectional half-section of the homopolar motor-battery com-
     bination
(b) Cross-sectional part-section of another example of the homopolar
     motor-battery combination

Source: U.S. Patent 4,024,422

Magnets **14** may be attached to the yoke with clearance for the shaft, or may be
replaced by soft pole pieces or extensions of the yoke **16** as is illustrated in Figure 5.6b. A conductor **20**, connected at one end to the shroud member **18**, and

a brush **22**, connected to the other end of the conductor and in contact with the shaft **12**, completes the electrical circuit. An insulating coating **24** covers (a) the entire surface of the shroud member **18**, except for the active surface immediately adjacent to the rims of the disks **10**, (b) the conductor **20**, (c) the brush **22**, except for the contact surface with shaft **12**, and (d) the shaft to the extent possible. Figure 5.6b shows a method of supporting magnets **14a** from the yoke **16** by means of pins **17** connected to the yoke and extending through holes **19** in the shroud **18**. Also shown, is an embodiment in which the yoke **16** is extended to eliminate the end magnets **14**.

When the entire assembly is immersed in or filled with an electrolyte, such as seawater or the like, the disks **10** form the anode and the shroud member **18** forms the cathode of the battery which sends electric current, I, through the body of the cathode, through the conductor **20**, the brush **22**, shaft **12** and radially outward to the rims of the disks **10** as shown by the arrows in Figure 5.6a. The resultant current from the disks causes a force reaction with the axially magnetic field produced by the magnets **14, 14a** whereby a torque is generated to cause rotation of the disks and shaft **12**. The brush **22** continues to transmit current to the shaft **12** as it rotates, resulting in a continuously rotating electric motor.

The motor will stop when it is removed from the electrolyte, but it may also be stopped by opening the conductor **20** with a switch. In this case, however, the interior of the shroud **18** may be wasted away unless it is processed in accordance with the usual practices used in battery manufacture.

The homopolar motor-battery combination provides a self-contained, light-weight power source for use in seawater such as for scuba assists. If such a device is desired for other environments, a housing with appropriate seals, etc., could be used to enclose the homopolar motor-battery combination, such housing being filled with an appropriate electrolyte. The unit may be switched off by draining away the electrolyte, or a switch can also be provided to turn the combination on and off. Alternatively, the electrolyte might also be displaced by a nonpolar fluid to stop the motor.

Pressure Compensated Battery

A process described by *A.R. Waltz; U.S. Patent 3,940,286; February 24, 1976; assigned to U.S. Secretary of the Navy* provides an improved deep ocean power source by using a battery employing seawater as the electrolyte and flexible walls which contain an inert substance during periods of shelf life. Upon immersion in the water, the pressure imposed by ambient seawater compresses the battery walls directly and, although a small amount of the inert fluid may escape, the entrance of the electrolyte is postponed until the battery is activated.

Activation is accomplished by uncovering apertures in the bottom and top of the battery and permitting the difference in the specific gravities of the inert filler and the seawater to displace the inert filler with the active electrolyte. In this fashion, a plurality of batteries may be electrically connected in parallel and sequentially activated such that a uniform source of power may be provided for prolonged periods keeping the expendable buoy operative in the on-station position for extended periods.

RESERVE BATTERIES

Anodized Lead Cathode for Leclanche Cell

J.R. Coleman and T. Valand; U.S. Patent 3,939,010; February 17, 1976; assigned to The Minister of National Defence, Canada have found that anodized lead is satisfactory material for the cathode collector in manganese dioxide-zinc electrolytic cells, and is moreover a relatively cheap material. Unanodized lead has been found to cause considerable voltage irregularities in the cells upon discharge.

Thus according to the process, there is provided a cathode suitable for use in acid Leclanche cells which comprise an anodized lead current collector having adhered to the anodic film a conductive compressed cathodic layer of manganese dioxide and carbon. The thickness of the anodic film on the current collector is not greater than that which would be produced by anodization in 8 normal sulfuric acid with the passage of 18,000 millicoulombs of electricity per square inch of lead surface. The process also includes a reserve primary cell of the acid Leclanche type comprising a cathode as defined above and an amalgamated zinc anode mounted in a container, the cell being constructed to be activatable by addition of sulfuric acid as electrolyte.

Example 1: In this and in the following examples, except where otherwise indicated, all work was performed with 1 inch by 1 inch cathodes. A dry blend was made of 85 parts by weight of an electrolytic manganese dioxide used in the dry battery industry and 15 parts by weight of a commercial graphite also used in the industry. A solution of 5 grams of carboxymethylcellulose (CMC) in 300 ml H_2O (1.67% w/v CMC) was prepared.

1.5 grams of the dry mix was blended with a suitable amount (0.65 to 0.75 ml) of the CMC solution, and this was spread on both sides of a 1 inch by 1 inch metal current collector with a tab attached at one corner for electrical connection. The current collector was held in a small pasting-frame. After drying in air until a suitable amount of water had evaporated (about 1 hour), the material was compressed, preferably at 7,000 lb load (7,000 psi) or above, and then after removal from the frame, dried completely. The cathode material, under compression, must still be moist enough to flow and consolidate but not so wet as to squash, and this is a matter of trial and error.

Anodes were prepared of commercial grade sheet zinc used in the dry battery industry, cleaned in aqueous sodium hydroxide, washed with water and then amalgamated by brief exposure to a dilute solution of mercuric chloride in water or in 8N H_2SO_4. Cathodes were prepared as described above on five metal current collectors: 80 mesh platinum gauze; 0.002 inch lead foil perforated with many small holes so that the cathode material would adhere firmly; and expanded nickel, silver and stainless steel grids.

Each cathode, in turn, was placed between two zinc anodes, with expanded polypropylene mesh as separator, activated by immersion in 8N H_2SO_4, and discharged at a constant current of 120 mA. The voltage of each cell, under load, was plotted against the discharge time in hours and revealed the run-out times (the time taken from the beginning of the discharge to the point at which the cell's voltage fell to 1.5 volts) to be clearly superior for those cells in which a platinum or lead cathode current collector was used. However, the cells employing the lead cathode current collectors show considerable voltage irregularities during discharge, and this was attributed to partial temporary breakdowns of the surface oxide or sulfate film on the lead. When anodized lead foils were substituted for the un-anodized foils, these voltage irregularities on discharge largely or completely disappeared.

Example 2: To show the effects of anodization for different times, cathodes were made on unanodized lead foils or foils anodized at 120 mA for various times up to 5 minutes, employing 8N H_2SO_4 for anodization. The cathodes prepared as described previously were discharged against zinc anodes in 8N sulfuric acid at 120 mA, with the following results:

	. . . . . . Anodization Time, minutes					
	0	1	2	2.5	3	4	5
Average run-out time (hours to 1.5 V)	2.62	3.4	3.3	3.37	2.95	2.9	2.5

Example 3: Employing lead foil anodized for 2 to 2½ minutes at 120 mA and preparing the cathodes as described above, the effect of sulfuric acid concentration on the discharge characteristics of the cells was studied, varying the sulfuric acid concentration from 2N to 10N. Run-out times at 2N were only about 1.5 hours; at 3N about 2.3 hours and at 4N to 6N between about 2.5 and 3 hours, with the best results being close to 3.2 hours or about theoretical for a one electron discharge. At 8N, the run-out times were only very slightly lower and at 10N were still satisfactory at more than 2 hours.

In the MnO_2 cathode discharge in sulfuric acid, an adequate supply of hydrogen ion from the acid must be available to secure reduction of the MnO_2 at the high voltage characteristic of this ion system. If an inadequate quantity of acid is present, the run-out time to the cutoff voltage adopted here (1.5 V) will be lowered. The lower run-out for the 2N acid referred to above is attributed to this effect.

Electrolyte Reservoir

D.K. Szidon; U.S. Patent 3,986,895; October 19, 1976; assigned to Honeywell, Inc. describes a reserve activated electrochemical cell in which a reservoir containing a quantity of electrolyte is placed within an outer cylinder and is biased to force the reservoir to a predetermined active position. Restraining means are provided to restrain the reservoir in a second or reserve position remote from the active position. Upon removal of the restraining means, the reservoir is moved to an active position and cell activation begins. Also included is an electrolytic cell to operably connect with the reservoir, and valve means are provided between the cell and the reservoir to permit flow of electrolyte from the reservoir into the cell under pressure.

As shown in Figure 5.7a, a reserve activated electrochemical cell is shown generally by the numeral 10. This cell includes an outer cylinder 12 having a central axis 13. Positioned within the cylinder 12 is a reservoir 14 which contains a quantity of electrolyte. Restraining means 20 are provided to maintain the reservoir 14 in a reserve position and prevent movement of the reservoir 14 toward the cell 22. Biasing means 18, in the form of a spring, is provided to force against the end 17 and the reservoir 14, serving to force the reservoir 14 towards the cell 22 to an active position upon removal of the restraining means 20. A valve means shown generally as 24 is provided to permit entry of the nozzle 21 into the cell 22 upon activation.

FIGURE 5.7:　RESERVE CELL

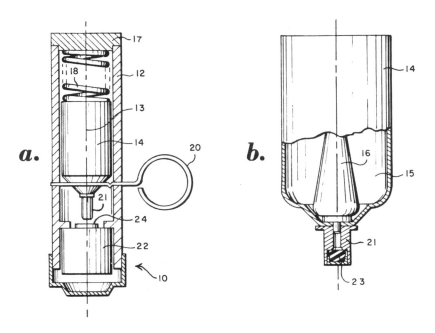

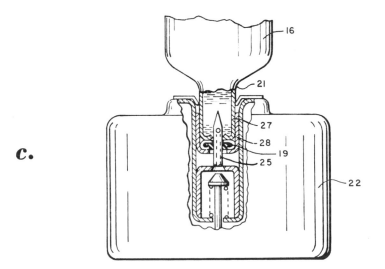

(continued)

FIGURE 5.7: (continued)

d.

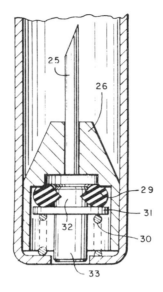

(a)   Partially sectioned view of a reserve activated electrochemical cell
(b)   Partially sectioned view of one portion of a cell
(c)   Partially sectioned view of another portion of a cell
(d)   Partially sectioned view showing a preferred cell

Source: U.S. Patent 3,986,895

As shown in Figure 5.7b, the nozzle is connected to an inner chamber **16** of the reservoir **14**. The inner chamber **16** and the outer wall **14** of the reservoir together define an outer chamber **15** which contains a quantity of inert gas such as argon under pressure. Rupturable means **23**, including a seal **19**, is provided to block the passageway through the nozzle **21** until activation. Rupture of the rupturable member **23**, which may be manufactured from glass, rubber, metals such as aluminum or other easily penetrated materials, or combinations thereof, may be accomplished by movement of the reservoir **14** to the predetermined active position and placement of the nozzle **21** in the valve means.

As shown in Figure 5.7c, the valve means contains a needle **25** which is positioned to penetrate the rupturable member **23** and the seal **19** upon insertion of the nozzle **21** and the reservoir of electrolyte into the predetermined active position. The tube end **27** of the inner chamber **16** and the sides of the valve **28**, which extend upward towards the reservoir, form a seal along with the rubber portion **19** of the rupturable means **23**, which prevents entry of the pressurized gas into the cell **22** from the outer reservoir **15** and allows the electrolyte to flow through the needle **25** in the cell **22**.

In Figure 5.7d, a preferred example is shown which is effective to seal the cell upon passage of electrolyte into the cell. Specifically, a hub **26** is attached to the needle 25 for support of the needle. A T-shaped member having a horizontal portion 31 and a vertical portion 32 is cooperatively mounted with an O ring 29. Spring 30 is positioned to bias the horizontal portion 31 against the O ring 29, thereby causing vertical portion 32 to prevent flow of fluid through the needle 25 and out port 33.

This spring 30 is set to give an effective pressure or resistance to flow in an amount tending to normally keep the valve closed, thereby preventing flow of fluid through the valve. Thus the cell 22 is effectively isolated from outside environments through its only opening, which is in conjunction with the port 33 of the valve shown in Figure 5.7d. Upon arrival of the reservoir to the predetermined position, followed by rupture of the barrier by the needle 25, the pressure contained in the outer chamber, which in one embodiment may be in the order of 80 to 100 psi, acts upon the electrolyte contained in the inner chamber to force it through the needle 25 into the cell 22. The spring 30 in the valve 24 is adjusted to give a nominal back pressure in the order of 14 to 15 psi tending to keep the valve closed.

As the fluid is under a significantly greater pressure than the spring is exerting, the spring is compressed and a quantity of electrolyte flows rapidly from the inner chamber 16 to the cell 22. Upon equalization of this pressure after the required amount of electrolyte has passed into the cell, the spring 30 then exerts its nominal 15 psi of pressure against the horizontal and vertical portions of the plug 31 and 32 and the O ring 29 to effectively seal the cell from the remaining portion of the device. The electrolyte is in the cell at this point and has activated the battery, and it is effectively prevented from escaping from the cell, regardless of the position of the battery during use.

A number of cells were built according to the process and were evaluated for various performance characteristics. A conventional, nonaqueous electrochemical cell, 7-plate (4 anodes and 3 cathodes) 750-mAh reserve battery, was employed. Several of these cells were stored for periods of time ranging up to two weeks at temperatures ranging from –40° to 160°F.

After storage, several of these cells were activated at equilibrium temperatures ranging from 0° to 125°F. The cells were operated against a constant resistance of 60 ohms and the time was measured for discharge from an initial voltage of approximately 3 volts to the discharge voltage of 2 volts. All of the cells activated upon removal of the restraining means and a number of them were successfully capable of discharging for a period of approximately 17 hours under these above-described conditions, thus clearly demonstrating that a useful reserve activated electrochemical cell has been constructed.

Dry Electrolyte Storage

S.F. Schiffer and R. DiPasquale; U.S. Patent 4,005,246; January 25, 1977; assigned to Yardney Electric Corporation describe a reserve-type cell which incorporates dry electrolyte in an electrolyte storage space with the electrolyte disposed in or on bibulous material. The storage space is located behind one of the electrodes and out of contact with the other of the electrodes. However, the electrode nearest the storage space is porous so that liquid electrolyte obtained by dissolving the dry electrolyte in the storage space with water can be passed through the porous electrode into contact with the other electrode in order to activate the cell. An amount of dry electrolyte, well in excess of that which is consumable during the life of the cell electrodes, can easily be incorporated into the storage space or added periodically. One or both of the electrodes may also be consumable and various types of electrode pairs can be utilized efficiently.

Referring to Figure 5.8, a cell or battery **10** is shown which comprises an outer cylindrical casing **12** having an upper end **14** fitted with a removable upper end cap **16**. Cap **12** may be of electrically insulative material, such as nylon, tetrafluoroethylene or polyethylene, and preferably snap fits into end **14**. An electrical contact **18** of steel or the like is disposed on and through cap **16** and into contact with the first or a pair of internally disposed electrodes described more particularly hereinafter.

FIGURE 5.8: RESERVE CELL

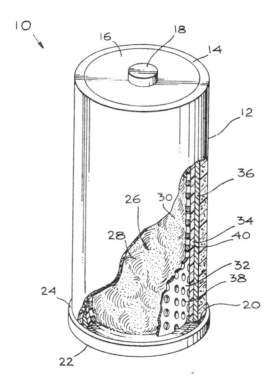

Source: U.S. Patent 4,005,246

Casing **12** also includes an open bottom end **20** slidably received within an electrically conductive cover **22** of copper, steel or the like having a raised rim **24**, the inner surfaces of which grip the outer surfaces of casing **12** at end **20**. Cover **22** is in electrical contact with the other of the electrode pair internally disposed within casing **12**. Cap **16** provides access to a central cavity **26** within which is disposed to a bibulous or water absorbent material such as fibrous material, particularly fibrous cellulosic material such as cotton wadding, paper wadding or the like. Further suitable examples are the following: asbestos, fiberglass, porous foams such as cellulose, polyurethane or polystyrene foams.

Disposed on or in the bibulous material 28 is a dry electrolyte 30 in particulate form as shown in Figure 5.8. Cavity 26 is defined by upper end 14 and cap 16 of casing 12, bottom cover 22 and by a porous cylindrical electrode 32 forming the sidewall thereof. Electrode 32 can be porous by virtue of a plurality of small openings 38 punched or otherwise made therein. Alternatively, electrode 32 can be fabricated of metal screening or of honeycombed or otherwise expanded metal material or material rendered porous in any other suitable manner. This electrode does not contain a dry electrolyte incorporated therein by fabrication or otherwise and, accordingly, can be very simply made. It need only transfer or permit the passage therethrough of a liquid electrolyte from cavity 26 into contact with a separator 34 concentrically aligned with electrode 32 and abutting the outer surface thereof.

Separator 34 acts as a means for spacing electrode 32 from a second outer electrode 36 concentrically aligned with electrode 32, the inner surface of electrode 36 abutting the outer surface of separator 34. Separator 34 can be fabricated of any suitable porous stable electrically insulative material. Thus, this material should be capable of withstanding, over a period of time, any corrosive effects which liquid electrolyte in the cell might otherwise exert. In this regard, various types of permeable absorbent materials have been utilized with success, for example, absorbent paper and the like.

Alternatively, the separator can be fabricated of, for example, fibrous felts and battings. Electrode 36 forming the other of the pair of electrodes in cell 10 need not be porous but can be if it is desired. However, if electrode 36 is of the air depolarized type, then casing 12 must permit the passage of air thereto. In such event, casing 12 can either be porous or perforated, as by slots. The particular electrodes involved can be selected from a wide variety of those which are made in a standard manner and which do not incorporate dry electrolyte therein.

Where consumable electrodes are used, replacement of one or both electrodes can be made by removal of cap 16 and sliding or otherwise removing one or both electrodes 32 and 36, separator 34, material 28 and electrolyte 30 from casing 12, all of which items can then be replaced as desired. If just the anode is consumable, electrode 36 is left in place. Electrode 36 abuts the inner surface of the sidewall of casing 12 and can be in slidable relation thereto for easy removal. Cover 22 forms a subunit with casing 12 and electrode 36.

It will be noted that electrodes 32 and 36 may be relatively thin and closely spaced by separator 34 for maximum efficiency. If the electrolyte were to be stored between the two electrodes or if the two electrodes were to incorporate the electrolyte therein, either the spacing between the two electrodes would be excessive and, accordingly, would change the electrical characteristics of the cell or the electrodes would be bulky and would also exhibit different electrical characteristics than those provided by the present cell 10.

In order to operate cell 10, water is added to cavity 26, as through the open top thereof after removal of cap 16. Cap 16 is replaced and, after the water dissolves electrolyte **30**, the resulting liquid electrolyte passes through openings **38** in inner electrode **32**, pores **40** in separator **34** and into contact with electrode **36** so that both electrodes are in electrical contact with each other through the aqueous electrolyte and the cell is activated. As the liquid electrolyte is consumed in the reaction, capillary action draws fresh liquid electrolyte from bibulous material **28** in the cavity through electrode **32** and into communication with separator **34** and electrode **36** to sustain the electrical output of the cell. Cell **10** continues to operate until one or both electrodes are depleted or until the liquid electrolyte is depleted. Additional liquid electrolyte can easily be provided by addition of further amounts of water to cavity **26** for dissolving of excess dry electrolyte **30** carried therein.

Thus, cell **10** has the advantage of being storable in a dry inactivated condition for an indefinite period of time without deterioration. Cell **10** provides means for storing an excess amount of dry electrolyte for activation of cell **10** when desired.

Thus, when it is desired to use cell **10**, all that need be done is to add water to dry electrolyte **30** therein or to add aqueous electrolyte solution to cavity **26** in the event that dry electrolyte **30** is not already disposed therein.

The amount of aqueous electrolyte provided or generated and made available to the electrodes determines whether the cell runs to exhaustion or shuts off beforehand and so is available for reuse. Cell **10** has substantial advantages over conventional cells in compactness, flexibility of use, simplicity of construction, low cost and replaceability of components.

Organic Electrolyte

A.N. Dey; U.S. Patent 3,930,885; January 6, 1976; assigned to P.R. Mallory & Co., Inc. describes a reserve cell for providing electrochemical energy and which is capable of long shelf life in the reserve state. This cell comprises a combined prismatic electrode chamber overlayed with a compartment containing the electrolyte.

The electrolyte is prevented from access to the electrode chamber by a frangible sealing means. The reserve cell is designed to be rapidly actuated by a separable actuator means of which the reserve cell actuator means described in U.S. Patent 3,484,297 is preferred. The electrolyte compartment is completely lined with an electrolyte impermeable lining forming the electrolyte container which is collapsible. Upon actuation, the applied pressure forcibly expels the electrolyte from the electrolyte container into the electrode chamber. This cell is particularly adapted for use with high energy electromotive couples including those dependent on active anodic metals above hydrogen in the electrochemical series and particularly such metals as lithium, calcium, sodium and potassium.

Referring to Figure 5.9a, the interior of the cell is shown in partial section. The prismatic cell **1** includes casing **1a**, which is preferably made of a nickel-plated steel or of molded polymers, such as polyolefins or phenolics, or other rigid materials capable of resisting atmospheric and chemical corrosion. Groove **2** placed just above electrode stack **3** is incorporated into the cell casing **1a** in order to provide a demarcation of the electrolyte compartment **6a** and to provide a support for a prismatic electrolyte containing capsule or reservoir **6**. The electrode stack **3** includes a parallel plate stack of rectangular anodes **3a** and cathodes **3b**. Between the electrodes is positioned bibulous separator material **3c** inert to the electrolyte and the electrodes. The cathode terminals are connected to a common tab **4a** which, in turn, is connected to a glass-metal seal terminal **4** attached to the cell top **5** to provide external terminals for the cell **1**. The anode terminals are similarly connected to the casing **1a** via a common tab.

The prismatic electrolyte capsule **6** in compartment **6a** provided with a circular opening on one side is molded from flexible polymeric material, such as phenolics, diallyl phthalates or high density polyolefins. On the side adjacent to the electrodes an opening is provided which is closed by a frangible thin walled sealing lid **8**. An impact transmitting bar **7**, consisting of three or more vertical impact bars **7a** with a flat circular disc **7b** at the top, is placed inside the electrolyte capsule **6** with the flat disc **7b** placed opposite to the lid **8** of the capsule **6**. The inner wall of the capsule **6** is sealed to this flat disc **7b**.

The capsule **6** is filled with electrolyte and a circular thin walled sealing lid **8** is placed over the circular opening of the capsule and sealed to the capsule. A layer of polyethylene may be used between the two pieces. The capsule **6** is then inverted so that the capsule lid end faces the electrode stack **3** and the entire capsule **6** is then positioned in the electrolyte compartment **6a** to rest on groove **2** of the cell case **1** directly above the electrode stack **3**.

Cell top **5** consists of a nickel-plated steel lid with a central orifice **9** for the contact positioning of the spring loaded plunger **10** of the standard activation mechanism according to U.S. Patent 3,484,297 in contact with bellows **11**. The glass-to-metal seal cathode terminal **4** is attached to cell top **5**. The inner side of the nickel-plated steel cell top **5** is hermetically heat sealed to a flexible polyethylene aluminum foil laminate bellows **11** of capsule **6**. Cell top **5** is sealed to the cell case **1a** after incorporation of the electrode stack **3** and the positioning of the electrolyte capsule **6** in the cell case **1a**.

Bellows **11** is free at the central part where it rests within the electrolyte compartment **6a** and in contact with capsule **6** inside the cell casing **1a** and the spring loaded plunger **10** outside of the cell. A metal protective disc may be protectively placed below the plunger head **10a** and the bellows **11** in order to protect the capsule **6** from puncture when it is impacted by the spring loaded plunger **10** during activation. Cell top **5** is also fitted with a pair of clasps **12** for retaining the activation mechanism housing **14**. These clasps permit the positioning of the separable activation mechanism **15** just prior to activation and the removal of the activating mechanism **15** after activation. The activation mechanism **15** is capable of interchangeable reuse for cells of this type.

The disposable activation mechanism **15** includes a spring loaded plunger **10** which is restrained by a pin **13**. Twisting of the activation mechanism housing **14** causes pin **13** to slip through a hole in the top of housing **14** thus releasing the spring and the plunger **10** which impacts the electrolyte capsule **6** through bellows **11** and the impact transmitting bar **7** which thereupon breaks the thin frangible lid **8** and compresses the electrolyte capsule **6** thus releasing and expelling the electrolyte into the electrode stack **3** and thus activating the cell **1**. Pin **13** is prevented from slipping through the hole in the housing **14** by safety pin **15** which has to be removed prior to cell activation by twisting off of the housing **14**. The activation mechanism housing **14** along with the spring and the plunger **10** is automatically released from cell top clasp **12** after activation.

FIGURE 5.9: ORGANIC ELECTROLYTE RESERVE CELL

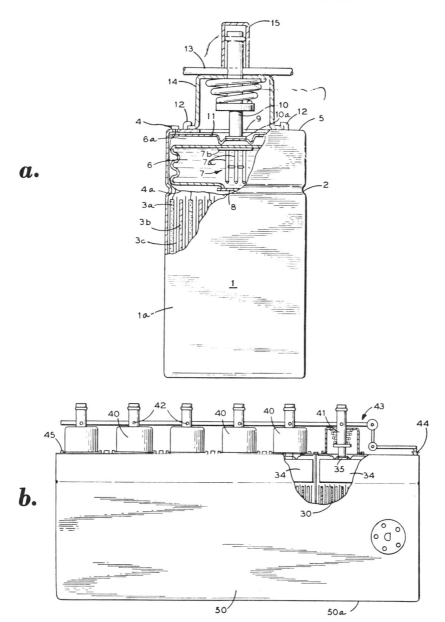

(a) Schematic diagram of a reserve cell
(b) Multicell reserve battery construction with parts cut away to show the arrangement of the cells to form a battery and showing the safety devices to prevent accidental activation or actuation of the reserve battery

Source: U.S. Patent 3,930,885

The electrode stacks **3** may be connected in series or parallel and by packaging them in a battery case **50a** as shown by Figure 5.9b. The battery **50** shown in Figure 5.9b consists of 6 cells **30**.

The simultaneous activation of all the cells **30** of battery **50** of Figure 5.9b is achieved by slight modifications of the activation mechanism **14** of Figure 5.9a. This is accomplished by removal of all of the restraining pins **42** of the spring-loaded plungers **41** of activators **40** of all the cells by means of ripcord **43** interconnected to all of the pins **42**. The handle of the ripcord is secured at one end of the battery in safety catch **44**. The activation mechanism housings **40** may be removed after use by pushing them from one side to dislodge them from cell top clasps **45**. The reserve cell structures of this process are applicable to a variety of organic electrolyte batteries and generally to all hermetically sealed batteries.

Electrolyte Stored in Frangible Envelope

R.I. Sarbacher and J.C. Bogue; U.S. Patent 4,031,296; June 21, 1977 describe energy cells which are basically reserve-type batteries in which the electrolyte is separated from the electrodes until the battery is ready for use. In one example, the battery is a cylindrical battery in which the electrolyte is stored in a frangible envelope inside the battery case and above the electrode structure. Means are provided to break the envelope holding the electrolyte to permit the electrolyte to flow into the electrode cavity. The anode and cathode are designed to permit free flow of the electrolyte.

In a second example, the battery is also cylindrically shaped but is formed in a very thin or short cylinder. In this second case, the electrolyte is stored in a frangible envelope in the center of the battery casing. Again, means are provided to break the frangible envelope so that the electrolyte can flow into the electrode cavity and activate the battery. A plurality of anodes surround an anode cylinder and are electrically secured thereto and a plurality of cathodes surround the anode cylinder and are electrically connected to the battery case. The battery may also be fabricated as a rectangular, flat pack. The electrolyte is stored in a frangible envelope which is broken by means of a pair of hinged plates which are folded over the battery and squeezed to release the electrolyte.

Dashpot Electrolyte Ampule for Projectile Firing

A process described by *M.I. Morganstein; U.S. Patent 3,945,845; March 23, 1976; assigned to The U.S. Secretary of the Army* relates to an improved dashpot type of electrolyte ampule for deferred action batteries for rockets, missiles, or other projectile firing means. More specifically, the process relates to an improved dashpot for an ampule of the type described in U.S. Patent 3,754,996.

Ampules of the type described in the above-mentioned U.S. Patent 3,754,996 employ a cutter mechanism to puncture the bottom of the ampule under linear gun acceleration (setback) so as to permit the subsequent release of the electrolyte under radial (spin) acceleration.

Referring in detail to Figures 5.10a to 5.10c, there is illustrated a first example of an ampule generally indicated as **10** including a cylindrical container **12** having a lid **12A** and a rupturable bottom **12B**. The ampule cylinder **12** is substantially filled with an electrolyte **14** which is to be subsequently released through rupturable bottom **12B**.

FIGURE 5.10: DASHPOT TYPE ELECTROLYTE AMPULE

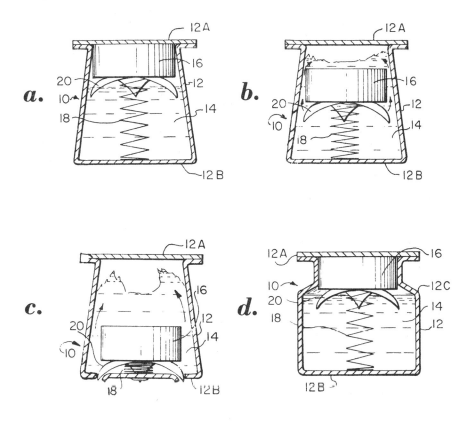

(a)-(c)   Ampule design and the stages of motion it experiences in response to
           setback forces
(d)   Ampule design

Source: U.S. Patent 3,945,845

The electrolyte within cylinder **12** is at a level which substantially coincides with the bottom of a piston **16** as shown in Figure 5.10a. Piston **16**, under static conditions of the ampule is held in the position shown in Figure 5.10a by a spring **18** and the buoyant force generated by electrolyte **14**. Cutter blades **20** are provided on the bottom of piston **16** and are adapted to engage and rupture bottom **12B** in response to setback forces to be described hereinafter. Three cutter blades are illustrated in Figures 5.10a to 5.10c.

The entire ampule **10** is adapted for insertion into a deferred action battery of the type illustrated in U.S. Patent 3,754,996. As described in that patent, the battery is mounted in a missile or projectile and the setback forces generated by the firing of the projectile cause displacement of the piston in the direction of the rupturable bottom of the ampule, whereby the cutter blades rupture the bottom and release the electrolyte into contact with the battery plates.

Figure 5.10a shows the ampule in a static condition before firing of the projectile. As the projectile is fired, setback forces in the direction of the ampule container bottom **12** cause piston **16** to downwardly move as shown in Figure 5.10b against the force of spring **18** and the buoyant force of electrolyte **14**. The piston continues to move downwardly until the cutter blades **20** rupture or cut through bottom **12B**, as shown in Figure 5.10c, thus releasing electrolyte **14** to the battery of the type described.

In Figures 5.10a to 5.10c, the container walls of container **12** taper or slope outwardly from the top to the bottom of the container. The piston **16**, container **12**, and electrolyte **14** function as a dashpot with a variable size orifice. The size of the orifice between piston **16** and container **12** increases as the piston travels through the container. The hydraulic damping of electrolyte **14** is a maximum with the piston **16** in the position of Figure 5.10a and is a minimum with the piston in a position **16** shown in Figure 5.10c. The net effect is to provide maximum safety from accidental forces and maximum efficiency for low *g* activation of the deferred action battery.

The side walls of container **12** in Figures 5.10a to 5.10c are shown as being gradually and continuously tapered from top to bottom. Tapering the can to a greater diameter at the bottom than at the top significantly increases the orifice clearance and reduces damping at the end of the cutting stroke to a large extent.

In another example as illustrated in Figure 5.10d, the side walls of container **12** may be stepped as shown at **12C**. Thus, the diameter of container **12** increases abruptly at **12C** with a concomitant decrease in hydraulic damping. The distance of the step at **12C** is so chosen that accidental forces such as dropping will not create an excursion of piston **16** beyond point **12C**. Thus, the safety factor of the ampule is not decreased.

Diaphragm Activation Technique

W.C. Merz; U.S. Patent 3,929,508; December 30, 1975; assigned to The U.S. Secretary of the Navy describes a reserve battery for providing electrical current when an unexpected need arises. The process provides a battery container including four dry porous cells, a bellow including a solvent to energize the cells, means for separating the solvent and the cells until electricity is needed and a means for puncturing the separating means and mixing the solvent with the dry porous portion.

According to Figure 5.11, the basic structure of the reserve battery consists of a housing or battery case **10**, a diaphragm plate **11**, and a collapsible bellows **12**. Within the bellows structure there is positioned a lance **13** mounted on a base **14** and supported by struts **15** and **16**. About the lance and strut structure at the lower end is a collar **17**.

There is mounted on the diaphragm a guide ring **18** centered about a reduced cross section or thinner portion of the diaphragm plate **11**. The lance **16**, collar **17**, guide ring **18** and reduced portion of the diaphragm **19** are all coaligned along a major axis **20**. As shown, the outer portion of the diaphragm ring **21** is threaded and is designed to accept the threaded portion **22** of a screw cap **23**. The upper portion **24** of the screw cap **23** engages the upper portion **25** of bellows **12**. Bellows **12** is welded to the upper surface of diaphragm **11** and defines therewith a space within which a solvent generally designated as **30** is confined.

FIGURE 5.11: RESERVE BATTERY

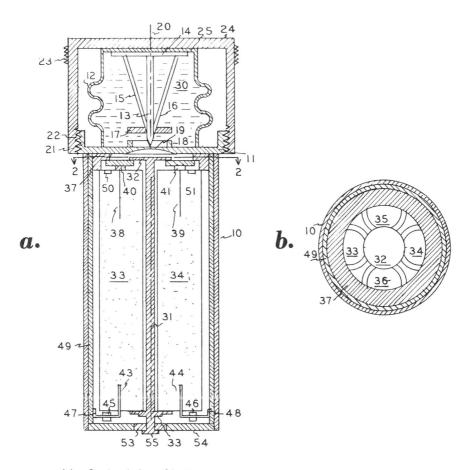

(a)   Sectional view of battery
(b)   Sectional view along lines **2–2** of Figure 5.11a

Source: U.S. Patent 3,929,508

Inside battery case **10** along the major axis is a support structure or spindle **13** whose upper portion **32** has an enlarged cap or portion. A base support plate **33** is affixed to spindle **31** and is designed to support four cells **33**, **34**, **35** and **36**. An insulated support ring **37** is designed to position the four cells within the battery case and to retain them in fixed relationship. Wires **38**, **39** are connected to electrical caps **40**, **41**. Wires **43**, **44** in the lower end of cells **33**, **34** connect to diodes **45** and **46** respectively. Electrical connection is made through wires **43**, **44** to diodes **47**, **48** and to an anode structure **49**, a cylindrical shaped member within battery case **10**. Additionally, valves **50**, **51** are provided in the cells **33**, **34**. Base plate **33** is connected through an insulated portion **53** of terminal plate **54** to an external cathode plate terminal **55**. In operation, these reserve batteries are provided in a survivor kit for energizing a radio transmitter when needed. Should electrical energy be necessary, a person would take the battery and screw the cap **23** in a downward direction along axis **20**. Lance **19** advances encountering the thin portion of the diaphragm plate and puncturing it. As the cap continues in its downward direction under the pressure of cap **23**, the lance penetrates until collar **17** encounters the guide line **18** at which time the lance and structure begin to collapse.

In one successful case, the pressure to penetrate the diaphragm was 2 to 3 pounds per square inch. Continued advancement of the cap drives the electrolyte solvent **30** into the space within the upper portion of the battery housing. The valves **50**, **51** open and the solvent electrolyte penetrates into the dry porous cell structure **33**, **34** at which time the chemical reaction starts which produces electrical current upon demand. The diode structure **45**, **46** is provided to prevent current reversal.

Multicell Mine Battery

P. Bro, R.H. Kelsey and N. Marincic; U.S. Patent 3,929,507; December 30, 1975; assigned to P.R. Mallory & Co., Inc. describe a mine battery having a smaller volume and a lower weight than current mine batteries. The battery is based on a high energy density lithium/sulfur dioxide system, utilizing a reserve structure for holding the corrosive electrolyte separate from the battery cell components until activation is desired, whereby storeability may be achieved in excess of 10 years. The battery consists of a plurality of cells.

One modification contains the electrolyte in a sealed internal reservoir container in each cell. Another modification contains a reserve sealed common reservoir container of the electrolyte, which may be activated when use of the battery is desired by transferring the electrolyte from the reserve common reservoir to each of the cells for activation and immediate operation. In each modification, each reserve container remains sealed until activation of the battery is desired. The container is then suitably opened to release the electrolyte to the related cell or cells.

As shown in Figure 5.12a, a battery **10** consists of five cells **12-1** to **12-5**, arranged in circular disposition, with an outer battery container **14** around a central axis **16**. The five cells are normally kept in inactive condition, each cell being provided with a restraining pin **20-1**, with a similar pin in each of the other cells through a series to **20-5**. As will be shown in connection with the description in Figure 5.12b, the restraining pins **20-1** to **20-5** are normally positioned in their detent positions, as shown in Figures 5.12a and 5.12b, as long as the battery is not in active use and is waiting in storage until it is to be activated. In Figure 5.12b, the detent pins **20-1** and **20-3** are shown on cells **12-1** and **12-3**.

The activating system is as schematically indicated in the parts shown in Figure 5.12a, provided with a pressure pad **22** and reaction ring **24**, which are appropriately operated, as shown later in Figure 5.12b, to develop pulling forces to pull radially on the detent pins **20-1** to **20-5**, through corresponding connecting cables **26-1** through the series to **26-5**, so all of the five detent pins **20-1** through **20-5** will be withdrawn from their detent positions to release the operating mechanism associated with each cell. The cells will thereupon be activated and placed in condition to function as electrical sources of energy, ready to supply such energy through their terminals **18-1** and **18-2** to an external circuit, upon demand. In order to retain the cells against casual displacement in response to any external forces, when the container shell **14** might be accidentally bumped or jarred, the cells within the container **14** are preferably potted. As may be seen in Figure 5.12a, there is considerable available empty space about the cells within the battery container **14**, and in cases where the use of the cell is predetermined, and such use may include connection to other circuit elements, those other circuit elements may be disposed in the space within the container **14** to the extent that the various spaces are available and adequate for accommodation of those other elements. As shown in Figure 5.12b, the battery container is a round box structure **14** with a closing bottom **30** and a similar closing top **32** to provide an otherwise hermetically sealed box container for the cells.

FIGURE 5.12: RESERVE MINE BATTERY

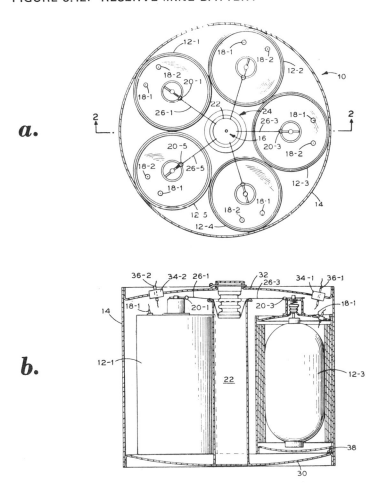

(continued)

FIGURE 5.12: (continued)

c.

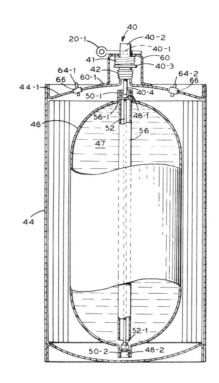

(a)   Plan view of battery showing five cells disposed in a circular disposi-
       tion around the central axis of the battery
(b)   Sectional view through the battery of Figure 5.12a taken along the
       vertical plane indicated by the line 2–2 in Figure 5.12a, and illus-
       trates the general internal construction of one cell with an internal
       reservoir of electrolyte
(c)   Enlarged schematic view of the cells in which the reservoir for the
       electrolyte is disposed as an internal container within the cell

Source: U.S. Patent 3,929,507

Provision is made in the top cover **32** of the battery box **14** for certain elements,
in a way that will maintain the hermetic seal unbroken. There are two glass
beads **34-1** and **34-2** providing a metal-to-glass seal in the top metal cover **32**, and
those glass beads serve as supports for two metal feed-through terminals **36-1** and
36-2, for connection to an external circuit. The connections from the terminals
36-1 and **36-2** to the internal cells **12-1** through **12-5** are not shown in Figure
5.12b, but their disposition will depend upon whether the cells are connected in
series circuit connection or in parallel circuit connection. Since the individual
cells **12-1** through **12-5** will be contained in metallic cans, appropriate insulation
between the cans and the battery container **14** may be provided by a suitable
lining of insulating material **38** on the internal surface of the battery container
14 and between the cells. The manner in which the withdrawal of the detent

pins **20-1** through **20-5** permits the activation operation of each cell in the modification of Figures 5.12a and 5.12b is shown in more detail in Figure 5.12b and in the enlarged view of Figure 5.12c.

As better shown in Figure 5.12c, the detent pin **20-1** is normally in its detent position and extends through a hole **40-1** in the head **40-2** of an impact pin **40**. The impact pin **40** comprises the head **40-2** and a coaxial reaction pressure flange **40-3** from which depends the impact shank **40-4**. The detent pin **20** holds the impact pin **40** stationary to compress a spring **41** against the reaction pressure flange **40-3** to store energy in the spring **41** until release. The flange **40-3** seats on the top of and is hermetically attached to a compressible bellows **42** and serves to close the bellows that serves to hermetically seal the cell casing **44**, which contains the operating elements of the cell, including the anode, the cathode and the vial **46** for containing and isolating the electrolyte **47**.

The vial **46** is made of a corrosion resistant material, such as a metal or a glass-lined metal or an equivalent, and is constructed to have two frangible end pieces **48-1** and **48-2** that are seated in, and sealed to, two hollow tubular extensions **50-1** and **50-2**, at the two ends of the vial **46**. The upper frangible piece **48-1** is of cup shape, in order to receive and accommodate and guide the lower end **40-4** of the fracture pin **40**. A rod **52** depends from the cup shaped fracturable element **48-1** and embodies an enlarged head **52-1** at the other end that normally rests on or near the equivalent bottom surface of the bottom fracturable element **48-2**, in order that downward movement of the fracturing pin **40-4** will fracture and break the cup shaped sealing element **48-1** and strike the rod **52** to compel the enlarged head **52-1** on the rod to engage and break the cup shaped bottom sealing element **48-2**.

In order to stabilize the rod **52** during the movement of the battery in handling and in order to control the transfer of the electrolyte, a hollow tube **56** is disposed in the vial **46** to concentrically surround the push rod **52**, and is appropriately supported from the upper end of the vial, in the manner schematically illustrated at **56-1** at the top of the hollow tube **56**, where it may be suitably joined to the inside of the vial **46**.

In order to actuate the push rod **52**, when the detent pin **20-1** is removed from its detent position in the fracture pin **40**, the compressed spring **41** is employed. Normally, the spring **41**, as a helical compression spring, is compressed to its maximum compressed condition between the flange **40-3** and the inside of a top wall of a tower cap **60** which is suitably anchored, preferably by welding, to the top cover **44-1** of the cell body **44**, along an annular peripheral flange **60-1** that sits coaxially on the top cover **44-1**.

When the detent pin **20-1** is withdrawn from the head **40-1** of the fracturing pin **40**, the pin is released to permit the stored energy in the compressed spring **41** to force the pin suddenly downward to cause the bottom end of the pin shank **40-4** to strike the sealing cup **48-1** of the vial **46** and simultaneously to act through the pressure rod **52** to press or strike the bottom sealing cup **48-2**, to thereby fracture both cups **48-1** and **48-2**.

The electrolyte **47** is thereupon expelled from the vial **46**, under pressure of the driver gas and of the internal vapor pressure of the electrolyte within the vial **46**, and the electrolyte is caused to move out into the operating space surrounding the vial **46** and into the spaces between the anode and cathode components that are wrapped around the vial. The cell is thus activated, and is immediately ready to supply energy to an external circuit to which the cell is connected. For the purpose of connecting the cells of the battery in appropriate circuit arrangement, as previously selected, each cell is provided with two terminals, shown in Figure 5.12c as terminals **64-1** and **64-2**, which may preferably be fed through electrical conductors sealed in glass-to-metal beads **66** anchored in the top cover **44-1** of each cell casing **44**.

Automatic Pressure Control and Venting

H.-K. Köthe and R. Schneider; U.S. Patent 3,928,078; December 23, 1975; assigned to Varta Batterie Aktiengesellschaft, Germany describe an automatically activated battery assembly with electrolyte chambers separated from the individual cells, and a pressure gas container.

Referring to Figure 5.13, there is shown a system wherein electrolyte is simultaneously pressed out of several chambers or pockets and fed to the electrochemical cells of a battery. The pressing-out operation is preferably effected by pressure gas from a reservoir **2** of limited volume, the gas being conducted into an expansion space **3** located between a stationary wall **4** and the pocket **1** or a movable wall of one of the aforementioned chambers. When the thus pressurized electrolyte flows out of the pockets **1** or the chambers and into the cells **5**, the gas in the cells **5** is displaced and vented from the cells **5** through a common venting pipe **6**. This venting pipe **6** ends in an electrolyte trap or nonreturn container **7** from which the gas escapes through a valve structure or vent **8** and an outlet **21**.

The valve structure **8** is a slide valve controlled by a driving gas pipe **22**. In a modification of Figure 5.13, a safety valve, such as a spring-loaded valve, can be substituted therefor, in which case the return line **11, 23** is omitted. Generally, after pressing the electrolyte out of the chambers or pockets, the expansion chamber **3** and the activation gas line **19** are vented. This venting takes place by means of an outlet that terminates preferably in an adjustable vent or capillary tube **9**, providing the advantage that a relatively precise adjustment of the venting time is possible.

After the venting process is completed, the slides of valves **8** and **12** are restored to their initial position. Although the venting period can be selected within a wide range, the length thereof has been established, according to the process, by the fact that the venting of the expansion chamber **3** or of the line **19** is completed when the cells develop gas during discharge. What is attained thereby is that with the use of the slide valve control according to Figure 5.13, either the gas is conducted along the path **11** and **23** into the expansion chamber of the pressure gas or, when a safety valve is used instead of a slide valve, the gas can return along a path through the electrolyte canal **10** into the previously emptied electrolyte pocket **1**.

Leftover electrolyte entrained by the gas is separated from the gas in both of the foregoing cases in the container **4**. The cleaned gas can then be conducted through the activation gas pipe **19** and can be discharged through the outlet **17**. In the first of the foregoing cases, the inner cell pressure does not exceed the ambient pressure of the battery. In the second case, naturally, the gas can be discharged only after the electrolyte pocket has been caused to burst by a given superpressure. In the latter case, the pressure exceeds the ambient pressure only by the amount necessary for bursting of the electrolyte pocket.

FIGURE 5.13: AUTOMATICALLY ACTIVATED BATTERY ASSEMBLY

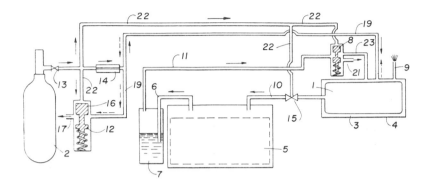

Source: U.S. Patent 3,928,078

According to the process, the vent or valve **12** is included in the system, as described below. Since the gas produced in the cells during the discharge cannot escape completely through the adjustable vent or capillary tube **9**, pressure will increase in the expansion chamber **3** and thereby in the cells **5** within a short time and would soon considerably exceed the ambient pressure. In order to avoid this, the expansion chamber **3** is opened by the vent or valve **12**, according to the process, when the pressure in the expansion chamber **3** of the gas that has caused the pressing out of the electrolyte pocket during the activation process has preferably become about equal to the ambient pressure.

The outlet in the expansion chamber **3** is situated in such a manner that only cleaned electrolyte-free gas can escape therethrough. The vent or valve **12** is closed by the highly compressed pressure gas at the start of the activation process against the bias of a spring and opens again under this spring pressure as soon as the pressure in the expansion chamber falls below a given value. At this instant, the pressure in the expansion chamber is abruptly reduced to that of the ambient pressure of the battery.

With the automatically activated battery constructed in accordance with the process, an inner pressure is produced in the individual cells during the activation as well as during the discharge of the cells, that inner pressure being only slightly higher than the ambient pressure of the battery so that bursting of the battery and escape of electrolyte is reliably avoided.

Aluminum-Silver Oxide Battery for Underwater Propulsion

G.E. Anderson; U.S. Patent 3,953,239; April 27, 1976; assigned to The U.S. Secretary of the Navy describes a primary battery system for providing a quiet high energy source for a limited period of time for use in underwater propulsion systems.

The primary battery comprises aluminum and silver oxide and utilizes a preheated potassium hydroxide electrolyte for the generation of power. The battery with its electrolyte preheated to a minimum of 120°F is capable of providing a very high electrical energy source of minimum weight and volume for underwater propulsion systems.

The electrolyte may be brought up to temperature by either conventional heating apparatus or a special mixing procedure for providing an instantaneous temperature use.

THERMAL BATTERIES

Alkali Metal Salt Solvate Electrolyte

D.C. Luehrs; U.S. Patent 3,977,900; August 31, 1976; assigned to Board of Control of Michigan Technological University describes a thermally activated electrochemical cell including a solid electrolyte which is capable of melting and becoming an ionically conductive liquid within a relatively short time after being heated to a temperature above its melting point.

The thermally activated electrochemical cell or battery includes a cathode, an anode and a normally solid electrolyte which is an organic solvent solvate of an alkali metal salt and is formed by coordinating the salt with a polar organic solvent.

These solvates, which are crystalline coordination compounds and nonconductive at ambient temperatures and become an ionically conductive liquid upon being heated above their melting points, have sharp and relatively low melting points and relatively low latent heats of fusion permitting them to become an ionically conductive liquid within a short time after being heated above their melting points, are compatible with most common electrode materials and have relatively low resistivities.

The organic solvent solvates used as the electrolyte in the thermally activated electrochemical cell or battery are formed by combining an alkali metal salt having a relatively small cation and a larger anion with a polar organic solvent to form a crystalline coordination compound. For any particular application, the alkali metal salt must be compatible with the particular electrodes used in the cell and soluble in the general class of polar organic solvents described below.

The preferred alkali metal salts are lithium chloride and lithium perchlorate with lithium chloride being the most preferred because of its high stability.

The polar organic solvents used to form the crystalline solvates used as the electrolyte preferably have a Gutmann donor number greater than that of water, i.e., 18. Solvents having a Gutmann donor number of at least 20 are more preferred and those having a Gutmann donor number of at least 25 are the most preferred. Also, the polar organic solvents preferably have a dielectric constant greater than 20. The preferred solvents are dimethylsulfoxide, dimethylacetamide, hexamethylphosphoramide and dimethylformamide.

The crystalline organic solvent solvates used as the electrolyte can be formed by dissolving the anhydrous alkali metal salt (which preferably has been dried under a vacuum at an elevated temperature of about 100°C or higher) in the pure organic solvent, heated to an elevated temperature, until the solution becomes saturated.

After the solution is cooled to room temperature, the alkali metal salt solvate crystallizes out and the solvate is thereafter separated by filtering, washed with an inert hydrocarbon, such as benzene, to remove excess solvent and then dried, such as under a vacuum. For example, 20 g (0.47 mol) of an anhydrous lithium chloride is dissolved in 90 g (0.50 mol) of hexamethylphosphoramide (HMPA) heated to a temperature of 100°C. When the resultant solution is cooled to room temperature, white crystals of LiCl·HMPA crystallize out and these crystals are separated from the solution by filtering, washed with benzene and then dried under a vacuum.

The table below lists representative examples of solid electrolytes within the scope of the process which were prepared by the methods outlined above. These alkali metal salt solvates were characterized by elemental analysis, melting point shown below and IR spectra. TMU is tetramethylurea.

Solvate	Melting Point, °C
LiCl·HMPA	150
LiBr·4HMPA	74
LiSCN·HMPA	180
LiSCN·2HMPA	82
LiClO$_4$·HMPA	128
LiClO$_4$·4HMPA	137
LiNO$_3$·HMPA	65
LiBF$_4$·HMPA	116
NaI·HMPA	154
NaSCN·HMPA	151
NaSCN·2HMPA	77
NaClO$_4$·HMPA	176
NaClO$_4$·2DMF	70
NaSCN·2DMF	49
LiNO$_3$·DMF	72
LiClO$_4$·2TMU	109
LiCl·TMU	63
LiBr·TMU	123
LiSCN·TMU	49
NaClO$_4$·2TMU	46
2NaI·3TMU	69

Of the solvates listed, LiBF$_4$·HMPA, NaI·HMPA, NaSCN·HMPA, NaSCN·2HMPA, NaClO$_4$·HMPA, NaSCN·2DMF, LiNO$_3$·DMF, LiCl·TMU, and 2NaI·3TMU while operable as an electrolyte, are presently considered to be less desirable because of their tendency to melt incongruently. It should be understood that the alkali metal salt solvates listed are only representative, and not all inclusive, of those which can be used in accordance with the process. However, from the melting points of those listed, it can be seen that electrochemical cells or batteries including such alkali metal salt solvates as the electrolyte can be thermally activated at relatively low temperatures and the activation temperature for a particular application can be varied as required by selecting the appropriate solvate.

Copper Iodide and an Alkali Metal Iodide Electrolyte

S. Senderoff and G.W. Mellors; U.S. Patent 3,963,518; June 15; 1976; assigned to Union Carbide Corporation have found that certain complex copper-containing compositions while very poor ionic conductors at ordinary room temperatures undergo a sharp transition at elevated temperatures to become good ionic conductors at such temperatures. This sharp change in conductivity results from a sudden phase change in the material at a specific temperature and is fundamentally different and more useful for most applications than the gradual increases in conductivity with increasing temperature that characterize the usual type of solid ionic conductor. The process comprises binary compositions of copper iodide and an alkali metal iodide or cyanide, the molar proportion of copper iodide with respect to the alkali metal iodide or cyanide being about 4 to 1. Alkali metals that may be used include potassium, rubidium and cesium, potassium being preferred.

Specific examples of compositions of the process include: 4CuI-KI and 4CuI-KCN. The former compound changes from poor ionic conductor to good ionic conductor at about 250°C. The latter shows this transition at about 270°C. Both materials retain good ionic conductivity up to and beyond their melting points, these being about 450°C for both. The materials are compounds which may be represented by the formulas KCu_4I_5 and KCu_4I_4CN. Both have a specific conductance at 300°C of above 0.2 $ohm^{-1} cm^{-1}$.

The materials may be prepared without difficulty simply by melting appropriate starting materials, e.g., CuI and KI, in a sealed vessel under a protective inert atmosphere, for example, of argon or helium. After melting and thorough mixing, the resultant molten mass is rapidly quenched to room temperature. The resulting solid may be crushed and pelleted according to ordinary techniques, care being taken to avoid moisture pickup.

Samples prepared as just described were tested for specific conductances using a standard 1,000 cycle conductance bridge and have been shown to have specific conductance of above 0.2 $ohm^{-1} cm^{-1}$ as indicated above. The specific conductance of 4CuI-KCN at room temperature, on the other hand is about 1×10^{-7} $ohm^{-1} cm^{-1}$ and that of 4CuI-KI is substantially greater than 1×10^{-3} at room temperature. The composition of 4CuI-RbI, another specific example of this process, is not an ionic conductor at room temperature but at about 252°F has a specific conductance of about 2.3×10^{-3} $ohm^{-1} cm^{-1}$. The specific conductance of this material at 343°C was determined to be 1.7×10^{-1} $ohm^{-1} cm^{-1}$.

The data indicate the suitability of the material to serve as electrolytes in devices activatable at temperatures in the range above 250° up to 350° to 400°C or thereabout. Examples of such devices include thermal batteries for fuses or missiles, warning devices activatable at elevated temperatures, and devices adapted to perform certain functions when the ambient atmosphere in which such device is placed reaches a specific and predetermined elevated temperature.

Compressed Powder Layers

G.F. Zellhoefer; U.S. Patents 3,954,504; May 4, 1976 and 3,954,503; May 4, 1976; both assigned to National Union Electric Corporation describes a fusible salt electrolyte cell system including a cathode, an anode, and a mass of nonfused electrolyte. The cathode and electrolyte components essentially consist of a series of layers of powders compressed under a pressure of the order of 50,000 to 60,000 psi into a single pellet with adjacent layers intimately united. The electrolyte is mixed with dehydrated kaolinite in a sheet layer crystalline structure form, which is present throughout the mass of electrolyte in an amount sufficient to impart nonflow characteristics to the fused electrolyte so as to maintain the electrolyte against substantial displacement under pressure when the electrolyte is fused thereby to retain the fused electrolyte in intimate contact with the anode.

Figure 5.14a illustrates a battery 10 consisting of a series of cells 12. As illustrated in Figure 5.14a, the cells 12 may be arranged in series in a stack which may employ any desired number of cells and with heat source material 14 surrounding the cells 12 for heating the same to an elevated temperature for activating the cells. The heat source material 14 may be covered by a layer 16 of suitable insulation which in turn is housed in a suitable container 18. Conventional ignition means may be used for igniting the material 14. A terminal 20 operatively connected with the end cell of the stack projects from the casing 18 through a suitable hermetic seal 22 while the other terminal may project through the opposite end of the case 18.

As shown in Figure 5.14b, each cell preferably comprises a cathode 24 and an anode 26 separated by and in contact with an electrolyte, the electrolyte preferably consisting of a cathode layer 28 and an anode layer 30. The cathode layer of the electrolyte 28 preferably consists of equal parts of KCl and LiCl mixed with an adsorbing agent such as kaolin, and a depolarizing agent. For example, the cathode layer 28 may consist of 19% KCl, 19% LiCl and 62% of V_2O_5 plus kaolin, when required, in an amount sufficient to retard and/or prevent the flow of electrolyte when the same fuses without interfering with the wetting of the cathode 24 by the electrolyte. Good results have been obtained when the adsorbent comprises up to about 25% of the cathode layer 28.

The anode layer 30 of electrolyte preferably consists of equal parts of KCl and LiCl and an adsorbing agent of the character previously described admixed therewith. The adsorbent may comprise approximately 40 to 50% by weight of the anode layer 30. Preferably the anode layer is substantially free of a depolarizing agent. The percentage of the adsorbent or nonflow agent used in any case depends upon the design of the cell and the amount required to retard or prevent the flow of electrolyte when the cell is activated and subjected to conditions of use and for some applications it may be feasible or desirable to omit the adsorbing agents in the cathode layer 28.

The adsorbing material may be prepared by employing a commercial grade of kaolin ($Al_2O_3 \cdot 2H_2O \cdot 2SiO_2$, a native aluminum silicate) which is spread in a shallow tray to a uniform depth of about 1 cm and then heated to 585° to 600°C for 4 hours to remove moisture and water of hydration.

For the anode layer 30 dry KCl-LiCl and dehydrated kaolin, prepared as above indicated and mixed in weight proportions as above indicated is ball milled. For the cathode layer 28 dry KCl-LiCl-V_2O_5 and dehydrated kaolin are mixed in the weight proportions as heretofore stated and ball milled. As an aid in securing uniformity of results, it has been found that it is desirable to compress the powder used in forming the cathode layer 28 of electrolyte into a slug of desired density and then granulate such material before pressing the layers of powder to form the pellet. It is also desirable to employ the same technique for the powdered material used to form the anode layer **30** and the anode **26**.

For the cathode **24** B_2O_3 and V_2O_5 in powdered form should be fused and solidified, then granulated and sifted to obtain a powder suitable for use in forming the cathode **24**.

FIGURE 5.14: FUSED SALT ELECTROLYTE CELL

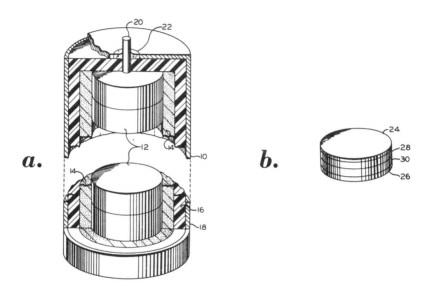

(a)   Fragmentary partial sectional view illustrating
        an assembly of a series of cells.
(b)   Elevation view of a single cell.

Source: U.S. Patent 3,954,504

In forming the pellet the powdered material forming the components of the cell and prepared as above indicated is stacked in a suitable die with the material arranged so as to form an anode **26**, an anode layer **30** of electrolyte, a cathode layer **28** of electrolyte and a cathode **24**. The thickness of each layer should be uniform. The layers of stacked powdered material are then subjected to a pressure of the order of 50,000 to 60,000 psi to form a four-layer pellet with each of the layers intimately united with the adjacent layers.

A pellet of 0.25 inch diameter and the length of 0.080 inch has been found to give good results where the cathode layer **24** is 0.010 inch, the cathode layer **28** of electrolyte is 0.028 inch, the anode layer of electrolyte is 0.027 inch and the anode is 0.015 inch. The optimum ratio of the layers of the cell depends upon the operating conditions of the cell. The use of two layers **28** and **30** of electrolyte is for the purpose of improving the electrochemical properties of the electrolyte and permits the use of a higher concentration of depolarizing agent in the cathode layer **28** than is the case when a single layer of electrolyte is used.

The KCl-LiCl cell will produce a no load voltage of about 2.75 to 2.90 volts throughout a temperature range of 360° to 520°C. The heating of the cell is effected by the ignition of the heat source material **14**.

Sodium Carbide and Halogen Gas as Activators

A process described by *R.A. Sutula; U.S. Patent 4,026,725; May 31, 1977; assigned to the U.S. Secretary of the Navy* relates to methods of activating molten salt electrolyte batteries. The process involves placing a powdered alkali metal carbide selected from the group consisting of lithium carbide, sodium carbide and mixtures into contact with the solid alkali halide electrolyte and then contacting the alkali metal carbide with a halogen gas selected from the group consisting of fluorine, chlorine, and mixtures.

Sodium carbide offers the advantage of reacting spontaneously with chlorine at low temperatures (0°C). Moreover, sodium carbide can be used in combination with other alkali metal carbides or lithium metal to provide mixtures which will also spontaneously ignite with chlorine gas at low temperatures (0°C). Further, these carbides also react with fluorine.

The proportion of alkali metal carbide to electrolyte must be great enough to melt the electrolyte and heat the electrolyte and certain other parts of the battery to the operating temperature of the battery. The proportion of carbide to electrolyte needed can be calculated from the heat of reaction of the alkali metal carbide with the halogen, the heats of fusion and the specific heats of the products and of the electrolyte, and specific heat and the heat losses of the particular battery used.

The proportion of alkali metal carbide which can be used is limited by the fact that when the carbon content of the electrolyte exceeds about 27% by weight the cells are likely to short out. Preferably the proportion of alkali metal carbide to electrolyte should be selected to produce, on combustion of the carbide, a carbon content of less than 10% by weight.

Metal-Metal Oxide Activator Composition

According to a process described by *W.H. Collins; U.S. Patent 4,041,217; August 9, 1977; assigned to Catalyst Research Corporation* thermal batteries are formed of two or more stacked cells, each cell being spaced from the adjacent cell by a combustible composition that undergoes exothermic reaction without the liberation of any substantial amount of gas and that forms an electrically conductive ash when burned. Thus, when the battery is activated by burning the combustible material to provide heat to melt or fuse the electrolyte, the ash formed thereby provides an electrical connection between the adjacent cells.

Suitable combustible compositions for use in the batteries include intimate mixtures of a finely divided metal oxide with a finely divided reducing metal that will exothermically react to form an electrically conductive metal oxide that is dispersed throughout the composition, as by melting or subliming. It is especially desirable to use combustible compositions that originally have an extremely low conductivity but form, on burning, an ash with extremely high conductivity; such compositions comprise higher oxides of metals having at least two valence states and that form on partial reduction an electrically conductive lower oxide, for example, tungsten trioxide and molybdenum trioxide. Any metal that reduces the oxide may be used, that is, any metal that forms oxides having a lower free energy of formation than the metal oxide to be reduced. Suitable reducing metals for use with the tungsten trioxide and molybdenum trioxide include iron, cobalt, nickel, chromium, molybdenum, aluminum, boron, magnesium, titanium and zirconium and tantalum.

The proportion of metal in the combustible composition is not more than about the stoichiometric amount required to reduce the higher metal oxide to the desired lower oxidation state. Somewhat more metal may be used for applications where atmospheric oxygen will be available to the reaction, e.g., to account for air entrapped in batteries. Some metals form a number of oxides, and the metal constituent may be used in any amount up to stoichiometric required by reaction with the oxide constituent to form the lowest suitable oxide.

For example, tungsten forms a number of oxides lower than tungsten trioxide, e.g., W_2O_5, W_4O_{11}, W_2O_3, and WO, all of which are conductive, thus the stoichiometric amount of metal may be based on a reaction to yield any of these lower oxides. Proportions of metal as low as about 20% of stoichiometric may be used as desired, the proportions being varied to provide the amount of heat and burning rate desired. Inert diluents, such as sand, may also be added to adjust the amount of heat produced; but the total amount of such diluents, including fibers used in forming pads, should not exceed about 40% of active combustible composition to avoid a noticeable effect on the electrical conductivity of the produced ash.

Pads of combustible material suitable for convenient use in batteries may be formed from thoroughly mixed slurries of metal oxide powder, metal powder, inert diluent if desired, and inorganic fibers, such as asbestos fibers, glass fibers, ceramic fibers or the like. It is preferred that the fibers be fine and flexible, since pads produced therefrom are more flexible and more dense. Pads may be formed from such slurries in a number of conventional ways, for example, by forming and drying in a sheet mold. Also, a mat may be formed by laying down the slurry on a screen, to form essentially a fiber paper filled with the combustible composition, using conventional papermaking equipment, such as a cylinder paper machine or Fourdrinier machine.

At least about 12% fibers are required to form a coherent pad and up to 25% or more fibers can generally be used without any noticeable effect on the conductivity of the ash formed by burning the pad. Drying temperatures must, of course, be kept below the ignition temperature of the combustible composition.

Alternatively, finely divided and well mixed constituents of the combustible composition, with or without fibers, may be compressed, either in the wet or dry state, to form a coherent compact pellet or wafer. The temperature during compression must, of course, be kept below the ignition temperature of the combustible composition and in some instances this may require that the mixture be cooled. Heating pads of the desired shape for use in batteries may be cut or punched from larger pads formed as described above, and the pelletized wafers are formed in the desired shape.

Example: A mat was formed in a sheet mold from a thoroughly mixed aqueous slurry containing 19.0 grams of powdered zirconium, 42.5 grams of tungsten trioxide, 6.5 grams of glass fibers and 6.5 grams of ceramic fibers. After being dried, the mat was positioned between two metal plates in one arm of a Wheatstone bridge. The mat was ignited and the resistance, determined in the usual manner, was found to be less than 0.1 ohm per square inch. Such pads have a resistance on the order of 100,000 ohms per square inch before ignition. Similarly, low conductivity ashes are obtained using other reducing metals or when using molybdenum trioxide with zirconium or other reducing metals.

Pyrotechnic Activation

According to a process described by *F.M. Bowers, J.H. Ambrus, G.S. Briggs, M.E. DeGraba, G.L. Green, A.S. Kushner, and D.L. Warburton; U.S. Patent 3,972,730; August 3, 1976; assigned to the U.S. Secretary of the Navy* a pyrotechnically activated lithium-chlorine cell possessing high voltage, high energy density and a high rate of discharge is composed of a porous graphite gas diffuser, a fused alkali metal salt electrolyte, a lithium anode in contact with the electrolyte, and a cartridge containing a pyrotechnic material disposed within the anode, all of the components being stacked inside a container that serves as the cell case. The porous gas diffuser is positioned within the cell case to provide a gas receiving chamber. Chlorine is introduced into the gas receiving chamber as the pyrotechnic material is ignited to activate the cell.

Example: A lithium-chlorine fused salt electrolyte cell of the process was fabricated in accordance with the following description. The cell case was a Poco AXF-5Q dense graphite cup 5.1 cm in diameter by 4.6 cm high. A 1.8 mm thick wafer of Poco AX porous graphite served as the gas diffuser and was fastened into the cup by means of a screw thread thereby forming a gas receiving chamber 4.0 mm high. The electrolyte consisted of a 1.5 mm thick wafer of lithium chloride-potassium chloride eutectic salt which was precast prior to assembly. The anode, lithium, was held in a cup-shaped nickel 200 Feltmetal matrix. A nickel 200 shield surrounded the matrix. The anode was insulated from the cell case by a boron nitride insulator which extended downward to serve as a lining for the cell case.

The pyrotechnic material used for activation was Pyronol No. 2, a mixture of aluminum, nickel and iron oxide. The complete cell including the pyrotechnic material, the lithium and the salt (but not the chlorine) weighed about 200 grams.

Cells of this construction were successfully and reproducibly activated in less than 30 seconds and discharged at 60 amperes and 2.1 to 2.5 volts for 18 minutes. The electrochemical efficiency of the lithium anode was approximately 50%.

Molten Metal Anode

According to a process described by *G.C. Bowser and J.R. Moser; U.S. Patent 3,930,888; January 6, 1976; assigned to Catalyst Research Corporation* the anode comprises a foraminous inert metal substrate wettable by and filled with an electrochemically active anode metal that melts at a temperature below the cell operating temperature and a housing having an impervious metal portion in electrical contact with the active anode metal and a porous refractory fibrous portion.

The porous portion is in sealing engagement against the periphery of the metal portion. It is preferred for use in long-life batteries for the housing to completely envelop the active anode with the porous portion at least partly covering the anode surface that abuts the cell components with which it electrochemically reacts, which for purposes of convenience will be designated as the inner surface. The inner anode surface may be left uncovered for use in cells or batteries designed for short-lived operation.

The batteries comprise a plurality of cells, each comprising the anode of this process, an electrolyte, and a depolarizer cathode. The preferred batteries comprise a stack of cells in which there are stacked in recurring sequence, an anode, a wafer containing electrolyte, depolarizer and binder, and a combustible composition that serves as a heat source and a cathode current collector. The batteries are activated by heating to a temperature above the melting point of the anode metal and the melting point of the electrolyte.

The active anode metals suitable for use in this process include alkali metals, alkaline earth metals or alloys thereof that melt below the cell operating temperature, or, for most purposes, below about $400°C$. It is preferred to use lithium or an alloy of lithium and calcium as cells with such anodes provide higher voltage, power density and energy density.

Vanadium Pentoxide Wafer Electrode

A process described by *D. Shevchenko; U.S. Patent 4,027,077; May 31, 1977; assigned to Catalyst Research Corporation* relates to thermal-type deferred action batteries. In this process, a properly prepared nickel strip is moved continuously lengthwise from one spool to another and, while it is traveling, one of its surfaces is coated across substantially half of its width with a mixture of V_2O_5 and H_3BO_3, or V_2O_5 and $(NaPO_3)_6$, and a liquid carrier to form a band along the strip. This band then is heated to melt it to form a glaze, and then the strip is cooled. The remainder of the same surface of the nickel strip then is coated with magnesium or calcium to form a band of that metal beside the glazing.

The two bands are then cut up into battery plates of the desired size and shape. One of the glazed plates and one of the bimetal plates then are placed against the opposite sides of a wafer having a V_2O_5 layer and an electrolyte layer. The V_2O_5 layer engages the glazed side of the glazed plate, and the coated side of the bimetal plate engages the electrolyte layer. This method of making a battery is very fast compared with the way they have been made before, and the two-layer wafer remains flat.

Using the process, a 500-foot strip of nickel can be provided with the glaze and metal coatings in a day and a half or two days. There likewise is a great saving in the etching time. The glaze is smooth and the same thickness throughout its area, which makes battery performance more consistent and reproducible. The time required to make a wafer is reduced at least a third and a much better bond is provided between the magnesium or calcium and the nickel than when the nickel plate was merely pressed against the magnesium or calcium layer of a wafer. Also a considerable amount of room is saved in a battery case by using double instead of triple layer wafers since Mg, or Ca bimetal is about 0.005 inch thick whereas a Mg layer is about 0.015 inch thick. This saved battery room is used for additional cells or for the extra insulation which preserves the heat, thus prolonging the battery life.

Depressions at Anode-Electrolyte Interface

*A.R. Baldwin and T.A. Reinhardt; U.S. Patent 3,914,133; October 21, 1975;
assigned to the U.S. Energy Research & Development Administration* describe
an improved thermal battery which includes a depression at the interface of
electrolyte pellet and anode for trapping of molten material.

Referring to Figure 5.15, the thermal battery may include a plurality of electrochemical cells **10** stacked one upon the other in electrical series within a suitable
casing **12** and thermal insulating barrier **14**. Electrical connections may be made
in an appropriate manner by suitable electrical leads and terminals **16,17** and **18**
to the respective positive and negative terminals of the upper and lower battery
cells in the stack. The heat or thermal generating elements for the battery, which
are generally positioned as a part of each battery cell with or without additional
heat generating elements at each end of the battery, may be ignited to activate
the battery by a suitable electrical match or detonator **20** and heat powder or
fuse **22** which is coupled between the match **20** and the heating generating elements in each cell.

The battery is normally formed by first stacking the individual cell elements to
form separate cells and then the cells are stacked together in the form shown in
Figure 5.15a and emplaced within the casing **12** and insulator **14** under suitable
pressure, such as by a compression force applied by a bolt **23** passing through
the center of the cells, or other suitable mechanisms. The so-stacked battery
cells may then be covered with an end cap insulator **24** and a casing cover **25**
in an appropriate manner. The battery is operated by initiating the electrical
match **20** and in turn the heat powder **22** and the individual heat generating
elements of the cell stack and the electrical current drawn off through appropriate leads **16,17** and **18**.

The individual cells, in accordance with this process, and as shown in Figure
5.15b, may include an anode electrode **26** and a heat generating disc **28** separated
by an electrolyte pellet **30**. Each of the discs or pellets is formed in a generally
circular or annular shape of similar or the same diameters and may be provided
with a central opening or bore **32** for receipt of the battery cell compression
mechanism. The individual cell **10** elements are stacked in the manner shown
to provide a sandwich of the anode electrode **26**, electrolyte pellet **30** and heat
generating disc **28**. The anode **26** may be formed of calcium or other like material as a solid disc or as a layer on a supporting conductive disc or plate, depending upon the electrochemical reaction utilized by the battery.

The electrolyte **30** may be formed from a mixture of a depolarizer and binder
with a normally solid fusible electrolyte which becomes conductive upon being
heated to above its melting point. For example, the electrolyte may be a eutectic mixture of potassium chloride and lithium chloride. The heat generating
disc **28** may be formed from such as iron-potassium perchlorate or the like which,
when ignited, is electrically conductive and produces a minimum of gas or other
deleterious products.

In accordance with this process, depressed portions or recesses are formed at the
interface between the anode electrode **26** and the electrolyte pellet **30** in one or
the other or both the electrode **26** and pellet **30**. The recesses may be formed
as an essentially annular or ring-shaped depression, groove or notch in one or

many concentric rings or a spiral or as a regular pattern of circular or other shape depressions. It is generally preferred for ease of manufacture and the like that the recesses, such as shown by depressions **34**, be formed in the surface **36** of pellet **30** which is adjacent to the anode electrode **26**.

FIGURE 5.15: THERMAL BATTERY

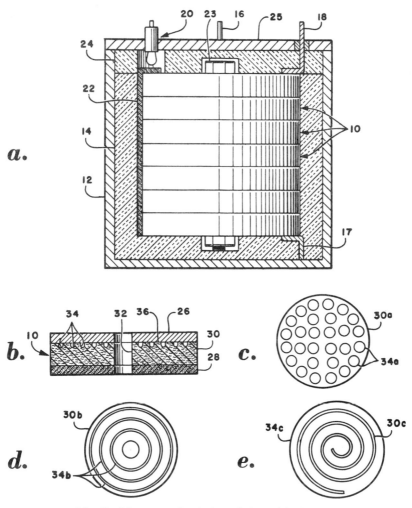

(a)   Partial cross-sectional view of thermal battery.
(b)   Cross-sectional view of battery cell which may be utilized
            in the battery of Figure 5.15a.
(c)   Top view of one arrangement of an electrolyte pellet for
            use in the cell of Figure 5.15b and battery of Figure 5.15a.
(d)(e)   Top views of other forms of the electrolyte pellet.

Source: U.S. Patent 3,914,133

The depressions 34 may take the form of a plurality of circular depressions 34a in a regular pattern on a surface of the electrolyte pellet 30a as in Figure 5.15c, one or more concentric rings 34b on a surface of the electrolyte pellet 30b in Figure 5.15d, or of a spiral 34c on a surface of the electrolyte pellet 30c in Figure 5.15e. The circular depressions 34a may be positioned, as shown, so as to fall along concentric rings about pellet 30a with the depressions in adjacent rings offset or staggered from those in adjoining rings. The depressions 34a,34b and 34c are preferably evenly distributed across the surface of the electrolyte pellets 30a,30b and 30c to insure the most effective distribution of the molten and active anode material over a maximum area of the electrolyte pellet. The depressions are also located so as to intersect all radii of the electrolyte pellet to maximize trapping of molten material wherever it may form at the interface.

As molten active anode materials are formed between the interface of anode electrode 26 and the electrolyte pellet 30, the molten materials may move about the surface or interface and reach a portion of the depression 34. The molten material will be effectively trapped in these depressions without degradation of the electrochemical cell reaction and battery outputs.

In a typical application, the depressions 34 may be from about 0.004 to 0.015 inch deep and from about 0.25 to 0.030 inch wide and include a depression volume of from about 0.001 to 0.002 cubic inch per square inch of pellet interface surface area of the cell 10. By positioning the depressions, as shown, in an essentially annular shape or regular pattern, the depressions will intersect all potential paths of molten material movement toward the outer edge of the electrolyte pellet 30. This effectively prevents formation of sufficient molten material at outer edges of the cell which may cause bridging of cell electrodes from molten materials formed at peripheral portions of the electrolyte pellet which are not traversed by a trapping depression.

The depressions 34 may be placed in a face of the electrolyte pellet 30 by any appropriate molding, cutting or other forming operation and may, in fact, be molded into the electrolyte pellet during its formation, as the pellet is commonly formed by compacting a mixture of powders in a mold. Thus, the electrolyte pellet may be formed in the desired shape in a common manner without any modification of prior formation processes, other than provision of appropriate mold shapes to provide the depressions.

Thermal batteries formed in accordance with this process can be operated at low current drains and produce electrical outputs for time periods of as long as 60 minutes or longer and generally for no less than 30 minutes using the arrangement shown in Figure 5.15 without shorting between cell electrodes from molten materials formed during the active discharge period. In addition, operating thermal batteries formed have greatly increased resistance to shorting due to movement of molten material resulting from mechanical environments of shock and vibration.

Spin Stabilized Battery

G.C. Bowser, Jr. and R.L. Blucher; U.S. Patent 4,044,192; August 23, 1977 and R.L. Blucher; U.S. Patent 4,002,498; January 11, 1977; both assigned to Catalyst Research Corporation describe a thermal battery which is suitable for use under high spin conditions. The battery comprises an aligned stack of annular cell elements, including combustible heating elements, forming an assembly, referred to as a cell stack, having a central hole extending vertically through it substantially on the axis of spin when in use. The central hole is filled with a combustible fuze material that when ignited will in turn ignite the combustible heating elements in the cell stack. The outer periphery of the cell stack is tightly enclosed by a sheet of nonporous electrical insulating material. The enclosed cell stack is contained in a hermetically sealed container and spaced from the container walls parallel to the axis of spin by a rigid noncompressible thermal insulation. Insulation between the cell stack and container walls perpendicular to the axis of spin may be compressible or noncompressible.

Referring to Figures 5.16a and 5.16b, which illustrate a preferred battery assembly, the cell stack is made up of a stack of annular cell elements each being in direct contact with adjacent cell elements. Each cell is made up of a metal anode **1,1a,1b,1c**, and **1d**, an electrolyte depolarizer pellet, **2,2a,2b,2c**, and **2d** and a combustible composition pellet that also serves as a cathode current collector **3,3a,3b,3c** and **3d**. A combustible composition pellet **4** provides heat to the outside of the bottom cell of the cell stack. Salt pellets **5** and **5a**, positioned on the ends of the cell stack, suitably $LiCl-Li_2SO_4$ or $KCl-Li_2SO_4$ serve as a heat sink and temperature regulator. Additional heating is provided by combustible composition **6**, and **6a**.

The central cavity of the cell stack is filled with a solid fuze material **7**, that is electrically insulating and forms a solid electrically insulating ash when it is burned. The entire outer surface of the cell stack is in tight engagement with mica sheet **8**. Rigid thermal insulation **9** is tightly fit between the mica sheet and metal container **10**. Thermal insulation **11** and **12**, spacing the cell stack from the end of the container, may be rigid insulation, the same as insulation **9**, or a compressible insulation such as packed asbestos fibers. Positive terminal **13** connected to metal current collector **14** extends through the insulation and container wall, suitably hermetically sealed by Kovar glass to metal seal **15**.

Leads **16** and **17**, hermetically sealed by Kovar seal **18**, are connected to an electrical source for firing the electrical match or squib **19** to activate the battery. Negative terminal **20** is secured to the housing. The electrolyte-depolarizer pellets of the cells of this process are single layer wafers containing all the cell electrochemical elements except the electrodes as described.

Such wafers or pellets are a compact of an intimate mixture of electrolyte, depolarizer and inert absorbent, the amount of absorbent being sufficient to absorb the electrolyte when it is melted. Generally, the depolarizer is present within the range of about 30 to 70 weight percent, the electrolyte within the range of about 20 to 45% and the absorbent in the range of about 5 to 30 weight percent, with the total of these three major components being essentially 100%. Suitable electrolytes include fusible alkali metal salts and mixtures thereof, such as, for example, lithium chloride-potassium chloride, lithium bromide-potassium bromide, lithium hydroxide-lithium chloride and other alkali metal salt mixtures, preferably eutectics.

A variety of easily reducible oxidizing materials are known and can be used as depolarizers, suitably including potassium dichromate ($K_2Cr_2O_7$), calcium chromate ($CaCrO_4$), barium chromate ($BaCrO_4$), molybdic oxide (MoO_3), manganese dioxide (MnO_2), and tungstic oxide (WO_3). Illustrative inert absorbent materials include bentonite, kaolin, magnesium oxide and finely divided SiO_2 such as Cab-O-Sil.

FIGURE 5.16: THERMAL BATTERY

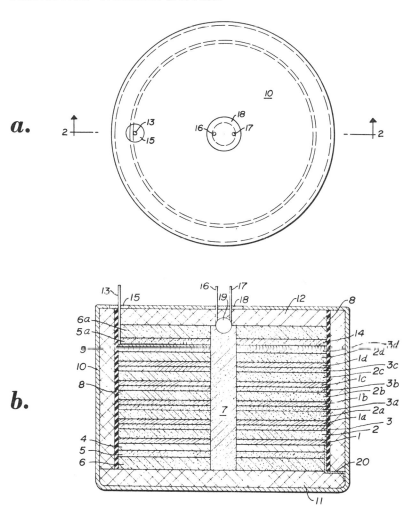

(a) Top plan view of battery.
(b) Vertical section of the battery of Figure 5.16a taken on line **2—2**.

Source: U.S. Patent 4,044,192

A variety of metals may be used to provide the electrodes in thermal cells in accordance with well-known electrochemical practice. For many purposes, it is preferred to use calcium as the anode and iron as the cathode, as this combination is productive of a high emf and calcium has a high melting point. However, various other metals may be used, such as, for example, magnesium anodes and cathodes of nickel, silver, or copper, The anode is conveniently formed of a laminar bimetal sheet having the lamina of active anode metal such as Ca on a lamina of metal with more structural rigidity, such as nickel or iron. The cathode may take various forms, suitably sheet metal or a pellet of compacted and/or sintered powdered metal. The metal component of the combustible composition, comprising a reducible metal and an oxidizer therefore, may serve as the cathode. In such cases, the reducing metal, suitably iron, of the composition is used in excess of stoichiometric amount that will react with the oxidizer component.

It is essential to obtain the full benefit of this process to tightly surround the outer periphery of the cell stack with a nonporous electrical insulation to form a seal with each cell. The preferred material is mica sheet as it forms an exceptionally effective seal, possibly by some limited reaction with the cell components when the battery is heated to activation temperature.

Other nonporous materials can be used to somewhat less advantage such as glass board sealed with a high-temperature cement or asbestos board sealed with water glass. Other rigid insulations, such as glass or asbestos board, do not provide this edge sealing action. To illustrate, several batteries were made in accordance with Figure 5.16a and 5.16b except that the mica sheet was not used and the rigid insulation, Glass Rock No. 50, tightly engaged the outer periphery of the cell stack. The cell stack was made up to 10 cells having a calcium-iron bimetal anode, the calcium being the active anode metal, a depolarizer-electrolyte pellet consisting of an intimate mixture of 30.9% KCl, 25.1% LiCl, 37.0% $CaCrO_4$, and 7% Cab-O-Sil compacted at 15 tons per square inch, and a cathode-heating composition consisting of an intimate mixture of 90% iron and 10% $KClO_4$ compacted at 8 tons per square inch.

The fuze material was 28% Zr and 72% $BaCrO_4$. The batteries were tested at various spins and the time, under load, for the output voltage to reach 12, 11 and 10 volts was measured, the 10 volt output representing the useful life of the battery in a particular application. The results set forth in the table below show that the performance of the battery at a spin of 100 rps was substantially the same as at zero spin, but that at a spin of 200 rps the performance had deteriorated to give less than ⅓ the zero spin life.

Spin (r.p.s.)	12 Volts	Seconds to 11 Volts	10 Volts
0	74.5	124.2	174
100	70.0	110.0	155.0
200	29.0	45.0	54.0

When a mica sheet was wrapped around the cell stack, as shown in Figures 5.16a and 5.16b, an otherwise identical battery gave substantially the same performance at 200 rps as at zero spin.

Spin (r.p.s.)	12 Volts	Seconds to 11 Volts	10 Volts
200	95.0	130.0	166.0

Another group of batteries was made as in Figures 5.16a and 5.16b and the previous example, except that the amount of combustible composition was slightly decreased. The performance of these batteries at various spin conditions was as follows:

Spin (r.p.s.)	12 Volts	Seconds to 11 Volts	10 Volts
0	79.0	122.0	156.0
150	75.0	120.0	157.0
300	75.0	120.0	157.0

Battery Design

According to a process described by *E.J. King; U.S. Patent 3,972,734; August 3, 1976; assigned to Catalyst Research Corporation* thermal cells are formed of an electrode of one metal that is enclosed within an envelope or receptacle of another metal serving as a case for the cell, with an appropriate electrolyte disposed between the two. In the preferred case a metallic electrode blank is coated, in part, with a different metal to provide positive and negative electrode areas on the same blank.

Such a blank is then enclosed by folding an electrode of, for example, the base metal in such manner that the coating metal is opposite the inside surfaces of the enclosing electrode. The electrolyte is disposed between the coated surface of the blank and the inner surface of the enclosing or receptacle electrode. In this general manner, a very compact and rigid cell structure is provided that may be developed into batteries of desired emf with a high current capacity.

Figure 5.17a includes three main sections **10,11** and **12**. Suitably the blank is formed of very light gauge (0.006 inch thick) sheet nickel. Sections **10** and **11** are substantially equal in size and are rectangular in shape. Section or extension **12** is of size and shape adapted to fit within an envelope formed by folding section **10** onto section **11**. While section **12** therefore must be sufficiently small to go within the resulting envelope, it is desirable, in the interests of obtaining as high a current discharge as possible, to make that portion as large as is consistent with the foregoing requirement. In other words, the opposing faces of the electrodes are as nearly equal as possible to maximize the rate of discharge in use.

The blank of Figure 5.17a actually functions to provide two separate electrodes. The receptacle formed from sections **10** and **11** constitutes one of the electrodes, for example, if made from nickel it will be the cathode where a nickel-calcium couple is used. The anode is formed from section **12** by treating it with a second metal, such as by providing a layer of calcium **14**, that covers substantially the entire section. The calcium (or other second metal) may be placed on **12** in any way desired; however, an advantageous method of application involves perforating section **12**, much in the manner that a cheese grater is perforated, to provide oppositely extending projection points **15**, and then rolling calcium sheets on both sides of the resulting structure.

It will be observed from the cutaway area of section **11** that the blank serving as the enveloping or receptacle electrode is provided with a screen or wire mesh **16** affixed thereto. The wire mesh suitably is made of the same metal as the blank.

FIGURE 5.17: THERMAL DEFERRED ACTION BATTERY

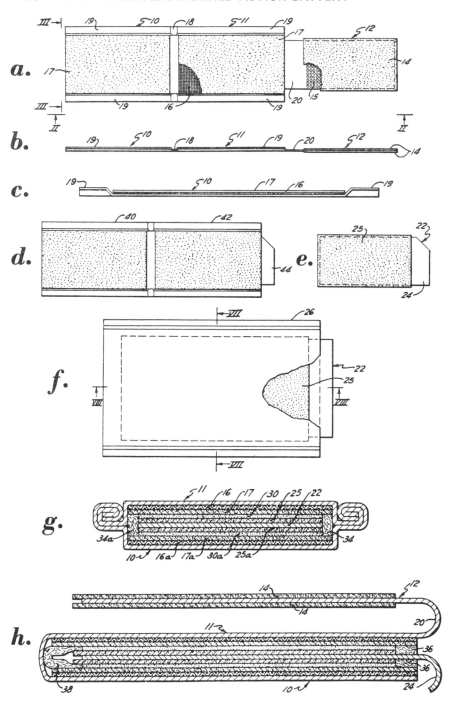

(continued)

FIGURE 5.17: (continued)

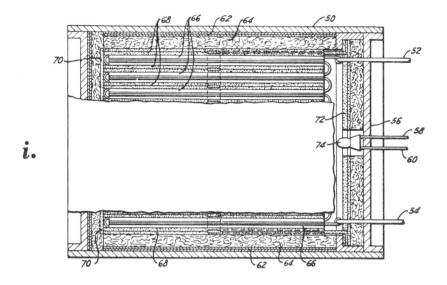

i.

(a)   Plan view of a cell blank that is used to provide a cathode of one
        cell and an anode of a second cell.
(b)   Side view of the blank of Figure 5.17a taken along line II—II.
(c)   End view of the blank of Figure 5.17a, but to a larger scale, taken
        along the line III—III.
(d)   Plan view of a blank that is used as an outer or receptacle electrode
        of an end cell.
(e)   Plan view of a blank used as a central or inner electrode of an end
        cell.
(f)   Plan view of a complete cell showing the relative disposition with
        respect to one another of its two electrodes.
(g)   Cross section of the cell of Figure 5.17f taken along line VII—VII.
(h)   Cross section of the cell of Figure 5.17f taken along line VIII—VIII.
(i)   View, partly in elevation, of a complete battery structure.

Source: U.S. Patent 3,972,734

Where a high rate of discharge is needed in a cell, it is advantageous to provide
a depolarizer adjacent the cathode. The screen **16** serves as a carrier for a depolarizer **17**; such screens are provided on each of the sections **10** and **11** of the
cathode. The relative disposition of the depolarizer and the screen is evident
from the end view of the blank shown in Figure 5.17c. The use of a depolarizer
and screen adds some thickness to the unit. Consequently, and for the additional
purpose of providing sufficient space to receive the central electrode, the blank
most suitably is stamped so that the central area thereof is recessed longitudi-
nally (Figure 5.17c).

In forming an electrode with a blank as shown in Figures 5.17a, 5.17b, and 5.17c,
the blank is prepared with a depolarizer and a second metal in place as just de-

scribed. Section **10** is then folded along its transverse axis, as along folding trough **18**, a depression stamped into the blank, until the depolarizer on section **10** is in face-to-face relationship with the depolarizer of section **11**. The side edges **19** of sections **10** and **11** are then crimped or welded together to provide a seam through its length. The seamed area is then folded upon itself (see the end areas of the cell in Figure 5.17g) to complete the seal and to reduce the width of the unit. The resulting structure constitutes an envelope-shaped electrode, or a receptacle-shaped electrode with an open end, adapted to receive a central electrode, and a second electrode integral with and extending from an end edge of one of the sides of the enveloping or receptacle electrode.

In constructing a cell, such as a first end cell of a battery, with an electrode receptacle such as just described, a central electrode with the second metal thereon (calcium) is prepared. Such an electrode **22** covered with the second metal **25** is shown in Figure 5.17e. Since this electrode blank serves as an electrode and as a means to provide one lead for the battery of cells, it is provided with an uncoated neck area **24** to facilitate connection of a battery lead. Prepared electrolyte pads are placed on each surface of the electrode **22** and the resultant unit then is placed on section **11** of the blank of Figure 5.17a. The blank is then folded to form the receptacle in the manner just described.

In Figure 5.17f is shown a completed cell, in plan showing the relative disposition of a central electrode **22** with respect to the side walls of the receptacle electrode **26**. A cross section of this cell is shown in Figure 5.17g. Considering Figure 5.17g, from top to bottom the elements in a completed cell are the outside enveloping section **11** of the receptacle electrode **26**, screen **16** affixed to that section which holds a layer **17** of a depolarizer. The electrolyte that is used is shown as a pad of electrolyte **30**. Adjacent the electrolyte pad is the calcium deposit **25** that is on the central electrode **22**. From the middle to the bottom of the cell the order of the foregoing components is reversed. On the central electrode **22** is the calcium deposit **25a** which is in contact with the electrolyte pad **30a**.

Adjacent the electrolyte pad is depolarizer **17a** which is compressed into screen **16a**, which in turn is affixed to the surface of the outer electrode **10**. To prevent accidental contact between the electrodes of the cell, it is desirable to include insulating pads around the edges of the central electrode. In Figure 5.17g, felt pads **34** and **34a** represent the insulating means that are along the sides of the central electrode.

The first end cell in addition to all other cells except the final end cell provides an adjacent cell with an electrode. The relative disposition of the auxiliary electrode of the first end cell is shown in Figure 5.17h, which is a cross section of the cell of Figure 5.17f taken along line VIII—VIII. In Figure 5.17h the auxiliary electrode of the completed end cell is shown in the plane above and parallel to the completed cell. This is accomplished by bending the area **20** of section **12** as shown. Section **12**, of course, is coated with the second metal. The numerals **36** and **38** constitute the insulating members, for example, felt pads, that are used at the ends of the central electrode of the cell to prevent accidental contact between the electrodes and electrolyte leakage during active life. The neck area **24** of the central electrode is shown bent down. In use, an electrical lead would be attached.

In a battery of cells in accordance with this process, there is at least one central cell. Such a cell is formed in the same general manner as described for the first end cell. However, the central or inner electrode of a central cell is the extension from the end of a side wall of the receptacle electrode of the next adjacent cell. Accordingly, there is no need to provide a special central electrode as in the instance of forming the first end cell.

In assembling a central cell, a prepared blank, such as that shown in Figures 5.17a, 5.17b and 5.17c, is folded around an element **12** of an adjacent cell to serve as its central electrode, with electrolyte pads in place, to form the receptacle electrode. The extension from the resulting receptacle electrode of the central cell is then available to serve as the inner electrode of the next central cell or the second end cell, depending on the number of cells desired.

The final end cell of a battery of cells need provide no electrode for an adjacent cell. Consequently, the structure of the blank used for such a cell may differ from the blank shown in Figure 5.17a. Such an end cell blank is shown in Figure 5.17d. This blank has three main sections **40, 42** and **44**. Sections **40** and **42** serve the same function as do sections **10** and **11** of the blank of Figure 5.17a. The tab extension **44** is provided as a means to which is connected a terminal lead for the delivery. In forming this cell, the blank with the depolarizer on its electrode surfaces is folded about the extension electrode of the next adjacent central cell and the side edges are sealed as with the other cells.

The batteries are intended for use where a high current discharge, relatively low voltage, thermal battery is needed. For the uses now known compactness is particularly desirable. On the other hand, to obtain a high current from a battery requires a large electrode area, since the quantity of current from a battery is proportional to the electrode area. It can therefore be seen that whereas compactness is desired on the one hand, the companion desideratum of a large current discharge requiring large electrode area is in conflict therewith. In the process, this conflict is resolved by minimizing bulk attributable to anything other than the electrodes.

This may be observed upon consideration of Figures 5.17f, 5.17g and 5.17h. It will be noticed that nothing is present, other than is absolutely necessary, to increase the thickness of the cell. Similarly the length and width of the electrodes are substantially the length and width of the cell, actually being different only in that additional width used for the crimped or sealed side edges and the external portions **20** and **24** of the central electrodes. It may also be noted that the receptacle electrode functions as the case for the cell, avoiding the need to add anything for that purpose.

Batteries that are made in accordance with this process include the completed individual cells, the battery terminals, a combustible material to activate the cells, a means to actuate the combustible, and a suitable casing. The structure of a typical battery is shown in Figure 5.17i. A hermetically sealed canister or case **50**, suitably of light gauge sheet iron or steel, is provided to receive a plurality of cells or groups of cells according to the voltage desired in the completed battery. The positive and negative terminal leads **52** and **54**, respectively, extend through the side (or top) **56** of the battery case **50** for ready access upon use, as do the terminal leads **58** and **60** of an electric match combustion initiator. In the battery shown, the cell units are series connected in the manner already described. Consequently, the first end cell and the final end cell are at opposite ends of the battery. Within the limits of operability of the combustion material used and similar considerations, the cell units may be placed in the battery case in any manner desired. One arrangement found suitable involves lining the battery case with insulating and shock absorbing material, such as asbestos and felt pads **62** and **64**, respectively. Other materials that can be used include, by way of example, fiber glass cloth and mica strips. These linings serve both as electrical insulation and as heat insulation, the latter to isolate the cells from the influence of ambient conditions and to retain heat from the combustible material in the area of the cells once the unit is actuated.

Individual cells and groups of cells are arranged with heat pads, of any composition as described hereinbefore, interspersed among them, and are then placed in the case within the lining. The heat pads are about the same length and width as the cell, and have powder trains connecting them to one another. The cells are designated **66**, the heat pads are **68** and the powder train is **70**. On the end of the stack of cells is placed a pad **72** carrying the electric match **74** used to set off the powder train, followed by layers of insulation and shock absorbing pads as desired. The powder train ignition means, which is the electric match **74** in this embodiment, is placed adjacent the end of the powder train to insure ignition. The sealing member **56** closes the case **50**, with the battery leads and combustion initiator leads extending therethrough in sealed relation. Where necessary the closure **56** is hermetically sealed to the remainder of the case.

Multimode Battery

A.A. Benderly; U.S. Patent 3,904,435; September 9, 1975; assigned to the U.S. Secretary of the Army describes a multimode battery which is capable of providing low, but useful, electrical current for a long period of time followed by a demand for considerable power for a short period of time. The thermal cell comprises: (1) electrodes; (2) at least one salt disposed between the electrodes, the salt being solid and capable of providing a low, but useful, ionic conductivity, at ambient temperatures and becoming a strong ionic electrolyte upon thermal activation; and (3) means contained in the cell for supplying a sufficient amount of heat to either melt or cause a solid state phase change in the salt and thereby causing the salt to become a strong ionic conductor.

More specifically, the thermal cells utilize a particular type of salt in order to achieve the desired two-level, energy requirements of the cell. These salts exist initially in a solid state capable of conducting a small and useful current thereby enabling the cell to provide energy for a particular or desired function. Upon thermal activation, the cell is capable of satisfying considerably greater energy requirements. These greater energy requirements are satisfied, for example, by a significant temperature rise taking place within the cell and converting the solid inorganic salt into a molten or liquid state.

Instead of undergoing a change of phase from solid to liquid, another embodiment of this process relates to the increased conductivity of the electrolyte by a significant temperature increase due to a change whereby transition from one crystalline state to another crystalline state causes the final solid phase to be substantially conductive.

Particularly suitable salts useful as electrolytes for the thermal cells of this process and capable of undergoing substantial changes in conductivity are selected from the group consisting of silver iodide, silver-rubidium iodide, lithium iodide, lithium tungstate, alkyl ammonium iodides, and mixtures thereof. The alkyl-ammonium iodides contain one or more alkyl groups, and preferably 4 alkyl groups attached to the nitrogen atom, e.g., a quaternary ammonium compound wherein the alkyl groups are preferably, though not necessarily, identical. When a quarternary ammonium compound is employed, each of the alkyl groups contains predetermined amounts of carbon atoms although best results are achieved when each alkyl group contains from 1 to 4 carbon atoms, preferably tetramethyl or tetramethyl ammonium iodide.

The thermal cells are applicable to those ordnance applications having relatively low power requirements over a relatively long period of time followed by a demand for considerable power for a rather short period of time. In a related mode of operation, a solid electrolyte battery, capable of being stored for a long period of time on an open circuit, would, on demand, receive a thermal input from a built-in thermal source. This rise in temperature would allow for a higher drain capability for a few minutes during which time a large capacitor would be charged.

Upon cooling, the battery would revert to an ordinary solid electrolyte power supply once more capable of supplying lower power to: (a) maintain the charge on the capacitor for a considerable period of time, e.g., 30 days and/or (b) operate timing or sensing circuits of low power requirements. This type of power supply would be useful in scatterable mines where long term, inactive storage is coupled with the need for days or weeks of active life followed by a firing pulse.

OTHER BATTERIES

Thermoelectric Battery

M.H. Brown; U.S. Patent 4,002,497; January 11, 1977; assigned to the United Kingdom Atomic Energy Authority, England describes a thermoelectric battery which comprises a thermoelectric module in the form of a matrix of rods of alternately P- and N- type semiconductor material connected together in the manner of a thermopile. The thermoelectric module is suspended in a casing in contact at one end with a heat sink and having attached at the other end a radioisotope fuel capsule. The attachment of the fuel capsule is supported against shocks or rapid accelerations by a combination of spring mounting and a fiber spider.

Referring to Figure 5.18, a cylindrical casing of stainless steel **11** encloses an assembly comprising a heat source **12** secured to one end of a thermoelectric unit which, in this example, comprises a thermoelectric module **13** in the form of a rectangular assembly of a plurality of thermoelectric elements secured together and electrically connected at their ends in the manner of a thermopile. The other end of the thermoelectric module **13** is secured with adhesive to a platform **14** of good thermal conductivity. In this example, the platform is made of stainless steel and a layer of **15** epoxy resin is interposed between the thermoelectric module **13** and the platform **14** to provide electrical insulation while maintaining good heat conducting connection between the module and the platform. The platform **14** abuts an insert **16** welded into one end of the cylindrical casing **11**. Electrical leads **17,18** for making electrical connection to the thermoelectric module **13** pass through apertures **19,21** in the platform **14**.

FIGURE 5.18: THERMOELECTRIC BATTERY

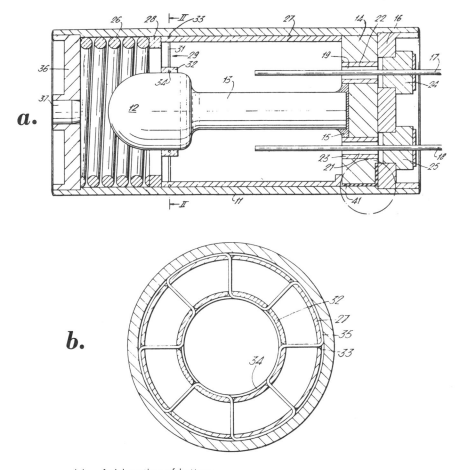

(a)   Axial section of battery.
(b)   Section on the line II—II of Figure 5.18a with some parts
      omitted.

Source: U.S. Patent 4,002,497

Each of the apertures is provided with an insulating sleeve **22,23**. An electrically
insulating gas-tight seal is provided at **24** and **25** between each of the electrical
leads **17** and **18** respectively and the insert **16**.

The platform **14** is a sliding fit within the cylindrical casing **11** and is biased
into contact with the insert **16** by the action of a spring **26** transmitted through
a cylindrical sleeve **27**, which is also a sliding fit within the casing **11**. The
sleeve **27** is of stainless steel but preferably of a different composition from that
of the casing **11** to avoid possible binding. The spreader ring **28** is interposed
between the spring **26** and the sleeve **27**. The end of the sleeve **27** adjacent the
spreader ring **28** supports a fiber spider **29**. In this example, the fiber spider **29**

is formed by a single fiber strand **31** which is wound in the manner best seen from Figure 5.18b between the sleeve **27** and a collar **32**, which is a loose slid-ing fit upon the heat source **12**. The collar **32** and the sleeve **27** are provided with a series of radially aligned apertures connected, in the sleeve **27**, by an ex-ternal circumferential groove **33** and, in the collar **32**, by an internal circum-ferential groove **34**. A portion **35** of the fiber strand extends in the groove **33** around the sleeve **27** between two apertures. The strand passes through the apertures across to the corresponding apertures in the collar **32** and extends around the groove **34** in the collar **32** to emerge from the next aperture and traverse to the sleeve **27**, and so on.

In this way a fiber spider support is provided for the heat source **12**. The por-tions of fiber extending in the grooves **33** and **34** are secured respectively to the sleeve **27** and the collar **32** by adhesive. The fiber material is chosen to provide optimum properties of strength and low thermal conductivity. An example of a fiber considered suitable is Du Pont's PRD 49 Type 1 described as an organic, low-density, high-modulus, high-strength fiber resistant to abrasion, cutting and tensile failure even after looping or knotting.

The enclosure **11** is sealed by an end cap **36** welded to the enclosure **11**. The end cap **36** is provided with a central pin **37** for final evacuation and gas filling after which the pin **37** is welded in place.

The thermoelectric module **13** is the form described in British Patent 1,303,834. Briefly, the module **13** comprises a plurality of semiconductor elements alter-nately of P- and N-type connected together to form a series of thermocouples by electrically conductive bridges.

The heat source **12** comprises a plutonium 238 radioisotope fuel contained in a high-temperature capsule of a tungsten/tantalum alloy with an oxide surface protective layer.

It will be seen that the configuration of the components provides for easy assem-bly into the casing **11**. The procedure is to bond the heat source **12** to the thermoelectric module **13** and the latter to the platform **14**. The module leads are then electrically connected to the leads **17** and **18** mounted in the seal hous-ing insert **16**. The assembly is inserted into the casing **11** and the insert **16** welded to the casing **11**. From the other end of the casing **11**, the sleeve **27** together with the fiber spider **29** is inserted, followed by spreader ring **28** and spring **26**. The collar **32** slides over and locates the heat source **12** as the sleeve **27** is inserted into the casing **11**. The end cap **36** is welded to the casing, the casing evacuated, filled with inert gas and the pin **37** welded to the end cap **36**.

If the battery is subjected to radial shocks, the effect of these upon the heat source **12** is absorbed by the fiber spider **29**, thereby reducing stress being im-posed upon the thermoelectric module **13** or its bonds to the heat source **12** and the platform **14**. Longitudinal shocks are cushioned by the spring **26**.

The interior of the sleeve **27** and the end cap **36** may be provided with radiation shielding material such as tantalum. The sleeve **27** itself may be made from tantalum and a tantalum shielding disc attached to the interior of the end cap **36** (welding difficulty making it impracticable to make the whole end cap **36** from tantalum). In such an arrangement, the spring **26** would be located at the other end of the sleeve **27**, so that the sleeve more effectively covers the source. Provision would then be required to restrict the longitudinal movement of the assembly. Improved resistance to high-frequency shocks may be achieved by providing a soft resilient material between the platform **14** and the casing **11** and end insert **16**. This provision is illustrated at **41** in the circular insert of Figure 5.18a.

Printed Circuit Board Battery Pack

A process described by *J.E. Sykes; U.S. Patent 3,992,225; November 16, 1976; assigned to Mauratron Incorporated* relates to printed circuit board battery packs for use in conjunction with batteries having snap connectors mounted on the opposite ends. The battery pack for batteries of the type having snap connectors mounted at the opposite ends comprises at least two circuit board members having snap connectors mounted thereon for mechanical engagement with and electrical connection to the snap connectors of the batteries.

Electrical connections are made between the snap connectors on the circuit boards and to terminals for the battery pack by means of printed circuit layers formed on the circuit boards. The circuit boards are positioned in a spaced apart relationship such that the batteries are retained by the snap connectors on the circuit boards while accommodating limited relative movement between the batteries and the circuit boards, whereby the batteries are floatingly supported and the difficulties that have been experienced with prior art battery packs due to shock loading are eliminated.

Referring to Figure 5.19a, there is shown a conventional battery **10**. The battery **10** comprises a cylindrical body **12** and flat, circular ends **14** and **16**. The battery **10** further includes a female snap connector **18** centrally disposed on the end **14** and defining a first terminal of the battery **10**, and a male snap connector **20** centrally disposed on the end **16** and defining a second terminal of the battery **10**. The battery **10** may further include indicia of terminal polarity **22**, which may be of any of the various well-known types.

In Figures 5.19b, 5.19c and 5.19e, there is shown a battery pack **30** comprising an example of the process. The particular battery pack **30** is utilized to receive and form electrical connections to four batteries **10** of the type shown in Figure 5.19a. The battery pack **30** includes a circuit board **32** comprising a thin, flat, substantially planar member formed from a dimensionally stable, electrically insulative material. The circuit board **32** may be formed in accordance with any of the various conventional techniques which are utilized in the electronic components manufacturing industry in the manufacture of printed circuit boards.

A plurality of male snap connectors **34** and a plurality of female snap connectors **36** are mounted on the circuit board **32**. All of the snap connectors are mounted on the same side of the circuit board **32** and are spaced predetermined distances apart, with the male and female snap connectors being alternately arranged. The snap connectors **34** and **36** are adapted to mechanically engage and form electrical contact with the female snap connectors **18** and the male snap connectors **20** of the batteries **10**, respectively. The snap connectors **34** and **36** of the circuit board **32** are positioned at spaced points along a straight line and may be considered as a linear battery pack. As is best shown in Figure 5.19c, the circuit board **32** has a plurality of printed circuit layers **38**, **40** and **42** formed thereon.

FIGURE 5.19   PRINTED CIRCUIT BOARD BATTERY PACK

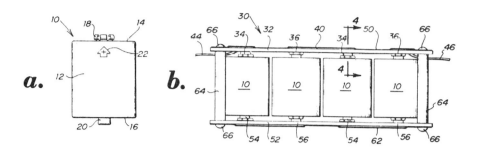

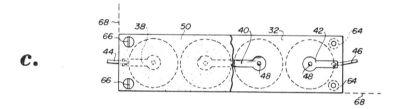

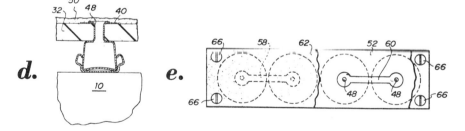

(a)   Conventional battery having snap connectors mounted at the opposite ends.

(b)   Side view of a battery pack comprising this process.

(c)   Top view of the battery pack of Figure 5.19b in which certain parts have been broken away.

(d)   Enlarged partial sectional view taken along the line 4—4 of Figure 5.19b in the direction of the arrows.

(e)   Bottom view of the battery pack of Figure 5.19b in which certain parts have been broken away.

Source: U.S. Patent 3,992,225

The printed circuit layers **38,40** and **42** may be formed on the circuit board **32** by means of any of the various techniques conventionally employed in the electronic components manufacturing industry in the manufacture of printed circuit boards. The printed circuit layer **38** is connected to one of the male snap connectors **34** of the printed circuit board **32** and defines a terminal for the battery pack **30**. A flexible lead **44** comprising a length of conventional insulated wire is connected to the printed circuit layer **38** by conventional means, such as a solder connection between the conductive portion of the wire and the printed circuit layer. The lead **44** extends from the printed circuit layer **38** through a hole formed in the circuit board **32** and hence into electrical contact with remote circuitry.

The printed circuit layer **42** also defines a terminal of the battery pack **30**. The printed circuit layer **42** is electrically connected to one of the female snap connectors **36** of the circuit board **32** and is in turn electrically connected to a flexible lead **46**. The lead **46** extends from the printed circuit layer **42** through a hole in the circuit board **32** and into electrical contact with remote circuitry.

Unlike the printed circuit layers **38** and **42**, the printed circuit layer **40** does not serve as a terminal for the battery pack **30**. Rather, the printed circuit layer **40** serves to electrically interconnect the male snap connector **34** and the female snap connector **36** which are not connected to the printed circuit layers **38** and **42**. Similar connections are formed between the terminals of the batteries **10** situated remotely from the circuit board **32**. In this manner the battery pack **30** functions to contact all of the batteries **10** received therein in series electrically.

The printed circuit layers **38,40** and **42** are each electrically connected to the underlying snap connectors by means of rivets **48**. Referring particularly to Figure 5.19d, the rivets **48** form electrical connections extending through the circuit board **30** whereby the printed circuit layers on one side and the snap connectors on the opposite sides are electrically interconnected. The rivets **48** also serve to secure the snap connectors to the circuit board **32**.

Referring again to Figures 5.19b and 5.19c, a layer **50** of electrically insulative material is formed on the same side of the circuit board **32** as the printed circuit layers **38,40** and **42**. The layer **50** preferably extends over the entire surface of the circuit board **32** remote from the positioning of the snap connectors thereon, and serves to electrically isolate the printed circuit layers **38,40** and **42**. The layer **50** may be formed from any of the various materials commonly employed as electrically insulative layers in the electronic components manufacturing industry. For example, the layer **50** may comprise polyethylene or any similar plastic material.

As is best shown in Figures 5.19b and 5.19e, the battery pack **30** further comprises a circuit board **52**. Like the circuit board **32**, the circuit board **52** preferably comprises a thin, flat, planar member formed from a dimensionally stable, electrically insulative material. The circuit board **52** may be manufactured utilizing any of the techniques commonly employed in the electronic components manufacturing industry in the manufacture of printed circuit boards.

The circuit board **52** has a plurality of female snap connectors **54** and a plurality of male snap connectors **56** mounted thereon. The female snap connectors **54** are adapted to mechanically engage and form electrical contacts with the male snap connectors **20** of the batteries **10**, and the male snap connectors **56** are adapted to mechanically engage and form electrical contacts with the female snap connectors **18** of the batteries **10**. The snap connectors **54** and **56** are all

mounted on the same side of the circuit board 52 and are positioned in a predetermined spaced-apart relationship which is identical to the spacing relationship of the snap connectors 34 and 36 of the circuit board 32.

Each of the female snap connectors 54 on the circuit board 52 faces and is positioned in alignment with one of the male snap connectors 34 on the circuit
board 32. Likewise, each of the male snap connectors 56 on the circuit board
52 faces and is in alignment with one of the female snap connectors of the circuit board 32. They are adapted to receive and retain the batteries 10 therebetween by engagement with the snap connectors 18 and 20 of the batteries.

As is best shown in Figure 5.19e, the circuit board 52 has a pair of printed circuit layers 58 and 60 formed on the side thereof opposite the positioning of the
snap connectors 54 and 56. The printed circuit layers 58 and 60 of the circuit
52 may be formed by means of the techniques conventionally employed in the
electric components manufacturing industry in the manufacture of printed circuit boards. The printed circuit layer 58 serves to electrically interconnect the
lefthand female snap connector 54 and the lefthand male snap connector 56
(Figure 5.19b) of the battery pack 30. The printed circuit layer 60 serves to
electrically interconnect the righthand female snap connector 54 and the righthand male snap connector 56 (Figure 5.19a) of the battery pack 30.

It will thus be understood that by means of the printed circuit layer 38 on the
printed circuit board 32, the male snap connector 34 electrically connected
thereto, the aligned female snap connector 54 on the circuit board 52, the
printed circuit layer 58, the male snap connector 56 connected thereto, the
aligned female snap connector 36 on the circuit board 52, the printed circuit
layer 40, the male snap connector 34 connected thereto, the aligned female snap
connector 54 on the circuit board 52, the printed layer 60, the male snap connector 56 connected thereto, the aligned female snap connector 36 on the circuit
board 32 and the printed circuit layer 42, all of the batteries 10 which are received in the battery pack 30 are connected in series electrically and serve to
establish a predetermined electrical potential across the terminals comprising
the printed circuit layers 38 and 42.

A layer of electrically insulative material 62 is formed on the side of the circuit
board 52 having the printed circuit layers 58 and 60 formed thereon. The insulative layer 62 is thus disposed on the side of the circuit board 52 remote
from the mounting of the snap connectors 54 and 56, and serves to electrically
isolate the printed circuit layers 58 and 60. The layer 62 may be formed from
any of the materials commonly employed in the electronic components manufacturing industry to form insulative layers. For example, the layer 62 may be
formed from polyethylene or similar plastic materials.

Referring again to Figure 5.19b, the circuit boards 32 and 52 are maintained in
a predetermined spacial relationship by means of a plurality of spacers 64. The
spacers 64 engage the engage surfaces of the circuit boards 32 and 52 having
the snap connectors 34, 36, 54, and 56 mounted thereon, and thereby maintain
a predetermined spacial relationship between aligned snap connectors. The circuit boards 32 and 52 are retained in engagement with the spacers 64 by means
of a plurality of fasteners 66. Aligned fasteners 66 may extend through the
spacers 64 into threaded engagement with one another. Alternatively, the fasteners 66 may be threadedly engaged directly with the spacers 64. The spacial

relationship between the circuit boards **32** and **52** is such that the batteries **10** are retained in the battery pack **30** by means of mechanical engagement between the snap connectors of the batteries and the snap connectors of the circuit boards. Thus, the snap connectors of the batteries are continuously maintained in mechanical engagement with an electrical contact with the snap connectors of the circuit boards. However, at least a limited amount of relative movement of the batteries **10** with respect to the circuit boards is permitted, whereby the batteries **10** are floatingly supported in the battery pack **30**. This has been found to be highly advantageous in preventing damage to the component parts of the battery pack due to shock loads. Likewise, the snap connectors of the batteries **10** do not become disengaged from the snap connectors of the circuit boards **32** and **52** even under conditions of extremely heavy shock loading.

Referring particularly to Figure 5.19c, it will be noted that whereas the process has been illustrated and described in conjunction with a battery pack adapted to receive four batteries **10**, battery packs may be constructed so as to receive any desired number of batteries. Thus, as is illustrated by the perpendicular dashed lines **68**, the battery pack **30** may be expanded either lengthwise or width-wise to accommodate a greater number of batteries, it being understood that in the case of a widthwise expansion the battery pack would no longer comprise a purely linear battery array. Conversely, the battery pack **30** may be reduced in length so as to accommodate a lesser number of batteries, if it is desired. The battery pack **30** may be arranged to accommodate any of the various conventional battery sizes including D, half D, and F, all of which are of the same diameter; C and others of the same diameter; A, AA, and several other sizes of the same diameter, etc.

Heat-Activated, Electrically Generating Fire Alarm

A. Fujiwara; U.S. Patent 4,009,055; February 22, 1977 describes an apparatus for producing an electric power when a fire breaks out, in which the method and the apparatus can apply to all electrically actuated devices or systems such as bell alarms, fire escape lights, emergency lights, smoke exhausters, broadcasting equipment and the like. The apparatus includes a self-contained sensor element which can detect a rise in temperature as a result of a fire, and a self-contained power supply source or electric generator which can supply an electric current to external fire alarms, emergency lights and the like for triggering them. The apparatus can be actuated at the very moment a fire breaks out, so that it can supply an electric current without delay.

Referring to Figure 5.20a, a water-activated cell or battery **1** is provided at the lower portion of a housing **17**, and comprises an element formed by disposing a porous water-absorbent material **3** such as sponge of continuously foamed synthetic resin material between anode-activated substance **2** in the form of anode plate and cathode-activated substance **4** in the form of cathode plate. A conductor wire **5** has one end connected to the anode of the cell **1** and the other end connected to a conductor **5a** of a lead-out wire **7**. Similarly, a conductor wire **6** has one end connected to the cathode of the cell **1** and has the other end connected to a conductor **6a** of the lead-out wire **7**.

A vessel **8** of glass material and provided with a glass capillary tube **14** contains an amount of pressure gas generating material **10** and an amount of cell activating electrolyte **11**. A thermosensitive element **12** of bendable material is provided inside the glass capillary tube **14**. The tube **14** has an opening **13** at the tip thereof, which is closed for sealing the tube **14**. The vessel **8** is disposed in the housing **17** so that its capillary tube **14** can be placed as shown in the gap or space lett beween the cell **1**. As shown, a net **16** is provided at the upper portion of the vessel **8**, on which an amount of drying material **15** is arranged. Pressure gas relief safety means **27** is detachably provided as shown, keeping the housing **17** sealed during the normal time.

FIGURE 5.20: FIRE ACTIVATED CELL

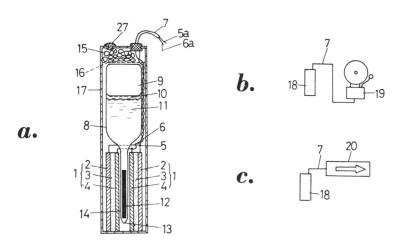

(a) Cross-sectional view of apparatus.
(b) Block diagram of an example in which the apparatus is used with the bell alarm.
(c) Block diagram of an example in which the apparatus is used with the fire escape light.

Source: U.S. Patent 4,009,055

In accordance with the apparatus constructed as illustrated above, if there is a rise in the ambient temperature as a result of a fire, the pressure gas generating material **10** starts to vaporize, causing the pressure chamber **9** of the vessel **8** to be filled due to increased pressure. In this case, provision may be made so that the vessel **8** can be broken when the internal pressure of the chamber **9** increases. However, this is not preferable since the temperature at which the vessel **8** can be broken must be extremely high. In the above embodiment, therefore, it is so arranged that the vessel **8** cannot be broken with only an increase in internal pressure.

If there is an increase in the internal pressure of the vessel **8**, the electrolyte **11** is always forced to go down, and the thermosensitive element **12** provided inside the tube **14** begins to bend with the rising temperature. When the element **12** is heated up to a given temperature, it breaks the tube **14**, bursting the electrolyte **11** under pressure out of the tube **14**. The water-activated cell **1** is then immersed in the electrolyte **11**, which activates the cell **1** to produce electricity. The electric power thus obtained is transmitted through the wires **5** and **6** to the external lead-out wire **7**.

Drying material **15** such as silica gel for example serves to absorb the moisture inside the housing **17**, and protect the cell **1** from the moisture. If the internal pressure of the housing **17** rises to excess when pressure gas is produced, the gas relief safety means **27** such as paraffin, detachably provided and normally sealing the housing **17**, can be broken away.

Figures 5.20b and 5.20c indicate block diagrams of examples in which the apparatus is used with other external devices. In Figure 5.20b, an example is shown in which the apparatus is used as power supply of a bell alarm **19**. In this example, the external lead-out wire **7** of the apparatus **18** is connected to the bell alarm **19**, triggering the alarm **19** when a fire occurs. Another example is shown in Figure 5.20c, in which the apparatus **18** is used with a fire escape light **20**. In this example, it is connected through the wire **7** to the light **20**, and can trigger the light **20** if a fire breaks out.

In the above examples, the apparatus **18** can be installed at any place and without limitations, and can also be connected through the wire **7** with external fire alarms and other devices located anywhere away from the apparatus **18**. In those examples, the apparatus **18** is used with a single fire alarm, but it may be used as plural-function power supply with a number of various external devices.

The following data were observed with reference to the uses described above: Note that nickel-titanium alloy (Nitinol) is used as thermosensitive element of bendable material. It was confirmedly observed that the apparatus can be actuated at the ambient temperature range of 53° to 58°C, and that it can start to produce electricity about 15 to 20 seconds after it is placed in a vessel kept to 60°C. Bimetal elements may also be used as thermosensitive elements. However, the bimetal element can only respond to the ambient temperature above 80°C, and its actuating temperature range is not definite.

The above data were obtained by using a cell **1** which consists of a copper plate and anode activation material composed of a mixture of silver chloride and pelletized silver pressed on the copper plate, and a plate of cathode activation material or magnesium, and by using a 15% solution of sodium chloride as electrolyte **11**. It should be noted that other electrolytes such as water, aqueous solution of potassium hydroxide, aqueous solution of sodium hydroxide and the like may also be used, and the choice of those electrolytes depends on the material of the cell used.

The pressure gas generating material **10** should desirably be composed of materials which are of nonsoluble and nonreactive character with regard to the electrolyte **11**, and are hard to ignite at the rising temperature. Halogenated hydrocarbons of high vapor pressure such as methylene chloride, chloroform and the like show good results. Other materials, though inflammable, may be used, including ethers such as ethyl ether or hydrocarbons such as pentane, isohexane and the like. A very small amount of any of those materials, namely 0.1 to 0.5 cc, can actuate the apparatus with certainty if a fire occurs, and can prevent the fire from developing into a serious problem or disaster. Furthermore, other materials which are of reactive character with regard to the electrolyte 11 and can produce pressure gas by reaction with the electrolyte 11 may be used, and include magnesium, zinc, aluminum and the like. Those materials also show good results if a small amount of any of them is added just before sealing the vessel 8.

Radioelectrochemical Converter

G. Duperray, C. Eyraud, G. Lecayon and J. Lenoir; U.S. Patent 3,971,671; July 27, 1976; assigned to Commissariat a l'Energie Atomique, France describe an energy converter which utilizes the products of radiolysis of an aqueous solution of an oxidation-reduction pair and comprises a leak-tight jacket initially filled with a pure gas which is identical with the gas evolved as a result of radiolysis. A porous electrode impregnated with the aqueous solution of the oxidation-reduction pair and specific to the reaction of this latter is placed within the jacket. The converter also comprises a gas electrode which is specific to the gas evolved as a result of radiolysis, a porous diaphragm for effecting the ionic junction between the two electrodes by gas/liquid surface conductivity and means for connecting the electrodes to an external circuit.

Figure 5.21a essentially comprises a jacket 1 of glass, ceramic material, vitrified or enamelled metal for example, in which are placed a porous electrode 2 impregnated with electrolyte (the electrode 2 can also be cylindrical) and a porous electrode 3 in the form of a cylindrical sleeve. The device also comprises a diaphragm 4 placed between the two electrodes and constituted by a porous wall in the form of a cylindrical cup, the base of which rests on the top face of the electrode 2; the porous wall is made of ceramic material or glass fiber. The cylindrical electrode 3 is applied by means of an insulating spacer guide 5 against the diaphragm 4 which is in turn applied against the internal wall of the jacket 1.

The connections of the electrodes 2 and 3 to an external circuit are provided by platinum lead-in wires 6 connected to the electrodes 2 and 3 by means of fine gold foil elements 7, the design function of which is to reduce contact resistances. The creation of a vacuum as well as the introduction of electrolyte and of gas which is identical with the gas subsequently evolved as a result of radiolysis is carried out through the exhaust tube 8.

In order to seal the element and to reduce the dead space, a cylinder 9 of glass for example is inserted into the interior of the cylindrical jacket 1 and the top edges of the concentric tubes are welded at a distance such that the different portions of the converter are not damaged by the heat evolved at the time of the sealing operation. As shown in Figure 5.21a, the source 10 which is intended to induce radiolysis is placed outside the jacket 1.

A better understanding of the advantages provided by the particular arrangements of the converter in accordance with this process will be gained from the following description of the application of the converter under consideration. The example relates to utilization of radiolysis of a Fe^{++}/Fe^{+++} system by a γ-ray source placed outside the converter.

Example: In this example, the electrolyte is constituted by a solution 0.8 N of sulfuric acid and 0.1 N to 0.5 N of ferrous ions. In this case the electrode 2 performs the function of cathode and is made of porous graphite. The electrode 3 performs the function of anode and is of porous material combined with pal-

ladium, rhodium or platinum. The cell is filled with hydrogen as a preliminary
step. The electrolyte is retained by capillarity, a very high proportion being retained in the cathode and a low proportion being retained in the diaphragm and
the anode. The electrolyte receives the γ-radiation from the source **10** consisting of a cobalt 60 or cesium 137 source placed externally of the converter.

FIGURE 5.21: RADIOELECTROCHEMICAL CONVERTER

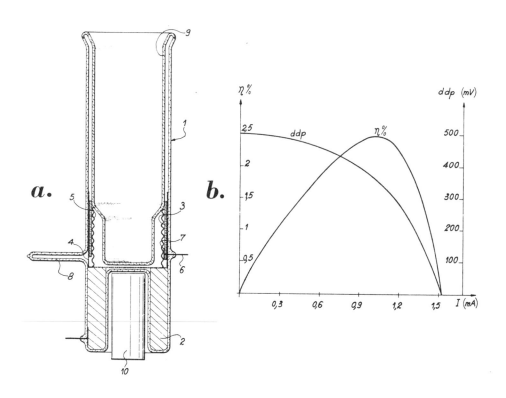

(a)   Diagrammatic cross-sectional view of the converter.
(b)   Graph indicating the current-voltage relationship utilizing the
         converter of Figure 5.21a.

Source: U.S. Patent 3,971,671

In the case of a weight of electrolyte of 15 grams and a dose rate of 0.12 Mrad
per hour, an intensity of 1.9 mA at 0.510 V is obtained. The energy efficiency
η obtained (ratio of electrical energy delivered to the energy dissipated by the
radiation within the mass of the electrolyte) is 2.2%. This value is deduced from
the I-V characteristic represented in the accompanying Figure 5.21b and obtained
by varying the load resistance. There is deduced from this latter the energy efficiency η as a function of the current intensity. The short-circuit intensity makes
it possible to define a maximum effective radioelectrochemical efficiency G (the
number of molecules converted in respect to 100 eV absorbed) of 10.3.

The converter's advantages are: (1) By virtue of the fact that the porous cathode is impregnated with electrolyte, practically the entire quantity of electrolyte is maintained in the vicinity of the cathode under the action of capillarity. The anode and cathode compartments are therefore well separated. This results in low cathode concentration of polarization; in fact, practically the whole quantity of ferric ions produced by radiolysis is formed in the immediate vicinity of the cathode. There is therefore no abrupt increase in the concentration of Fe^{+++} ions and consequently no voltage drop.

(2) The ionic conductivity between the anode and cathode compartments as ensured by means of the porous diaphragm permits surface conductivity without any capillary condensate or at least with a very limited volume of condensate. In the diaphragm, the ionic conductivity is mainly ensured by the protons, the ratio of the diffusion coefficients of the H^+ and Fe^{+++} ions being in the vicinity of 20. Thus the hydrogen produced by radiolysis is given off by the porous mass which constitutes the cathode and passes freely in the gas phase into the anode compartment in which it is consumed.

(3) The fact that the jacket which constitutes the converter is hermetically sealed makes it possible to prevent any out-leakage of hydrogen and any admission of inert gas or oxygen. This makes it possible in addition to ensure confinement of the radioactive source in the event that this latter is located inside the converter.

(4) By virtue of the fact that the converter is initially filled with hydrogen, the anode potential is very close to that of a hydrogen electrode, the more so as the hydrogen pressure is higher and the ferric ion concentration is lower at the outset. The electromotive force obtained is therefore close to the theoretical value of that of a reference pile of the same type.

(5) The two electrodes are not immersed in a solution of electrolyte. In the converter according to this process, this solution impregnates the first electrode and the diaphragm which are both porous; and if the converter is mounted on a moving body, displacements and disturbances of operation which are liable to arise need no longer be anticipated by reason of the adsorption of this solution. This original feature makes it possible to mount the converter on a moving body such as, for example, the carrier of a heart assistance device or a vehicle.

Variable Precision Multivoltage Step Battery

According to a process described by *F.G. Fagan, Jr. and S.J. Angelovich; U.S. Patent 4,025,700; May 24, 1977; assigned to P.R. Mallory & Co., Inc.* to provide an indication that a predetermined amount of the energy of a cell has been used up, the positive electrode is formulated with two or more ingredients having different electrochemical potentials relative to the negative electrode. As each ingredient is used up, the voltage between the electrodes drops to the next potential difference, thereby indicating a predetermined quantity of energy has been used, and that the remaining reserve is at or less than the amount at which the voltage is predesigned to drop.

The use of such a step voltage cell in a battery provides a measure of the condition of the battery when the voltate drops, due to the voltage drop in such pilot cell. To illustrate the nature and utility of the process, it is described in one mode of operation as applied in use with an alarm system. Conventional alarm system batteries consist of a number of similar cells connected in a series arrangement to achieve the battery terminal voltage required for a given application. This battery supplies continuous power to the alarm system sensing circuit; such as, for example, a smoke detection device, a temperature sensing device, or a breathing rate sensing device. Generally, the continuous power required to activate the sensing circuit is at a modest low level, and consequently, the current delivery capabilities required of the battery are also at a modest level, e.g., 20 to 75 microamperes.

However, an additional requirement imposed on the battery is that when the sensing circuit detects an emergency condition, the battery must supply adequate power to activate the alarm device, such as, for example, an audible horn, buzzer, etc. Generally, the power required to activate the alarm device is many times that required to activate the sensing circuit and the current delivery capabilities required of the battery are similarly increased, e.g, 50 to 75 milliamperes, a quantity that is one thousand times that for the sensing circuit. Therefore, in any given application, the selection of battery size, capacity, and current delivery capability is dictated by the greater requirements of the alarm device.

In the operation of such alarm systems, with continuous discharge of the battery powering the sensing circuit, a time is reached near the end of battery life, when, although still supplying the modest power required by the sensing circuit, the battery, due to its near exhaustion, is incapable, if called upon, of supplying adequate power to activate the alarm device. Thus, at that time, a malfunction could occur with potentially disastrous results, because of no warning of the battery condition that would have alerted a supervisor of the need for a replacement.

The process overcomes this possibility of malfunction without prior warning by producing in the battery a characteristic, which, when continuously monitored by the alarm system, will indicate that the battery is nearing its end of life, and that characteristic will activate the alarm device. A separate alarm device may be employed for this end-of-life warning, or the alarm mode, to indicate end of battery life, can be conveniently differentiated from the emergency condition mode, for example, by using an intermittent signal as contrasted with a continuous signal. Thus, the time and need for battery replacement will be indicated well in advance of battery failure due to exhaustion, and at such a time while there is still remaining in the battery sufficient reserve capacity even to power the alarm device, for an adequate time period, in the event of an emergency condition, to permit appropriate battery substitution to be made.

According to this process, the battery indicates the approach of its end of life condition, to the alarm system battery monitoring circuit, by an abrupt controlled decrease of the battery terminal voltage, and does so at a predetermined time well in advance of battery exhaustion, according to design. This voltage decrease is accomplished by modifying the positive electrode formulation of one or more of the cells of a multicell battery.

The process has been demonstrated, for example, by modifying a Mallory Battery Company cell, whose standard positive electrode formulation is 89% mercuric oxide (HgO), 6% manganese dioxide (MnO_2), and 5% graphite (C) and whose rated capacity is 900 milliampere-hours. The desired voltage step, according to the process, i.e., decrease of 0.75 volt from approximately 1.25 volts to approximately

0.50 volt, in one cell, or in more than one, where so desired, can be caused to occur after the discharge of 465 milliampere-hours of cell capacity, by substituting a new positive electrode whose formulation is 45% mercuric oxide, 3% manganese dioxide, 45% cadmium oxide (CdO), and 7% graphite.

The electrochemical principle established and employed here is to control the functioning of this special positive electrode formulation in the advantageous manner described, by intimately mixing two or more electrochemically active materials and compressing them into a common electrode, so that when such an electrode is discharged against an opposing electrode, the discharge proceeds first to exhaust completely the active material having the highest relative potential against the opposing electrode, and then, in stepwise fashion, the discharge proceeds to exhaust completely, the materials having successively lower potentials.

Thus, in the special modified cell of this process, the 3% manganese dioxide and the 45% mercuric oxide fractions of the positive electrode discharge against a zinc (Zn) negative electrode in the voltage range 1.36 to 1.25 volts, until they are substantially exhausted. Then the battery terminal voltage falls abruptly from approximately 1.25 volts to approximately 0.50 volt. Finally, the 45% cadmium oxide fraction discharges against the zinc negative electrode in the voltage range 0.50 to 0.40 volt, until the cadmium oxide is substantially exhausted.

In the example cited, the voltage step has been designed to occur at 465 milliampere-hours discharge capacity or 51.7% of the 900 milliampere-hours cell rated capacity. However, by readjusting the relative percentages of mercuric oxide, manganese dioxide, and cadmium oxide in the formulation, the voltage step can be designed to occur at any discharge capacity desired with an accuracy of plus or minus 5%. A discharge capacity in the range of 225 to 675 milliampere-hours, or 25 to 75% of rated capacity, appears to satisfy the majority of contemplated applications.

Deep Diode Atomic Battery

T.R. Anthony and H.E. Cline; U.S. Patents 4,024,420; May 17, 1977 and 4,010,534; March 8, 1977; both assigned to General Electric Company describe a deep diode, or semiconductor, atomic battery comprising a body of semiconductor material. The body has walls defining a central cavity to contain a radioactive gamma or x-ray emitter, a plurality of regions of first type conductivity and selective resistivity and a plurality of regions of second type conductivity and selected resistivity. The material of the second regions is recrystallized semiconductor material of the body and of the first regions and contains a substantially uniform level of a dopant impurity material throughout each second region and is sufficient to impart the second and opposite type conductivity thereto.

P-N junctions are formed by the contiguous surfaces of pairs of regions of opposite type conductivity. Electrical contacts are affixed to the respective regions of first type conductivity and to the regions of second type conductivity which become the anode and cathode of the battery.

The semiconductor atomic battery is energized by inserting a radioactive gamma or x-ray source into the central cavity in the semiconductor body. The dimensions of the semiconductor body are large enough so that a large proportion of the gamma or x-rays emitted from the central cavity in the semiconductor body are observed by the semiconductor body thereby enabling the battery to have a high efficiency and a low level radioactivity level at its external major surfaces. The dimensions and geometry of the regions of first type conductivity and regions of second type conductivity are chosen so that the distance from any point in the regions of first or second type conductivities to the nearest P-N junction formed by the contiguous surfaces of these regions is less than the minority carrier diffusion length in these regions. The radioactive source has a high specific activity and the energy level of the gamma or x-rays is selected to be less than the energy necessary to cause displacement of atoms in the semiconductor material of which the battery is comprised to avoid radiation damage to the semiconductor body.

The semiconductor atomic battery operates by the conversion of the gamma or x-rays emitted from the sources in the semiconductor body into electron-hole pairs on absorption by the surrounding semiconductor body. Because all points in the semiconductor body are within a minority carrier diffusion distance of a P-N junction, the majority of electron-hole pairs are separated by the built-in field of the P-N junction before they recombine. The separated electron-hole pairs forward bias the P-N junction and thus deliver power to an electrical load connected to the battery. Each absorbed gamma ray produces a plurality of electron-hole pairs to boost the current of the battery.

With reference to Figure 5.22a, there is shown a body 10 of semiconductor material having a selected resistivity and a first type conductivity. The body 10 has opposed major top and bottom surfaces 12 and 14, respectively, and opposed major side surfaces 30 and 32. The semiconductor material comprising body 10 may be silicon, germanium, silicon carbide, gallium arsenide, a semiconductor compound of a Group II element and a Group VI element and a semiconductor compound of a Group III element and a Group V element.

A central cavity 40 is first made in the semiconductor body 10 by grinding, ultrasonic drilling or sandblast drilling. The cavity 40 is preferably located midway between side surfaces 30 and 32 and extends between, and is perpendicular to, the opposed major surfaces 12 and 14. Alternatively, cavity 40 may terminate midway between the opposed major surfaces 12 and 14. The body 10 is mechanically polished, chemically etched to remove any damaged surfaces, rinsed in deionized water and dried in air. A preferential acid-resistant mask 16 is disposed on surface 12 of the body 10. Preferably, the mask 16 is of silicon oxide which is either thermally grown or vapor-deposited on the surface 12 by any of the methods known to those skilled in the art.

Employing well-known photolithographic techniques, a photoresist such, as for example, Kodak Metal Etch Resist, is disposed on surface of the silicon oxide layer 16. The resist is dried by baking at a temperature of about 80°C. A suitable mask of an array of spaced lines or dots of a predetermined dimension and spaced a predetermined distance apart is disposed on the layer of photoresist and exposed to ultraviolet light. After exposure the layer of photoresist is washed in xylene to open windows in the mask where the lines or dots are desired so as to be able to selectively etch the silicon oxide layer 16 exposed in the windows. The separation distances between the dots on lines of the array are all equal to or less than the minority carrier diffusion length within the semiconductor material of body 10. The thickness of the lines of the array or the width of the dots of the array are all equal to or less than the minority carrier diffusion length in the recrystallized semiconductor material of the body **10**.

Selective etching of the layer **16** of silicon oxide is accomplished with a buffered hydrofluoric acid solution (NH_3-HF). The etching is continued until a second set of windows corresponding to the windows of the photoresist mask are opened in layer **16** of the silicon oxide to expose selective portions of the surface **12** of the body **10** of silicon. The processed body **10** is rinsed in deionized water and dried. The remainder of the photoresist mask is removed by immersion in concentrated sulfuric acid at 180°C or immersion in a mixture of one part by volume hydrogen peroxide and one part by volume concentrated sulfuric acid.

Selective etching of the exposed surface area of body **10** is accomplished with a mixed acid solution. The mixed acid solution is 10 parts by volume 70% nitric acid, 4 parts by volume 100% acetic acid and one part by volume 48% hydrofluoric acid. At a temperature of from 20° to 30°C, the mixed solution selectively etches the silicon of body **10** at a rate of approximately 5 microns per minute. A depression **18** is etched in the surface **12** of the body **10** beneath each window of the oxide layer **16**. The selective etching is continued until the depth of the depression **18** is approximately equal to the width of the windows in the silicon oxide layer **16**.

However, it has been found that the depression **18** should not be greater than approximately 100 microns in depth because undercutting of the silicon oxide layer **16** will occur. Undercutting of the layer **16** of the silicon oxide has a detrimental effect on the surface penetration ability of the metal to be placed in the depressions **18** by vapor deposition. Etching for approximately 5 minutes at a temperature of 25°C will result in a depression of from 25 to 30 microns. The etched body **10** is rinsed in distilled water and blown dry. Preferably, a gas such for example, as Freon, argon and the like is suitable for drying the processed body **10**.

The processed body **10** is disposed in a metal evaporation chamber. A metal layer **20** is deposited on the remaining portions of the layer **16** of silicon oxide and on the exposed silicon in the depressions **18**. The metal of the layer **20** comprises a material, either substantially pure in itself or suitably doped by one or more materials to impart a second and opposite type conductivity to the material of body **10** through which it migrates.

FIGURE 5.22: DEEP DIODE ATOMIC BATTERY

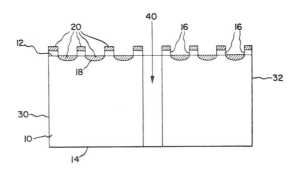

a.

(continued)

FIGURE 5.22: (continued)

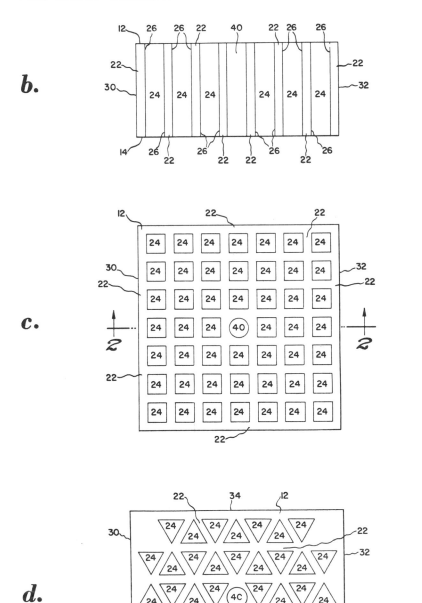

(continued)

FIGURE 5.22: (continued)

e.

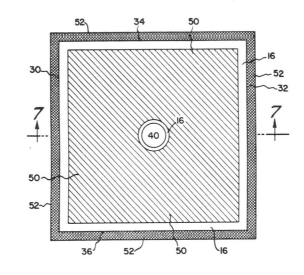

f.

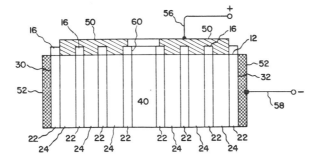

(a)   Side elevation view, in cross section, of a semiconductor body
       being processed.

(b)   Side elevation view, in cross section, of the same semiconductor
       body after temperature gradient zone melting processing.

(c)   Top planar view of the semiconductor body of Figure 5.22b
       incorporating an array of square columnar P-N junctions.

(d)   Top planar view of the semiconductor body of Figure 5.22b with
       a close packed array of triangular columnar P-N junctions.

(e)   Top planar view of the metallized anode and cathode of a deep
       diode atomic battery with all individual cells connected in
       parallel.

(f)   Side elevation view, in cross section, of the metallized anode and
       cathode of a deep diode atomic battery, wherein all of its indi-
       vidual cells are connected in parallel.

Source: U.S. Patent 4,024,420

The thickness of layer **20** is approximately equal to the depth of the depression **18**. Therefore, if the depression **18** is 20 microns deep, the layer **20** is approximately 20 microns in thickness. A suitable material for the metal layer **20** is aluminum to obtain P-type regions in N-type silicon semiconductor material. Prior to migrating the metal in depressions **18** through the body of silicon **10**, the excess metal layer **20** is removed from the silicon oxide layer **16** by such suitable means as grinding away the excess metal with a 600 grit carbide paper.

It has been found that the vapor deposition of the layer **20** of the aluminum metal should be performed at a pressure of approximately 1×10^5 torrs but not greater than 5×10^5 torrs. When the pressure is greater than 5×10^5 torrs, it has been found that in the case of aluminum metal vapor deposited in depressions **18**, the aluminum does not penetrate into the silicon and migrate through the body **10**. It is believed that the layer of aluminum is saturated with oxygen and prevents reduction by the aluminum metal of the very thin silicon oxide layer between the deposited aluminum and the silicon, that was formed in the air shortly after etching the troughs **18**.

Thus, the initial melt of aluminum and silicon required for migration is not obtained because of the inability of the aluminum layer to wet and alloy with the underlying silicon. In a similar manner, the aluminum deposited by sputtering is not as desirable, as sputtered aluminum appears to be saturated with oxygen from the sputtering process, thereby preventing the reduction of any intervening silicon oxide. The preferred methods of depositing aluminum on the silicon body **10** are by the electron beam method and the like wherein little, if any, oxygen can be trapped in the aluminum.

The processed body **10** is placed in a thermal migration apparatus, not shown, and the metal-rich liquid bodies in the depressions **18** are migrated through body **10** by thermal gradient zone melting process. A thermal gradient of approximately 50°C/cm between the bottom surface **14** which is the hot face and the surface **12**, which is the cold face, has been found to be appropriate at an apparatus operating temperature of from 800° to 1400°C. The process is practiced for a sufficient length of time to migrate all the metal-rich liquid bodies through the body **10**. For example, for aluminum-rich liquid bodies of 20 microns thickness, a thermal gradient of 50°C/cm, and a 1200°C mean temperature of body **10**, a furnace time of less than 12 hours is required to migrate the metal-rich liquid bodies through a silicon body **10** of one centimeter thickness.

For a more thorough understanding of the temperature gradient zone melting process employed in this process and for a more thorough description of the apparatus employed for this process, one is directed to U.S. Patents 3,901,736; 3,910,801; 3,898,106; 3,902,925 and 3,899,361.

It has been found that when the substrate 212 is of silicon, germanium, silicon carbide, gallium arsenide semiconductor material and the like, the migrating metal droplet has a preferred shape which also gives rise to the regions being formed having the same shape as the migrating droplet. In a crystal axis direction of $<111>$ of thermal migration, the droplet migrates as a triangular platelet lying in a (111) plane. The platelet is bounded on its edges by (112) planes. A droplet larger than 0.10 centimeter on an edge is unstable and breaks up into several droplets during migration. A droplet smaller than 0.0175 centimeter does not migrate into the substrate 212 because of a surface barrier problem. The ratio of the droplet migration rate over the imposed thermal gradient is a function of the temperature at which thermal migration of the droplet is practiced. At high temperatures, of the order of from 1000° to 1400°C, the droplet migration velocity increases rapidly with increasing temperature. A velocity of 10 centimeters per day or 1.2×10^{-4} centimeter per second is obtainable for aluminum droplets in silicon.

The droplet migration rate is also affected by the droplet volume. In an aluminum-silicon system, the droplet migration rate decreases by a factor of 2 when the droplet volume is decreased by a factor of 200. A droplet migrates in the <100> crystal axis direction as a pyramidal bounded by four forward (111) planes and a rear (100) plane. Careful control of the thermal gradient and migration rate is a necessity. Otherwise, a twisted region may result. It appears that there is a nonuniform dissolution of the four forward (111) facets in that they do not always dissolve at a uniform rate. Nonuniform dissolution of the four forward (111) facets may cause the regular pyramidal shape of the droplet to become distorted into a trapezoidal shape.

The migration of metal wires is preferably practiced in accordance with the planar orientations, migration directions, stable wire directions and stable wire sizes of the following table.

Wafer Plane	Migration Direction	Stable Wire Directions	Stable Wire Sizes (μ)
(100)	<100>	<011>*	<100
		<0$\bar{1}$1>*	<100
(110)	<110>	<1$\bar{1}$0>*	<100
(111)	<111>	(a) <01$\bar{1}$>	
		<10$\bar{1}$>	<500
		<1$\bar{1}$0>	
		(b) <11$\bar{2}$>*	
		<$\bar{2}$11>*	<500
		<1$\bar{2}$1>*	
		(c) Any other direction in (111) plane*	<500

*The stability of the migrating wire is sensitive to the alignment of the thermal gradient with the <100>, <110> and <111> axes, respectively. Group (a) is more stable than group (b) which is more stable than group (c).

The process has been described relative to practicing thermal gradient zone melting in a negative atmosphere. However, it has been found that when the body of semiconductor material in a thin wafer of the order of 10 mils thickness, the thermal gradient zone melting process may be practiced in an inert gaseous atmosphere in a furnace having a positive atmosphere. Upon completion of the temperature gradient zone melting process, the resulting processed body 10 is as shown in Figure 5.22b. The migration of the metal-rich liquid bodies in the depressions through the body produces a body having a plurality of first spaced regions 22 of a second and opposite type conductivity than the material of the body 10. Each region is recrystallized material of the body 10 suitably doped with a material comprising the metal layer 20 and having an impurity concentration sufficient to obtain the desired conductivity. The metal retained in the recrystallized region is substantially the maximum allowed by the solid solubility of the metal at the migration temperature practiced in the semiconductor material through which it has been migrated. It is recrystallized material of solid solubility of the metal therein. The region 22 has a substantially constant uniform level of impurity concentration throughout the entire planar region. The region has less crystal imperfections and extraneous impurities than the original material of the body. The thickness of the region 22 is substantially constant for the entire region. Depending upon the geometry of the original deposited array of metal-rich liquid bodies, the crystallographic plane of deposition and the crystallographic direction of migrations, regions 22 can be a lamellar array of planar zones, an array of square columnar zones, an array of triangular columnar zones, an array of hexagonal columnar zones, a square array of planar zones, a triangular array of planar zones, a diamond array of planar zones or a hexagonal array of planar zones.

The peripheral surface of each planar or columnar region 22 comprises in part the top surface 12 and the bottom surface 14 of the body 10. In addition, the peripheral surface of each planar zone comprises in part the peripheral side surfaces of the body 10. The body 10 is also divided into a plurality of spaced regions 24 having the same, or first, type conductivity as the body 10. A P-N junction 26 is formed by the contiguous surfaces of each pair of mutually adjacent regions 22 and 24 of opposite type conductivity. The P-N junction 26, as formed, is very abrupt and distinct, resulting in a step junction.

When regions 22 are planar regions, the regions are made so that the planar thickness does not exceed twice the minority carrier diffusion distance in the recrystallized material containing the dopant impurity that imparts the second and opposite type conductivity to regions 22. When regions 22 are columnar regions, the regions have a maximum cross-sectional dimension that does not exceed twice the minority carrier diffusion distance in the recrystallized material containing the dopant impurity that imparts the second and opposite type conductivity to regions 22. When regions 22 are recrystallized silicon containing the solid solubility limit of aluminum of 2×10^{19} atoms/cm^3 and the lifetime of this recrystallized material is 1 microsecond, then the critical cross-sectional dimension of regions 22 must not exceed 70 microns or twice the minority carrier diffusion length in the recrystallized P-type silicon.

In a similar fashion, the maximum cross-sectional width of regions 24 must not exceed twice the minority carrier diffusion length in the semiconductor material comprising regions 24. For N-type silicon of 5×10^{14} carriers/cm^3 and a minority carrier lifetime of 20×10^{-6} microsecond, this critical cross-sectional width for regions 24 must not exceed 150 microns or twice the minority carrier diffusion length in the N-type silicon. With all points in the semiconductor body 10 within a minority carrier diffusion length of a P-N junction 26, most of the electron-hole pairs created by the absorption of a gamma ray or x-ray from a source in the central cavity 40 will be collected and separated by the P-N junctions 26 before recombination.

Efficient collection of the generated electron-hole pairs by P-N junctions 26 will lead to an efficient battery. The resulting structure of body 10 after thermal gradient processing is shown in Figure 5.22b wherein the body 10 has a central cavity 40 to contain a gamma emitter or an x-ray emitter and regions 22 and 24 of opposite type conductivity forming P-N junctions 26 at interfaces between regions 22 and 24.

Referring to Figure 5.22c, a square columnar array of regions **24** is shown embedded in region **22**. This structure is obtainable by depositing an aluminum layer **20** in a square array of square depressions **18** on a (100) crystallographic plane of silicon and migrating the resulting aluminum-rich liquid droplets in a $<100>$ crystallographic direction. The square array and the square depressions are aligned so that their sides are parallel to the $<011>$ and $<0\bar{1}1>$ directions.

Referring to Figure 5.22d, a triangular columnar array is obtainable by depositing an aluminum layer **20** in a triangular array of triangular depressions **18** on a (111) crystallographic plane of silicon and migrating the resulting aluminum-rich liquid droplets in a $<111>$ crystallographic direction. The triangular shaped regions of the array are aligned so that their three sides are parallel to the $<011>$, the $<101>$, and the $<\bar{1}10>$ directions, respectively.

With reference to Figures 5.22e and 5.22f, a deep diode atomic battery with all collecting P-N junctions connected electrically in parallel is shown. These P-N junctions are of a columnar type as those shown in Figures 5.22c and 5.22d. An ohmic electrical contact **50** is affixed to and connects all regions **24** in parallel. A layer of electrical insulating material **16** such, for example, as silicon oxide, silicon nitride and the like, is disposed over regions **22** on top surface **12** to insulate regions **22** from contact **50**. A central aperture **60** in the ohmic contact **50** and insulating layer **16** is provided to allow access to the central cavity **40** containing the x-ray or gamma emitter.

With the columnar geometry for regions **24** shown in Figures 5.22c and 5.22d, regions **22** are interconnected and continuous so that an ohmic electrical contact on the side peripheral surfaces **30,32,34** and **36** can collect carriers separated by the field of the junction **26** and injected into region **22**. A deep diode atomic battery with ohmic electrical contacts connecting all regions **24** in parallel will develop a minimum voltage for the cell. This cell voltage is determined by the highest forward bias voltage that can be sustained by the P-N junction **26** before the forward current of the P-N junction becomes larger than the current generated by the junction field separation of electron-hole pairs produced by absorbed gamma rays since the gamma-ray-generated current is in a direction opposite to the forward current of the P-N junction. This voltage is about 0.5 volt for silicon, 0.1 volt for germanium and 1.0 volt for GaAs.

BATTERIES FOR PACEMAKERS
AND FILM PACKS

PACEMAKERS—GENERAL

Gas Absorbing Device for Alkaline Zinc Type

W.L. King and K.B. Stokes; U.S. Patent 3,943,937; March 16, 1976; assigned to Medtronic, Inc. describe an implantable electrically actuated medical device having gas storage characteristics. The device includes one or more electrochemical cells and an operative electric circuitry mounted in a metallic container, all of which are positioned in a mounting member made of a material which is highly permeable with respect to hydrogen gas, relatively impervious to liquid and which bonds well with an epoxy resin encapsulant coating over the device.

The mounting member may be foamed as it is molded to provide voids therein for storage of gas or the recesses in which the electrochemical cells are mounted may be supplied with sponge-like pads for absorbing excess gas until it can permeate through the encapsulant to reduce pressure buildup within the device and prevent cracking of the casing.

As shown in Figures 6.1a and 6.1b, the implantable medical device includes an encapsulation 15 of an epoxy resin covering a plurality of electrochemical cells 25 typically of the alkaline zinc-mercury type which are positioned in a mounting member 30 along with a sealed can or housing for the electronic circuitry of the device indicated generally at 20. The cells 25 are interconnected to one another and to the input terminals of the device, as indicated by the conductors 35, and the implantable device has a coupling member extending through the encapsulation as indicated at 11 with suitable output leads 12 extending therefrom and to electrodes or other apparatus associated with the device.

The details of the electrochemical cells and the electronic circuitry are omitted for simplicity. The cells are of the type which generate a gas, such as hydrogen, upon depletion. The epoxy resin covering or encapsulation provides a relatively liquid-tight seal for the device of a material which is biocompatible and liquid impervious.

FIGURE 6.1: GAS ABSORBING IMPLANTABLE ELECTRICAL MEDICAL
DEVICE

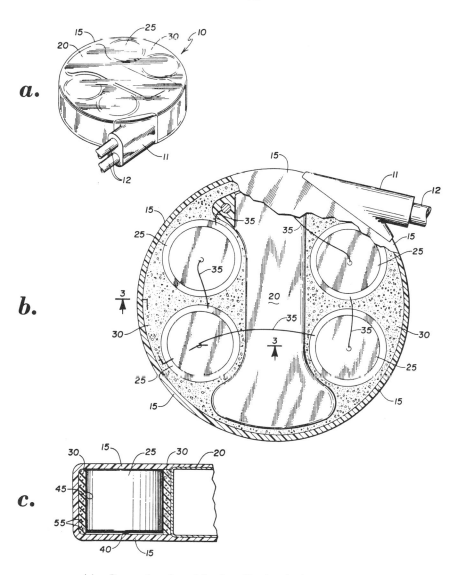

(a)   Perspective view of implantable electrically actuated medical
        device
(b)   Plan view of the device of Figure 6.1a with parts broken away
        to show the arrangement of parts
(c)   Sectional view of the device of Figure 6.1b taken along lines
        3-3 with parts broken away

Source: U.S. Patent 3,943,937

Where the words liquid-tight or liquid-impervious are used herein, it is meant that the resin will not pass fluid in liquid form while recognizing that the plastic resin will transmit small amounts of vapor. The epoxy resin covering is also relatively impermeable to any hydrogen gas generated in the cells. Such generation of gas takes place with cell depletion with the gas being vented from a suitable vent port, indicated at **40**, in the base of the cell and to some degree through the cell seal spacing the cell electrodes. Because the epoxy resin material is relatively impermeable to gas any sudden release of the same will create a pressure buildup within the encapsulation which may cause fracture or explosion of the same.

In the process, as indicated in Figure 6.1c, the mounting member **30** is formed of a highly hydrogen permeable plastic which may advantageously be a polyphenyleneoxide/polystyrene material (Noryl). The mounting member or spacer **30** has recesses **45** therein in which the cells and electronic circuit can **20** are positioned. The mounting member provides for location of the parts and rigidity to the package to support the encapsulation which bonds well with the mounting material to seal the same.

In the process, the material forming and mounting member is constructed of a structural foam providing voids or spaces in the mounting member, as indicated at **55**, which act as a reservoir for the hydrogen gas when released from the cells through the vent extremity **45** thereof. The mounting material of polyphenyleneoxide/polystyrene is of considerably higher permeability to hydrogen gas than is the epoxy resin and is also biocompatible with body fluids. The release of gas from the cells is absorbed in the voids or foamed spacer member to provide a very slow development of pressure within the casing permitting diffusion of the gas or release of the same through the encapsulant material or epoxy resin gradually without an excessive pressure buildup from the same.

W.L. King; U.S. Patent 3,970,479; July 20, 1976; assigned to Medtronic, Inc. describes an electrochemical cell in which the sealing material between the electrodes of the cell is impervious to liquid but permeable to hydrogen gas. The improvement in the cell construction is a shortened path through the sealing material to permit a continual and gradual discharge of the hydrogen gas generated in the cell and prevent a buildup of gas therein. The shortened path is effected by openings in the inner terminal member associated with the sealing material. The cell is particularly suitable for use in an implantable electrical device.

Insulation Grommet

A process described by *R.L. Cannon; U.S. Patent 3,986,514; October 19, 1976; assigned to American Optical Corporation* relates to an improvement in batteries used for powering implantable heart pacers. A grommet, which can be constructed from neoprene rubber, is located or inserted between positive and negative terminals of the battery. The grommet is designed to extend substantially beyond the surface enclosure of the battery for the purpose of providing substantial contact area between the neoprene surface and the epoxy which encapsulates both batteries and heart pacer circuitry. The improved bonding between epoxy and grommet forms a leak-proof covering over each electrode. Any liquid which forms in the area of the battery, as a result of external moisture seeping in or as a result of venting of the batteries themselves, will not provide a conductive path from one electrode to the other because of the barrier created by the bonding. In a particular example, a fluted grommet is utilized to further improve bonding action between grommet and epoxy. It is an advantage of the process to substantially extend battery life and hence pacer operational life.

Referring to Figure 6.2a, a negative electrode **103** is separated from positive electrode **101** by grommet **102**. The electrodes and grommet may have generally cylindrical shape and conform to each other in a concentric manner. However, grommet **102** extends above the exterior surface of the battery by only a small distance. This small distance does not permit sufficient surface area contact between the grommet walls and the encapsulating epoxy which is contiguous to the surface of this structure to form a leak-proof bond.

FIGURE 6.2: PACEMAKER

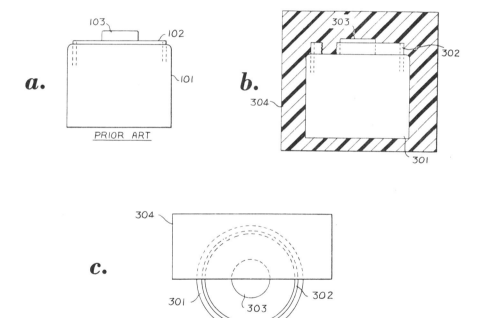

(a)   Elevational view of a prior art battery
(b)   Elevational view of an illustrative form of this process
(c)   Top view of Figure 6.2b

Source: U.S. Patent 3,986,514

Note that the internal portion of grommet **102** is hidden from view by the outer surface electrode **101** and is thus partially shown by dashed lines. The shape of the grommet internal to the cell may deviate from a cylindrical shape; thus, the dashed lines are not completely extended and are intended to represent the fact that a variety of shapes can be utilized. Referring to Figures 6.2b and 6.2c the negative electrode **303** is separated from positive electrode **301** by grommet **302** which is indicated to be substantially extended from the surface of the battery. Grommet **302** is selected to be nonconductive and to have the property of forming a leak-proof chemical bond with nonconductive, encapsulating epoxy **304**. Figures 6.2b and 6.2c depict epoxy **304** encapsulating only half of the battery or cell for purposes of clarity of illustration. Grommet **302** can be formed from neoprene rubber, for example.

Significantly increased surface contact area is achieved between grommet **302** and the surrounding epoxy. Epoxy **304** is contiguous with the exterior surface of the battery and grommet. This larger contact area results in a leak-proof bonding between epoxy and grommet. This bonding results in a leak-proof barrier or covering over each of the electrodes. The inside circumferential wall of the neoprene rubber grommet and the centrally located epoxy form a covering over centrally located negative electrode **303**. The outer periphery of the grommet and peripherally located epoxy form a second covering over positive electrode **301**.

It should be understood that venting of battery electrolytes described above creates high pressure. It is estimated that pressures on the order of thousands of pounds per square inch have developed in these pockets immediately external to the electrodes. In the event that pressure does get too high even for the improved bond of the process, the longer path length created by the larger grommet results in a higher resistance and a lower leakage rate in the event that conductive liquid does not interconnect the two electrodes. Consequently, the process solves the problem of pacers which fail due to an abrupt failure of one cell because of high leakage paths. It is estimated that 20 to 30% of pacers which fail have this failure mode.

Lithium-Iodine and Charge Transfer Complex

A process described by *R.T. Mead and W. Greatbatch; U.S. Patent 3,937,635; February 10, 1976; assigned to Wilson Greatbatch* provides a lithium-iodine cell including a lithium anode, a lithium-iodine electrolyte and a cathode comprising a source of iodine and a transport medium comprising iodine-containing depolarizer material connecting the iodine source to the electrolyte for transporting iodine ions from the source to the electrolyte.

The depolarizer material comprises a charge transfer complex of an organic donor component and iodine in the form of 2-vinyl pyridine iodide. The cathode is formed by providing substantially solid iodine having an electrical conductivity additive and coating a surface of the iodine element with a relatively thin layer of the depolarizer material. The cell is completed by coating a lithium surface of the anode with the depolarizer material, and placing the coated surfaces of the anode and iodine element against different surfaces of a barrier material which is penetrable by the depolarizer material and nonreactive with iodine.

Referring to Figures 6.3a, 6.3b and 6.3c, a lithium-iodine cell includes anode means comprising a pair of lithium members **12,14** having an anode current collector element **16** sandwiched or positioned between. The arrangement of lithium members **12,14** and collector **16** is fitted within an anode frame or holding means **18** which, in turn, can be fixed within a suitable cell casing or housing.

FIGURE 6.3: LITHIUM-IODINE BATTERY

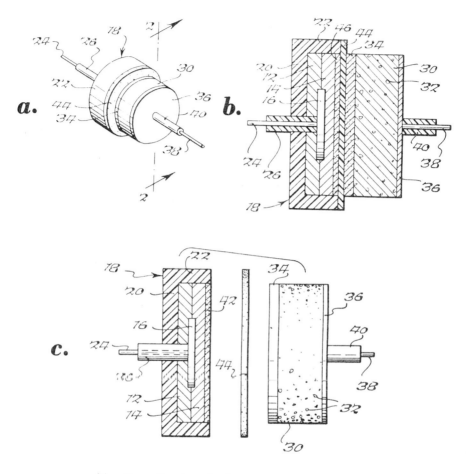

(a)   Perspective view of cell
(b)   Sectional view taken about on line **2-2** in Figure 6.3a
(c)   View illustrating the construction of the cell of Figure 6.3a

Source: U.S. Patent 3,937,635

In particular, holding means **18** is of synthetic plastic or similar electrical insulating material and formed to have a substantially planar face portion **20** and a continuous peripheral rim portion **22** extending from face portion **20**. Lithium members **12,14** can comprise a pair of disc-shaped or rectangular-shaped plates or foil members which are fitted face-to-face within rim portion **22** of holder **18**, the inner surface of rim **22** having a shape which conforms to the shape of the peripheral surfaces of members **12,14**.

Anode current collector member **16** is positioned or sandwiched between plates

12,14 in contact therewith, and can comprise a relatively thin sheet of expanded zirconium or nickel mesh. Face portion **20** of holder **18** and lithium plate **12** adjacent portion **20** are provided with small aligned slots or apertures for receiving an anode current collector lead **24** provided with a sheath or covering of electrical insulation **26** which lead **24** is connected at one end such as by welding to current collector **16** whereby an external electrical connection to the anode of the cell can be made. Anode frame or holding means **18** preferably is of a material which does not exhibit electronic conduction when exposed to iodine, e.g., the Halar or Tefzel fluoropolymers.

The anode assembly can be fabricated as follows. The material of anode frame or holder **18**, such as the above fluoropolymers, in addition to being of electrically insulating material, preferably also will be pressure-bondable to lithium. Lithium plate **12** is placed in holder **18** so as to be fitted within rim **22** wherein the aperture in plate **12** is in registry with the aperture in face portion **20**. The insulated lead **24** initially is welded or connected to current collector **16**, and then lead **24** is inserted through the aligned apertures in face **20** and plate **12** until current collector **16** is in contact with the exposed face of plate **12**.

Then plate **14** is placed in contact with current collector **16** and fitted within rim portion **22**. The anode assembly then is pressed together with a suitable force, e.g., about 3,000 pounds, causing the assembly to be bonded together. As a result, lithium plates **12,14** are bonded together in a manner sealing current collector **16** between the plates **12,14**, and the peripheral juncture at the edges of plates **12,14** is sealed by rim **22** of holding means **18**. If desired, the junction between the inner surface of rim portion **22** and the periphery of plates **12,14** can be sealed further by a suitable sealant or cement.

The cathode comprises a source of iodine in the form of a substantially solid block or pellet **30** of substantially pure iodine provided with an additive **32** of relatively high electrical conductivity in the form of graphite particles, metal fibers or the equivalent. The cathode comprises a transport medium comprising iodine-containing depolarizer material in the form of a charge transfer complex of an organic donor component and iodine connecting the iodine source to the electrolyte formed in the completed cell for transporting iodine ions from the source to the electrolyte.

In preferred form the transport medium comprises a layer or coating **34** of polyvinyl pyridine iodine material on a surface of iodine block **30**. The cathode is completed by a cathode current collector **36** in the form of a relatively thin plate of metal such as zirconium or nickel fixed or otherwise secured in contact with a surface of block **30**, in this instance the surface opposite that to which layer **34** is applied, and a cathode current collector lead **38** provided with suitable electrical insulation **40** is connected at one end to current collector **36**. Lead **38** enables an external electrical connection to be made to the cathode of the cell.

Before the anode and cathode components of the cell are assembled together, the arrangement prior to assembly being illustrated in Figure 6.3c, a layer or coating **42** of material identical to that of layer **34**, in particular a charge transfer complex of an organic donor component and iodine such as polyvinyl pyridine iodide, is applied to the exposed surface of lithium element **14** of the anode as shown in Figure 6.3c. A screen or barrier element **44** is placed between layers **34** and **42** in contact therewith when the anode and cathode elements are positioned together. In other words, the anode and iodine element **30** are placed against different surfaces of barrier element or screen **44** in a manner so that the surfaces of the anode and iodine element coated with the iodine-containing depolarizer material are in operative contact with the different surfaces of element **44**. Screen **44** is of a material penetrable by the material of layers **34** and **42** and nonreactive with iodine, and it prevents the occurrence of an internal electrical short circuit between iodine block **30** containing the additive **32** of electrically conductive material and the lithium element in the anode of the cell such as lithium member **14**.

The lithium-iodine cell operates in the following manner. When the anode and cathode elements are assembled together as shown in Figures 6.3a and 6.3b, the iodine-containing cathode material or depolarizer from layer **34** as well as from the layer **42** comes in contact with the exposed portion of lithium member **14**. A lithium-iodine electrolyte **46** begins to form at the interface and an electrical potential difference exists between the anode and cathode electrical leads **24** and **38**, respectively. The pellet or block **30** serves as a source of iodine ions to sustain or continue the reaction, and the layer or coating **34** of polyvinyl pyridine iodide serves as a vehicle to transport iodine ions from the reservoir **30** to the interface of lithium-iodine electrolyte **46** and lithium anode element **14**.

The relatively thin coating or layer **34** of depolarizer material on the iodine block **30** thus acts as an intermediary or a transport medium to diffuse iodine from the solid iodine reservoir or source **30** rather than having the cathode consist entirely of a charge transfer complex of an organic donor component and iodine, such as polyvinyl pyridine iodine. The polyvinyl pyridine iodine material of layer **34** and of layer **42** serves as a glue-like material to completely wet both the exposed surface of lithium element **14** and the surface of iodine block **30**. As fast as iodine is consumed from the polyvinyl pyridine iodine material more diffuses in from the iodine block or source **30**.

The cell according to the process is relatively economical to manufacture as compared to a lithium-iodine cell wherein the cathode material consists entirely of a charge transfer complex of an organic donor component and iodine such as polyvinyl pyridine iodine. This is because only a part or portion of the cathode consists of the polyvinyl pyridine iodine material and the remainder of the cathode comprises partly pure and substantially solid iodine.

The provision of relatively thin layers or coatings of depolarizer material, i.e., the charge transfer complex of an organic donor component and iodine in the form of polyvinyl pyridine iodine, contacting lithium element **14** in the anode of the cell and contacting the iodine block or reservoir **30**, insures complete contact between the cathode and the lithium anode and minimizes assembly problems encountered with handling the material such as polyvinyl pyridine iodine. With complete contact between cathode and lithium anode being provided, there are no problems of air gaps or abnormally high impedance at the anode-cathode interface.

The material of the coating or layer **34** and of coating or layer **42** is a charge transfer complex of an organic donor component and iodine. A preferred organic material is 2-vinyl pyridine polymer. The material is prepared by heating the organic material, i.e., 2-vinyl pyridine polymer, to a temperature greater than the crystallization temperature of iodine and then adding iodine to the heated mixture. The amount of iodine added should be greater than about 50 wt % of the resulting mixture so that enough iodine is available in the material to provide sufficient conductivity for proper cell operation. The amount of iodine added, however, should not be so excessive as to interfere with surface contact between layer 34 and block 30 and between layer 42 and lithium plate 14 by recrystallization. The material is applied with a brush or other suitable applicator to form a relatively thin coating or layer, in effect painting the material onto the surface, and the exact thickness of either layer 34 or 42 is not critical.

Alternatively, the layers may comprise fiber glass material coated with polyvinyl pyridine whereby formation of the iodide complex arises from a reaction with or a diffusion of iodine from the source or pellet 30 into the polyvinyl pyridine coating. As the cell operation causes the lithium iodide electrolyte 46 to form, the initially applied polyvinyl pyridine iodide layer 42 is displaced as the lithium iodide electrolyte layer 46 grows in thickness, the electrolyte 46 being shown in Figure 6.3b at a thickness which has completely displaced the initial polyvinyl pyridine iodide layer.

In the cell the iodine availability is enhanced by providing an iodine block of pellet 30 containing a high percentage of iodine. Since iodine in the pure state is a nonconductor of electricity, an additive 32 of high electrical conductivity such as graphite, metal fibers or the equivalent is incorporated in the block 30. The high conductivity additive could create an internal electrical short circuit if it were to contact the lithium metal of the cell anode. In effect, the charge transfer complex of the organic donor component and iodine separating the anode of the cell from the iodine pellet 30 acts as an electrolyte prior to the formation of the lithium iodide electrolyte in response to cell operation.

Screen or barrier 44 serves as a separator to prevent contact between the cell anode and the iodine pellet 30 provided with the high conductivity additive 32. In other words, barrier 44 is provided to prevent the graphite 32 in pellets 30 from contacting lithium element 14. Screen 44 must be of a material which is resistant to iodine in the sense of not becoming an electronic conductor in any appreciable degree when exposed to iodine. Furthermore, the material of screen 44 must be sufficiently porous to permit penetration of the screen by a flowable conducting material such as the iodine-polyvinyl pyridine complex without permitting physical contact between the lithium element of the cell anode and iodine block 30.

In this regard, screen 44 has characteristics similar to a membrane. By way of example, suitable materials for screen or barrier 44 are fiber glass cloth or fibrous or porous forms of fluoropolymer materials known as Teflon, Tefzel, Halar or polyester material. Screen 44 is necessary only early in the life of the cell, because as soon as a layer of the lithium iodide electrolyte 46 forms, iodine block 30 no longer can come in contact with or touch the lithium element 14. In Figures 6.3a and 6.3c, stippling of screen or barrier 44 indicates its porosity or permeability.

R.T. Mead, N.W. Frenz and F.W. Rudolph; U.S. Patent 3,996,066; December 7, 1976; assigned to Eleanor & Wilson Greatbatch Foundation describe a lithium-iodine cell which comprises a pair of cup-shaped container elements each having a peripheral flange and each containing a lithium anode element and iodine-containing cathode material. The container elements are of material which is heat sealable and which does not exhibit electronic conduction when exposed to iodine. The container elements are juxtaposed with the cathode material of each container being in operative contact with a cathode current collector, and the peripheral flanges of the container elements are heat-sealed together. The cell is encapsulated in potting material which is nonreactive with iodine, and the encapsulated cell is contained in an hermetically sealed metal casing having rounded surfaces. The lithium-iodine cell has an outer shape which causes little or no discomfort when implanted in a human body.

W. Greatbatch, R.T. Mead, F.W. Rudolph and N.W. Frenz; U.S. Patent 3,969,142; July 13, 1976; assigned to Wilson Greatbatch Ltd. describe a lithium cell comprising a cathode including a region of iodine-containing material having a pair of operative surfaces and a cathode current collector in the region between the surfaces, a pair of lithium anode elements operatively contacting corresponding cathode surfaces and each having a current collector, and electrical conductors connected to the cathode and anode current collectors.

Each anode element is fitted in a holder in a manner exposing a surface of each lithium element to the cathode material and sealing the anode current collector from exposure to the cathode material, the holders being of a material which does not exhibit electronic conduction when exposed to iodine. A pair of separator elements insulate the cathode conductor from the lithium anode elements. A pair of cells electrically connected in series and encapsulated in a single body provide a battery having an output of about 5 volts.

W. Greatbatch and R.T. Mead; U.S. Patent 3,981,744; September 21, 1976; assigned to Wilson Greatbatch Ltd. describe an enclosure for a lithium-iodine cell which includes a first casing containing the cell components and a second casing containing the first casing. The first casing is disposed so that the lid thereof is adjacent the bottom of the second casing. Both casings are of a material which is nonreactive with iodine such as epoxy material. The first casing is encapsulated in a polyester material for electrical insulation and sealing against iodine migration. The second casing is placed in an hermetically sealed outer casing of metal, the second casing being spaced from the lid of the outer casing.

Lithium-Iodine System

A.A. Schneider; U.S. Patent 4,010,043; March 1, 1977; and U.S. Patent 4,049,980; September 20, 1977; both assigned to Catalyst Research Corporation describes a lithium-iodine primary cell which is particularly well suited for use in prosthetic devices such as cardiac pacemakers. Generally, the primary cell comprises a lithium anode encasing member having an aperture therethrough. A cathode is positioned within the anode encasing member and consists essentially of an organic charge transfer complex and iodine. The cathode preferably includes a current collector having a lead portion with an insulating coating thereon that is positioned through the aperture.

The aperture may comprise a slit in the lithium or, alternatively, may be formed by enveloping the insulated cathode lead between folds in the lithium used in assembling the receiving vessel. In both cases, the lithium is bonded to the plastic insulator under pressure to form a seal. A lithium iodide electrolyte, preferably formed in situ, is coextensively positioned between and in contact with the inner surface of the anode-encasing member and the cathode. In a preferred case, the lithium anode-encasing member includes an anode current collector having a lead portion. The encasing member is provided with a layer of iodine-resistant cement and a thin plastic insulating coating, and the coated assembly is hermetically sealed in a metal protective case made, preferably, of stainless steel.

The cathode material is preferably a charge-transfer complex of organic material and iodine. Charge-transfer complexes are a well-known class of materials that have two components, one an electron donor, and the other an electron acceptor, that form weakly bonded complexes that exhibit electronic conductivity higher than either component. The charge-transfer complexes are in chemical equilibrium with small amounts of free iodine that is available for electrochemical reaction.

Cathodes containing intimate mixtures of low-conductivity complexes with powdered graphite or inert metal have high conductivities and can provide performance comparable to cells using high-conductivity complexes. Suitable charge complexes may be prepared using an organic donor component such as polycyclic-aromatic compounds, e.g., pyrene, anthracene, and the like; organic polymers, e.g., polyethylene, polypropylene, polyvinyls; or heterocyclic compounds containing nitrogen or sulfur, e.g., phenothiazine, phenazine, and the like. Preferably, the charge transfer complexes comprise a mixture of iodine and solid poly-2-vinyl pyridine$\cdot I_2$ or poly-2-vinyl quinoline$\cdot I_2$.

The electrolyte is preferably lithium iodide which may be formed in situ by contacting the anode and cathode surfaces wherein the lithium reacts with iodine in the cathode to form a solid lithium iodide electrolyte layer that contacts both the anode and the cathode. Alternatively, the electrolyte includes a coating of lithium iodide or lithium halide on the lithium anode formed by reaction of the lithium with iodine or other halogen.

The lithium-iodine cell is made by forming a sheet of lithium metal into a receiving vessel having an aperture therethrough and an opening. The vessel is formed to include at least one extending portion having a shape that conforms dimensionally to the opening and which is adapted to sealingly close the opening. A chemically resistant material, e.g., expanded zirconium metal or rigid foil, is welded to a plastic-coated lead wire to form a cathode current collector.

The cathode current collector is positioned, preferably, in the center of the formed vessel such that coated lead wire is securely located through the aperture. Alternative methods of accurately positioning the cathode lead comprise positioning the coated lead between the folds of lithium during the pressure forming of the lithium receiving vessel and/or heat sealing a plastic frame around the periphery of the cathode current collector where the frame is dimensioned to precisely fit the interior of the receiving vessel.

The cathode, preferably an organic charge transfer complex and iodine, is heated to a flowable consistency, e.g., to between 200° and 225°F. The formed vessel is then completely filled with the heated cathode material to provide intimate contact with all of the inner surfaces of the lithium vessel and cathode current collector. The vessel is chilled, e.g., to a temperature to between –130° and

−65°F, to solidify the cathode material. While the cathode material is in the solidified state, the extending portion of the vessel is positioned in abutting relationship with the cathode material and the peripheral edges of the opening are pressed against the vessel and cold-welded to form the anode encasing member.

Shock-Support System for Nuclear Battery

A process described by *H.C. Carney; U.S. Patent 4,026,726; May 31, 1977; assigned to General Atomic Company* relates generally to nuclear batteries which employ a radioisotopic heat source on one end, and a heat sink on the other end of a thermoelectric converter which is constructed of doped, semiconductor elements. In accordance with the known Seebeck Effect, the temperature gradient across the thermoelectric converter generates an electric potential which may be used to power a variety of electrical apparatus.

More particularly, this process relates to compact nuclear batteries of the type described above which may be implanted into the human body to power intracorporeal, life-assisting devices. The longevity of the radioisotopic heat source (87.8 years half-life) makes nuclear batteries especially useful for implantation into the human body. The long-term, relatively constant electrical output from such a battery reduces the need for repeated surgery or constant medical attention that is necessary with other types of power supplies, such as chemically reactive or rechargeable batteries. In addition, the compactness with which these nuclear batteries may be constructed makes them even more appropriate as a human implant.

This process provides a support system which will protect the thermoelectric converter in a nuclear battery from shock and vibration while cooperating with the battery insulation system and not detracting seriously from the thermal efficiency. In Figure 6.4a, a radioisotopic heat source 8 is independently supported within a battery housing 10 by a support tube 12. A thermoelectric converter 14 is secured between the heat source and a terminal cap 16, which functions as a heat sink, by thermally conductive alignment caps 18 and 20. Spring means 22 serve to isolate and cushion the converter from shock or vibration which may occur during manufacture, shipment or actual use, and a thin film 24 is bonded to the sides of the converter to further increase its resistance to fracture.

The nuclear battery housing 10 is preferably metallic, and it includes a barrel portion 26 which is closed at the upper end by the end cap 28 and at the lower or base end by the terminal cap 16, which is insertably positioned within the barrel and appropriately secured, as by welding. The terminal cap forms the base end of the battery housing and functions as the "cold" side to provide the temperature gradient across the thermoelectric converter. Glass-to-metal seals 30 are provided within the terminal cap to seal lead wires 32 which extend through the terminal cap for connection to the thermoelectric converter 14. Epoxy potting 34 on the underside of the terminal cap further seals the base of the housing and the lead wires.

To isolate the thermoelectric converter from axial and moment forces more effectively, the heat source 8 is independently supported within the battery by the upstanding support tube 12 which extends upwardly into the barrel 26. The base of the support tube includes a radially extending, peripheral flange 36 which is sandwiched tightly between an interior barrel flange **38** and the terminal cap **16**, to rigidly secure the support tube within the battery housing **10**. The support tube **12** tapers upwardly from its base to a smaller cylindrical portion which houses the heat source.

FIGURE 6.4: NUCLEAR BATTERY

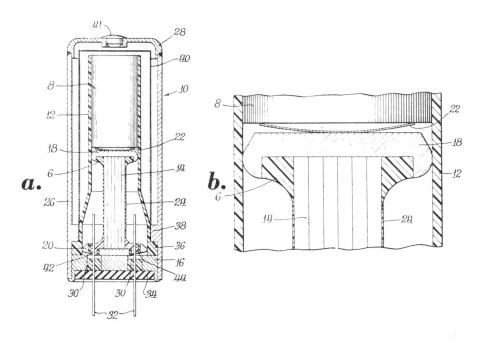

(a)   Vertical sectional view of a nuclear battery
(b)   Enlarged, fragmentary view of the portion of the shock sup-
      port system between the heat source and the thermo-
      electric converter

Source: U.S. Patent 4,026,726

The tapered-barrel shape of the support tube **12** is especially adapted to firmly carry the heat source while efficiently maintaining a temperature gradient across the thermoelectric converter **14**, which is positioned between the heat source **8** and thermal cap **16**. The support tube is rigidly secured to the housing at the base end, and it is near this connection that maximum forces are likely to occur in case of shock or vibration. The conical base of the support tube is designed to minimize total conduction and radiation parasitic heat loss by providing increased strength at the high stress, cantilevered end which is also the low temperature end. The wide base of the support tube allows it to be constructed of thinner material, while maintaining sufficient strength to absorb the stress or strain that results from shock or vibration.

Because the support tube is thus thinner than might be required otherwise, less heat is conducted along the support tube between the heat source and terminal cap. The smaller diameter portion of the support tube which encloses the heat source **8** also serves to reduce radiation and conduction heat losses which increase with increased diameter (surface area) of the support tube. And to further reduce radiation heat loss, by reducing the emissivity, the support tube is vacuum-coated with a layer of gold, usually less than 0.10 micron thick.

In the preferred case, the support tube **12** is 0.010 to 0.015 inch (0.0254 to 0.0381 cm) thick and is constructed of a polyimide polymer, such as that available under the name Vespel Sp-1, which is characterized by relatively high strength and comparatively low thermal conductivity. A support tube constructed of Vespel Sp-1 can adequately support the heat source **8** while limiting the transfer of heat between the heat source and the heat sink (terminal cap).

Another material which may be used for the support tube is a poly(amide-imide) resin which is available under the name Torlon 3000 and which has an even lower thermal conductivity than Vespel Sp-1. Although these represent the preferred materials for construction of the support tube, any material of sufficiently high strength, low thermal conductivity, and which is thermally stable at about 100°C, may be used.

The heat source **8** is housed within the smaller, upper cylinder portion of the support tube **12**. In the preferred case, the heat source is made of medical-grade plutonium (90% $^{238}PuO_2$), which is processed by well-known means to reduce undesirable radiation and chemical reaction and to make the material more compatible with container materials. The plutonium is then hot-pressed into a ceramic pellet and enclosed within a three-layer capsule for shielding against radiation and for the prevention of any accidental release of radioactive material during cremation. The capsule is described with more particularity in U.S. Patent 4,001,588. The heat source capsule is bonded by epoxy, into the upper end of the support tube, and the radioactive decay of the plutonium provides the necessary heat for the battery.

Because the battery is designed for human implantation, it is necessary that the thermal conversion efficiency be maximized to reduce nuclear radiation from the plutonium heat source. To reduce the parasitic transfer of heat to the battery housing, a reflective foil **40** radiation heat barrier is located between the support tube **12** and the battery housing **10** and the interior of the housing is filled with inert gas of low thermal conductivity inserted through seal plug **41**.

The thermoelectric converter **14** is also enclosed within the support tube **12** and is spaced between the heat source **8** and the terminal cap **16**. In the preferred case, the thermoelectric converter is made of elongated, alternating elements of bismuth-telluride N-type and P-type semiconductor material. These elements are assembled into a converter module, with gold contacts vacuum-deposited at each end and serially connecting the various elements. Terminal contacts for electrical lead wires **32** are provided in the form of gold strips bonded by epoxy to the last N and P elements in the series. The particular method for construction of this type of thermoelectric converter may be found in U.S. Patents 3,780,425 and 3,781,176. Because the fragile semiconductor elements are serially connected so that any break within the converter **14** will completely disrupt the operation of the battery, it is necessary to isolate the converter from shock, vibration and other external forces as fully as possible. To cushion the thermoelectric converter from the various bending and shock forces that may be encountered during the lifetime of the battery, alignment caps **18,20** and spring means **22** structurally isolate the converter from the heat source **8** and the thermal cap **16**.

The alignment cap **18**, which is more clearly shown in Figure 6.4b, is generally disc-shaped, with a generally flat upper surface and a recessed undersurface which is epoxy-bonded **6** to the top of the converter. The cap is preferably made of molybdenum for good thermal conduction from the heat source, and chemical compatibility with battery materials during any accidental fire or cremation.

The sides of the alignment cap which may contact the interior of the support tube are relieved so as to roll or slide when the support tube is flexed during shock or vibration. In particular, the side surface of the alignment cap has a radius of curvature of at least about 0.01 inch (0.0254 cm) and preferably about 0.015 inch (0.038 cm) in the region which is most likely to contact or rub the support tube. This curvature continues around to the underside of the cap. Upwardly of the region of contact, the side of the alignment cap slopes inwardly along a conical plane that is generally tangential to the curved portion of the side and that intersects the flat upper surface of the cap at an acute angle of about 80°.

The spring means **22** shown in Figure 6.4b, is preferably a saucer-shaped shell of molybdenum foil approximately 0.002 inch (0.0051 cm) thick which is seated on the alignment cap **18** and opens upwardly toward the heat source **8**. The spring is about 0.010 inch (0.0254 cm) high in the compressed position and about 0.015 inch high (0.0381 cm), with a radius of curvature of about 1.5 inches (3.81 cm), in the released position. The spring, in addition to cushioning the thermoelectric converter from shock also provides some allowance for variations in manufacturing tolerance. Heat is conducted across the gap between the heat source **8** and the top of the converter **14** directly via the spring and the gas-filled gap, which is not sufficiently large to materially impair thermal conduction, and by radiant heat transfer from the heat source.

Alignment cap **20** is also constructed of molybdenum and is epoxy-bonded to the base of the thermoelectric converter **14**. The alignment cap **20** rests upon the terminal cap **16** which functions as the heat sink or cold side of the temperature gradient across the thermoelectric converter. As with cap **18**, the side of the disc-shaped cap **20** is curved, the radius of curvature being at least about 0.01 inch (0.0254 cm) and preferably about 0.015 inch (0.0381 cm) in the region of contact with the support tube, to provide a rolling contact against the support tube and to reduce the transmission of moment forces to the converter upon shock or vibration to the support tube **12**. This is to be contrasted with a rigid, cantilever connection used in some prior art batteries.

The bottom of the alignment cap **20** is also relieved to allow a rolling contact with the terminal cap **16** upon flexure or movement of the support tube. In particular, the undersurface of the alignment cap tapers upwardly from a centrally flat portion at an angle of approximately 10° until it tangentially meets the curved contact portion. This small angle allows the alignment cap to rock upon flexure of the support tube. Upwardly of the region of contact, the side of the alignment cap slopes inwardly along a conical plane that is generally tangential to the contact region of the side and intersects the flat upper surface of the cap at an acute angle of about 80°.

To accommodate electrical connection to the thermoelectric converter, two holes are provided in the alignment cap 20 for the lead wires 32. Alumina insulator tubes 42,44 insulate the wire from possible grounding against the alignment cap. The lead wires may then be connected to the gold strip terminals provided on the converter module, as earlier described.

To further reinforce the thermoelectric converter against brittle cracking and to reduce radiation heat transfer from the converter, a thin film 24 is epoxy-bonded to the exposed sides of the thermoelectric converter. Preferably, the film is a laminate about 3 mils thick, including a layer of aluminum less than 0.1 micron thick, for low emissivity, vacuum deposited on a polymer film such as that available under the name Kapton. The preferred foil is bonded to the exposed sides of the thermoelectric converter by epoxy cement. This reinforcement serves to reduce failure of the converter which is caused by stress concentrations arising from surface irregularities on the sides of the converter. The very thin aluminum layer provides a lower emissivity which reduces radiant heat transfer from the converter.

Thus, the process provides a nuclear battery which is sufficiently shock resistant to withstand a variety of shocks, vibration or other stresses arising from external forces. Tests show that a battery constructed in accordance with this process is capable of absorbing shocks greater than 3,000 g without disrupting battery service.

Power-Source Canister

R.H. Comben and R.L. Doty; U.S. Patent 3,957,056; May 18, 1976; assigned to Medtronic, Inc. describe a power-source canister which can be easily incorporated into existing pacemaker designs. Generally speaking, a pacemaker unit should be as small as possible or, alternatively, of a shape which will be perceived as small by the biological environment. Since most pacemakers are implanted near the surface of the skin, a relatively flat shape is often selected. Within this constraint, the surface area to volume ratio is optimized in a squat cylinder.

This shape is a beginning point for many pacemaker designs with the edges of the cylinder being tapered to reduce the perception of thickness by the body. Thus, the larger pacemaker components are commonly located within the central area which is the position where the pacemaker thickness is the greatest. The power-source canister of the process is adapted to be incorporated within a pacemaker unit configured as described above. Of course, it may be modified in obvious manner, to be located outside the central region of the pacemaker, where appropriate.

Figure 6.5a illustrates a power-source canister in which the canister inner void or cavity is defined by a bottom wall 10 and a side wall composed of four generally straight portions 11 joined to each other by arcuate side wall portions 12.

FIGURE 6.5: POWER-SOURCE CANISTER

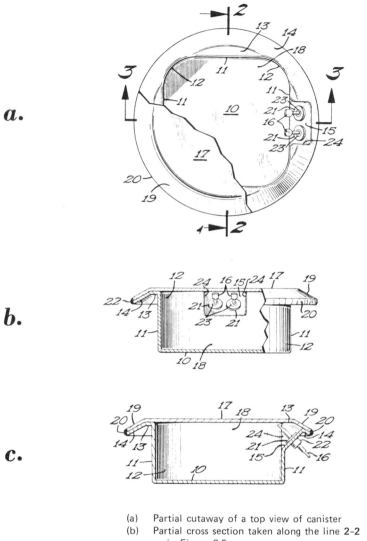

(a)   Partial cutaway of a top view of canister
(b)   Partial cross section taken along the line **2-2**
       in Figure 6.5a
(c)   Cross section taken along the line **3-3** in
       Figure 6.5a

Source: U.S. Patent 3,957,056

As illustrated, the straight portions are of approximately equal length, each being generally perpendicular to two of the others and generally parallel to a third. The arcuate side wall portions **12** have a radius of curvature approximately equal to one-fourth the distance between the parallel ones of the straight side wall portions **11**. A flange **13** having an arcuate flange portion **14** extends from the side wall and a bevelled portion **15** extends from a straight portion **11** of the side wall to the flange **13**. Walls **24** extend between the bevelled portion **16** and the side wall and feed-through terminals **16** pass through the bevelled portion **15**. A cover **17** overlies the canister cavity.

Referring to Figure 6.5b, there is shown a partial cross section of the preferred example of Figure 6.5a taken along the line 2-2 in Figure 6.5a. As seen in Figure 6.5b the inner void or cavity **18** of the canister of the process is defined by the bottom wall **10** and the continuous side wall formed by the straight portions **11** and arcuate portions **12**. The flange **13** extends from the side wall and terminates at the arcuate flange portion **14**. The cover **17** is coextensive with the flange and includes an arcuate cover portion **19** which is complementary to the arcuate flange portion **14**.

With the cover **17** in position on the canister of the process, the canister inner void or cavity **18** is closed and the cover **17** may be secured over the cavity **18** through the flange **13** without encroachment on the cavity **18**. Such attachment may be by any means known to the prior art such as bonding with any suitable adhesive. However, in those applications where the power sources to be positioned within the cavity **18** must be isolated from the device or devices to be powered, when a corrosive electrolyte such as lithium-iodide is used, e.g., the cavity **18** may be hermetically sealed.

A preferred manner of hermetically sealing the cavity **18** is through the use of a heat-generating sealing process such as welding. The bead resulting from such a process is illustrated at **20** in Figure 6.5b. Thus, through the cooperation of the flange **13** and a cover **17** which is coextensive with the flange **13**, the cavity **18** may be hermetically sealed through the use of a heat-generating sealing process at a point spaced from the cavity to minimize the effects of heat on the power sources contained therein.

Pacemakers are commonly cylindrically shaped with their sides tapering towards each other to reduce the perception of thickness by the body. The design of Figures 6.5a, 6.5b and 6.5c is uniquely adapted to fit within such a system, particularly within the central region of the pacemaker unit. The overall height of the power-source canister is selected to accommodate the thickness of the power source or sources to be contained therein while still falling within the thickness of the pacemaker at its central portion.

The arcuate portions **14,19** of the flange **13** and cover **17**, are similarly intended to accommodate the tapering sides of the pacemaker unit while providing an extension away from the cavity **18** for the purpose of allowing the use of a heat-generating sealing process without effect on any power sources which might be housed within the cavity **18**. The amount of curvature within the flange and cover is dependent upon the overall configuration of the pacemaker unit itself and is easily determinable for any such unit. It should be noted, that the central portion of the cover **17** need not be flat. Instead, it can be provided with a curvature, crown or other configuration which will accommodate itself in the overall design of the pacemaker unit. Of course, if the pacemaker design will permit, the entire cover **17** and flange **13** may be flat.

Referring to Figure 6.5c, there is shown a cross section taken along the line 3-3 in Figure 6.5a. In Figure 6.5c, the cavity is again illustrated as being formed by the bottom wall 10 and the continuous side wall. The bevelled portion 15 extends between one of the straight portions 11 of the side wall to the flange 13. As illustrated, the bevelled portion 15 engages the arcuate flange portion 14. However, the point at which the bevelled portion 15 engages the flange is dependent upon the dimensions of the flange and the existence or nonexistence of the arcuate flange portion 14, the particular configuration of the flange 13 at the point it is engaged by the bevelled portion 15 being noncritical in the process.

Feed-through terminals 16 are supported through the bevelled portion 15 of the side wall by a collar 21 which engages the inner face of the bevelled portion 15 and has a tubular portion 22 of a smaller diameter extending through the bevelled portion 15. The collar 21 and portion 22 are insulated from the terminals 16 by an insulating member 23 (Figure 6.5a) surrounding the terminal 16 and the entire assembly is spot welded to the bevelled portion 15. The assembly composed of the terminal 16, collar 21, portion 22 and insulation 23 are known to the art.

By passing the feed-through terminals 16 through the canister side wall, the power sources may be positioned within the cavity 18, connected to the terminals 16, and that connection tested. The cover may then be positioned over the cavity 18 and secured and/or sealed, as desired, without disturbing the connection between the terminals 16 and the power source or power sources within the cavity 18. The cover may also be removed without disturbing the connection. The inclination of the terminals 16 resulting from their passage through the bevelled portion 15 of the side wall makes the terminals 16 easily accessible within the cavity 18 by inclining them toward the canister opening. The terminal inclination also reduces the terminal intrusion into that portion of the cavity 18 intended to be occupied by the power source or power sources. Additionally, it is part of the process to space each of the terminals 16 from the others along the perimeter of the side wall, whether through a bevelled portion 15, or otherwise, such that no terminal 16 obscures the view or access to the others within or without the cavity 18.

PACEMAKERS—BIOFUEL CELLS

Stimulating Electrode

F. von Sturm and G. Richter; U.S. Patent 3,941,135; March 2, 1976; assigned to Siemens AG, Germany describe a heart pacemaker which comprises a stimulating electrode, a counterelectrode, a pulse generator and an implantable glucose-oxygen biofuel cell. The area of the glucose electrode of the biofuel cell is made larger than the stimulating electrode and the stimulating electrode is connected electrically with the glucose electrode so that they are both at the same potential. By so constructing the pacemaker the stimulating electrode's potential is reduced to a range only about 50 to 200 mV as measured with respect to a reversible hydrogen electrode in the same electrolyte. That is to say, it is reduced to a potential at which good blood compatibility and body compatibility will be expected. Since the geometric area of the stimulating electrode is smaller than that of the glucose electrode, local currents flowing between the stimulating electrode and glucose electrode due to the stimulating electrode acting as an oxygen electrode because of the material used cannot influence the potential of the stimulating electrode in an unfavorable manner. The dimensioning of these two electrodes according to the process is further advantageous since it is desirable to keep the stimulating electrode small in order to be able to supply a high current density at the point of stimulation with a relatively small power output. On the other hand, it is necessary that the glucose electrode be as large as possible so that the power delivered by the biofuel cell is in the required range from about 50 to 100 μW.

Figure 6.6 illustrates schematically on an enlarged scale the pacemaker of the process. The pacemaker designated generally as **10** and which will preferably be disposed in the heart is designed with a shape which forms a good flow profile. In essence, the housing of the pacemaker **10** comprises a biofuel cell **11** having diameter and height approximately 6 mm in the form of a hollow cylinder with plastic caps **12** and **13** closing the hollow cylinder. Materials used are blood compatible plastics, particularly epoxy resin and silicone resin. The shape of the plastic caps **12** and **13** is that of a dome, i.e., a hemisphere or spherical segment. A stimulating electrode **14** is mounted in the cap **12**. The stimulating electrode is a platinum/iridium alloy containing 10 to 20% iridium by weight.

FIGURE 6.6: PACEMAKER WITH BIOFUEL CELL

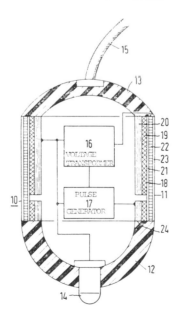

Source: U.S. Patent 3,941,135

The effective area of the stimulating electrode, i.e., the area projecting beyond the cap **12**, is essentially in the form of a hemisphere. The hemisphere has a a radius of approximately 1 mm resulting in an effective area for the stimulating electrode of approximately 6 mm². Attached to the component 13 is a connection 15 to permit removing the pacemaker from the blood stream. The pacemaker 10 includes conventional electronics including a voltage transformer 16 and a pulse generator 17 both of which are housed inside the hollow cylinder. The biofuel cell 11 is made up of a number of elements. The largest portion of the inner cylindrical shell is in the form of a glucose electrode (anode) 18. The glucose electrode is made up of two layers. An outer Raney platinum layer 19 is the active layer and is supported by an inner layer 20 of platinum/nickel alloy which is used both for support and for current take off.

This anode is prepared using the above-described method so that the nickel in a hollow cylinder consisting of a platinum/nickel alloy (atomic ratio Pt:Ni = 1:6) and having a wall thickness of a few tenths of a millimeter is dissolved out of the outer cylinder wall by chemical etching with an inorganic acid such as hydrochloric acid, sulfuric acid or nitric acid or an acid mixture, or by an electrochemical activation such as a potentiodynamic or potentiostatic treatment, such that there will be a Raney platinum layer (the outer cylinder wall) with a layer thickness of between 1 and 100 μM on top of a layer comprising a platinum/nickel alloy (the inner cylinder wall of the hollow cylinder).

The outer cylinder wall, i.e., the Raney platinum layer 19, is then covered by a hydrophilic membrane 21 for separating the two electrodes. The membrane 21 may comprise an ion exchange material but preferably will be a hydrogel. The arrangement comprising the glucose electrode 18 and membrane 21 is fitted into a hollow cylinder comprising a silver/aluminum alloy (approximate silver content 20 wt %) and having a wall thickness of approximately 100 μM. The aluminum is then dissolved out of this alloy using a potassium hydroxide solution so that a porous oxygen electrode 22 of Raney silver is obtained. The oxygen electrode 22 is then provided with a hydrophilic membrane 23.

The membrane 23, permeable to glucose and oxygen, is used for keeping proteins away from the electrode surface. As mentioned above, suitable materials for the membranes are particularly hydrogels, i.e., polymers produced by means of hydrophilic crosslinking agents; the polymerization takes place in the presence of water and the polymers assume their final swell state during the polymerization. Such polymers are preferably produced by the polymerization of glycol methacrylate (methacrylic acid-2-hydroxy ethyl ester) as a monomer with water-soluble tetraethylene glycol dimethacrylate as a crosslinking agent in the presence of more than 45% water. Copolymers containing up to 3 wt % of methacrylic acid or methacrylic acid-2-dimethylaminoethyl ester may also be used.

The membrane 23 may be applied to the oxygen electrode, e.g., by immersing the oxygen electrode in a solution of hydrophilic polymer and subsequently drying it. If applicable, a crosslinking process may follow to make the polymer insoluble. However, the membrane 23 may also be applied by first providing the oxygen electrode with a porous carrier material, in particular with a thin paper sleeve, and subsequently impregnating it with a solution containing the following ingredients: a monomer to form a polymer, a crosslinking agent and, if applicable, a polymerization catalyst.

The glucose electrode 18 or layer 20 is electrically connected with the stimulating electrode 14. The voltage transformer 16 is also connected to the layer 20 of the glucose electrode **18** and to the oxygen electrode **22**. A connection exists between the voltage transformer and pulse generator **17** with the pulse generator having a connection to the stimulation electrode **14** and to a counterelectrode **24**. Counterelectrode **24** is designed as a glucose electrode and is electrically insulated from the glucose electrode **18**. The electrodes have approximately the following areas: oxygen electrode 1.1 cm^2, glucose electrode 1 cm^2 and counterelectrode 0.1 cm^2. In operation, the oxygen electrode adjusts itself to a potential of approximately 640 mV each measured against a hydrogen electrode in the same electrolyte.

As a result the cell voltage is approximately 0.5 V. The potential of the stimulating electrode connected to the glucose electrode is approximately 145 mV, thereby being slightly more positive than the potential of the glucose electrode. As is evident from Figure 6.6, the stimulating and counterelectrode inside the heart are closely adjacent. This saves electrical energy since the current must travel only a short distance from the stimulating electrode to the counterelectrode. This aids even further in miniaturization.

As indicated in Figure 6.6, the inner hollow cylinder of the biofuel cell **11** is divided into two sections with the larger section forming the glucose electrode **18** and the smaller one a counterelectrode **24**. Such a division into sections which function as separate electrodes is also advantageously made but a plurality of stimulating electrodes are provided for the pacemaker, each electrode having a separate energy supply. In such a case each stimulating electrode is then connected to a separate glucose electrode.

Silicone Rubber Membrane Casing

E. Weidlich; U.S. Patent 3,915,749; October 28, 1975; assigned to Siemens AG, Germany describes a biogalvanic metal-oxygen cell which is encased in an oxygen permeable silicone rubber membrane having one metal anode, two oxygen cathodes disposed on both sides of the anode comprising silver screens to which a catalytically active material is applied, and electrolyte chambers between the anode and the cathodes.

The edge of the metal anode is joined to a plastic frame of greater width than the metal anode for the formation of the electrolyte chambers. The marginal areas of the silver screens of the cathodes are free of catalytically active material and are extended across the plastic frame and joined to each other. The silicone rubber membrane encases the cathodes and the marginal areas of the silver screens so as to make close contact, and a silicone rubber jacket is applied over the portion of the silicone rubber membrane which encases the marginal areas of the silver screens.

The process is explained in greater detail by way of Figure 6.7 showing a partial cross section of the cell. To produce a round metal-oxygen cell, an aluminum anode **2** approximately 2 mm thick is cemented into an epoxy resin ring **1** serving as an annular plastic frame. The ring is approximately 6 mm thick with an o.d. of 50 mm and an i.d. of approximately 36 mm. To simplify the assembly of the cell, the annular plastic frame **1** may be composed of three parts, each approximately 2 mm thick, a center part **12** and two outer parts **13**. In that case the metal anode **2** is connected in a suitable manner to a silver current take-off lug **11** and glued into the center part **12** of the frame. The two outer parts **13** are then cemented to the center part, defining in the finished cell the electrolyte chambers **4** located between the anode **2** and the cathodes **3**.

FIGURE 6.7: BIOGALVANIC METAL-OXYGEN CELL

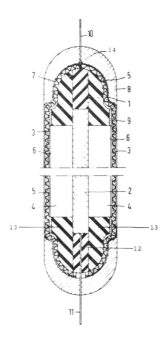

Source: U.S. Patent 3,915,749

Silver screens **5** (wire diameter: 0.12 mm; approximately 220 screen openings per cm²) approximately 60 mm in diameter are used to produce the cathodes **3**. A silver ring **9** (o.d.: 40 mm; thickness: 100 μm approximately) 2 mm wide is spotted concentrically on one side of each silver screen. The silver screen is soft-annealed for half an hour at 500°C and then pressed flat under approximately 40 N/mm² (approximately 400 kg/cm²). A mixture consisting of 190 mg activated carbon containing nitrogen and 190 mg of a 5% aqueous polyvinyl alcohol solution is brushed on the side of the silver screen **5** facing away from the silver ring **9** to cover the area corresponding to the area defined by the silver rings. The next operation is drying and crosslinking of the mixture at 150°C for 1 hour.

The charge amounts to approximately 15 mg charcoal per cm². A round polyvinyl alcohol membrane **6** (diameter: approximately 40 mm) approximately 12 μm thick is provided on the reverse side of the silver screen, e.g., on the side provided with the silver ring. This side of the cathode is subsequently glued, by means of an epoxy resin, to one side of the plastic frame containing the anode; correspondingly, a second cathode is glued to the other side of the plastic frame.

The two silver screens **5** are then interconnected by spot welding and a silver lug **10** is provided for current take-off. This silver lug goes through the plastic frame **1** and is electrically insulated against the silver screens **5** by providing the silver screens with a cut-out **14** as shown.

A common salt solution, i.e., a 0.9% NaCl solution, mixed with 100 mg/l sodium carbonate to stabilize the pH value, is filled into the two electrolyte chambers through holes in the plastic frame. The filling holes are subsequently closed by epoxy resin. Then the entire cell is enclosed in a silicone rubber membrane **7** approximately 19 μm thick. For this purpose, the cathodes as well as the silver screen areas free of catalyst are coated with about 20% silicone rubber solution in dichloromethane, whereupon the membrane is glued on and dried out. The anode **2** is also provided with a silver lug **11** for current take-off.

If the plastic frame is composed of three parts, the two outer parts may be provided first with a cathode each, enclosed unilaterally by a silicone rubber membrane and subsequently glued, by means of epoxy resin, to the center part containing the metal anode. Finally, a silicone rubber jacket **8** is cast around the plastic frame area of the cell produced in the described manner. The finished cell is approximately of pocket watch size with a volume of approximately 12.4 ml and a weight of approximately 31 g.

Aluminum Anode

A process described by *E. Weidlich; U.S. Patent 3,915,748; assigned to Siemens AG, Germany* provides an improved aluminum anode which resists premature passivation and pitting, reduces the development of hydrogen, reduces charge losses, and prevents premature corrosion of the outer zones of the electrode. This is accomplished by an aluminum anode comprising a metal screen having a layer of aluminum on both sides of the metal screen. The outer surfaces of the aluminum layer (i.e., the surfaces facing away from the metal screen) are lapped or sandblasted to form finely ground surfaces and are provided with an anodized layer at a portion of their outer surfaces.

By virtue of the surface treatment of the aluminum layers, i.e., by lapping or sandblasting, the aluminum surface is consumed uniformly and premature passivation of the electrode is prevented during operation of the cell. The surface treatment also prevents contamination which can lead to pitting. In addition, the development of hydrogen and the charge losses connected therewith can be reduced in this manner. By anodizing the outer surfaces premature corrosion is also prevented.

The use of a metal screen, to which aluminum is applied in layers on both sides, assures a reliable and durable contact with the anode. The metal screen preferably consists of silver; however, titanium and tantalum can also be used. Screens of titanium and tantalum can additionally be provided with a layer of a metal carbide or nitride. It can be insured in this manner that the mechanical stability of such screens is not affected by hydrogen that might be present. For the construction of the anode, the two aluminum layers are pressed on the metal screen and the anode is then cast over with epoxy resin at the edge.

The two aluminum layers of the anode according to the process can advantageously be provided with openings, which extend from one surface of the aluminum layer to the other. In this manner the possibility exists for the electrolyte to pass through the anode and thereby equalize shifts in pH value, since variations in the pH value can also lead to the development of hydrogen. This is especially true in so-called double-chamber cells, in which the anode is arranged as a partition in the center of a symmetrically designed, chamber-like cell and the two outside walls parallel to the anode are designed as cathodes. The openings provide the further advantage that pressure equalization is possible in the anode because a pressure difference is produced by uneven hydrogen development on both sides of the anode.

In order to ensure a uniform pH value at all points of the aluminum surface, the anode can additionally be enclosed with a membrane, preferably made of polyvinyl alcohol. The membrane may also consist of a polyvinyl alcohol and polyacrylic acid mixture.

FLAT CELLS FOR PHOTOGRAPHIC APPLICATIONS

Film Assembly with Hermetically Sealed Battery

A process described by *R.M. Delahunt; U.S. Patent 3,967,292; June 29, 1976; assigned to Polaroid Corporation* relates to photography and more precisely, to photographic film packs or assemblies comprising a container holding self-developing film unit(s) integrated with an electrical power supply system.

Referring to Figure 6.8a, a plurality of hermetically sealed batteries are shown sealed by way of a high-speed, continuous lamination process. Essentially, each battery is hermetically encapsulated between first laminar assembly **100** and second laminar assembly **200** by sealing the peripheral portions of the assemblies and about sealing area **210** containing apertures providing access to contacts **220**. Encapsulation is accomplished by way of specialized, high-speed equipment designed to continuously position a laminar battery between rolls of first and second laminar assemblies (**100** and **200**) so that the battery can be encapsulated between the superposed assemblies by applying the requisite heat or pressure to the portions thereof to be sealed.

FIGURE 6.8: FILM-BATTERY ASSEMBLY

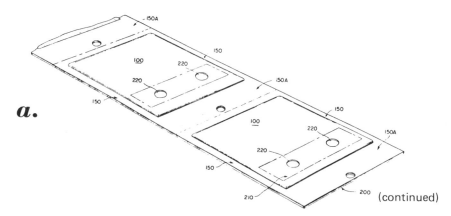

a.

(continued)

FIGURE 6.8: (continued)

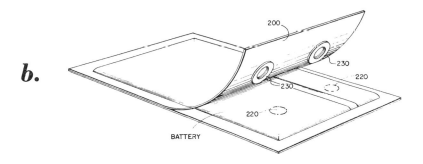

b.

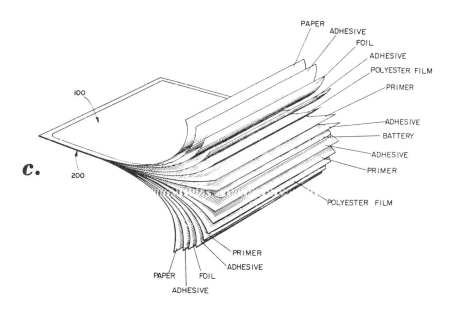

c.

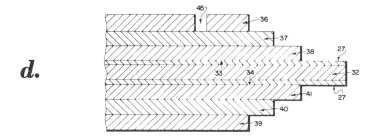

d.

(continued)

FIGURE 6.8: (continued)

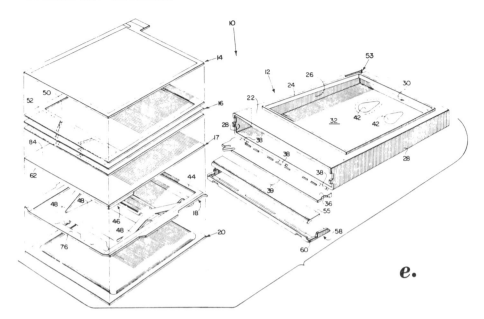

e.

(a)   Perspective view of a plurality of hermetically sealed batteries fabricated by
      way of continuous, high, speed production techniques
(b)   Perspective view of a partially encapsulated battery illustrating a particular
      manner for sealing particular portions of one of the laminar assemblies
      about portions of the battery
(c)   Exploded view of the elements of the preferred laminar assemblies employed
      to hermetically seal a battery
(d)   Partial cross-sectional view of a battery encapsulated within the sealing means
      of the process and illustrating the essential elements of batteries employed
      in the photographic film packs
(e)   Exploded view of elements of the photographic film pack showing the con-
      struction and arrangement of the elements of the film pack

Source: U.S. Patent 3,967,292

However, encapsulation of the battery positioned between assemblies **100** and
200 is achieved in a continuous fashion preferably with pairs of opposed rollers
employed to heat-seal peripheral portions **150** while portions **150A** and area
210 are preferably heat-sealed by heat-stamping techniques. As shown in Figure 6.8a, sealed area **210** is generally rectangular in shape and provides effective
sealing of portions of second laminar assembly **200** about contacts or terminals
220. The shape of sealed area **210** is not especially critical so long as effective
sealing about contacts **220** is achieved.

An especially effective method for assuring effective sealing of second laminar
assembly **200** about the area defining contacts **220** is shown in Figure 6.8b. The
second laminar assembly provides two prepunched apertures for defining the cathode and anode contacts which are shown disposed adjacent each other on one

flat surface of the battery. Circular-shaped sealing inserts 230 are shown arranged about the prepunched apertures on second laminar assembly 200 with one surface of sealing insert 230 securely sealed to second laminar assembly 200 while the other surface of sealing means 230 is available for sealing about portions of the surface providing contacts 220.

The respective surfaces of sealing insert 230 providing the adherent capability can have the same or different adhesive so that effective sealing between second assembly 200 and the surface of the battery providing contacts 220 can be achieved. Depending upon the particular circumstances, both surfaces of sealing means 220 can have the same or different type of adhesive, e.g., a heat activable or pressure sensitive adhesive. Alternatively, one surface of sealing means 220 can have one type of adhesive, e.g., a pressure sensitive adhesive while the other surface has another type of adhesive, e.g., a heat activable adhesive thereby providing a feature for assuring effective adhesion between the surface of the second laminar assembly and the surface of the battery should adhesion between these surfaces present peculiar adhesion problems.

Details of particular elements of preferred first and second laminar assemblies (100 and 200) will be better appreciated by reference to Figure 6.8c. As shown there, first laminar assembly 100 comprises a paper stock which can be, e.g., a 12 lb/3,000 ft^2 tissue stock. Coated on the inner surface of the tissue stock is a layer of adhesive preferably a low density polyolefin adhesive providing a layer of about 5 to 10 lb adhesive per 3,000 ft^2 to adhere the tissue stock to an aluminum foil sheet which can be in the order of about 0.00035 inch thick.

Another layer of polyolefin adhesive similar in construction and thickness to the layer discussed above provides adhesion between the foil sheet and an electrically nonconductive plastomeric film preferably a polyester film (Mylar) about 0.0005 inch thick. Finally, a layer of heat activable adhesive about 1 mil thick is adhered to the plastomeric film by a primer layer. In the especially preferred embodiment of the process, the heat activable adhesive is a commercially available ionomer sold under the name Surlyn.

Second laminar assembly 200 is preferably somewhat thicker than first laminar assembly 100. As shown in Figure 6.8b, laminar assembly 200 comprises a paper stock which can be a supercalendered bleached kraft paper stock of about 50 lb/3,000 ft^2. A layer of low density polyolefin adhesive of the composition and thickness described before is employed to adhere the paper stock to a foil sheet material preferably in the order of about 0.0005 inch thick. In turn, another layer of low density polyolefin adhesive overcoated with a primer provides adhesion between the foil sheet material and the electrically nonconductive plastomeric layer which as mentioned, is preferably a layer of polyester about 0.005 inch thick. Finally, a primer layer provides suitable adhesion between the plastomeric layer and the layer of heat activable adhesive, e.g., a layer of Surlyn about 1 ml thick.

Laminar assemblies of the type described above have been found to be especially advantageous in hermetically sealing laminar batteries to effectively protect film units included with the battery in the film packs of the process from any adverse effects of gaseous materials that may be evolved from the battery. Broadly, the laminar batteries involved in the process are flat, primary battery structures comprising anode materials, sheet-type separator elements, cathode materials and electrolyte combined with sheet-type electrically conductive oppositely disposed sheet-type current collector elements combined to form a multicell structure. The preferred battery of the film packs of the process comprise a Leclanche electrochemical system including a zinc anode **38** (Figure 6.8d) and a manganese dioxide cathode **41**. The aqueous electrolyte **27** generally comprises an aqueous ammonium chloride/zinc chloride (about 4:1) electrolyte and usually a small amount of mercuric chloride all dispersed in a polymeric binder. The dispersion of electrolyte **27** is applied to or impregnated in central portions of separator **32** and in contact with surface **33** of anode **38** and surface **34** of cathode **41**.

Essential elements of the cells of battery **20** further include metallic sheet current collector **36** having a plurality of gas ports **45** designed to permit efficient transmission of gas generated or liberated within the cell. Metallic current collector **36** is fabricated of aluminum, lead or steel and is arranged in contact with polymeric current collector **37** which comprises a sheet of electrically conductive carbon-impregnated, water-vapor-impermeable thermoplastic polyvinyl chloride film of the type sold under the name Condulon. Zinc anode **38** comprises active zinc material either as a paste or a zinc sheet secured to polymeric current collector **37**.

Cathode, metallic sheet current collector **39** can be similar in material and construction to current collector **36** (together with ports **45**) while cathode polymeric current collector **40** can be similar to polymeric current collector **37**. Cathode **41** comprises an active cathode material such as a manganese dioxide, conductive carbon paste deposited on current collector **40**.

The components of a photographic film pack **10** to which the process pertains are shown in exploded fashion in Figure 6.8e. They include a box-like film container **12** and its contents, a dark slide **14**, a plurality of self-developing film units **16**, a resilient pad member **17**, a film support member **18**, and electrical battery **20**. Film container **12** is molded of an opaque thermoplastic material, such as polystyrene, and includes relatively thin, substantially planar walls.

A forward wall **22** includes a generally rectangular upstanding rib **24** which defines the bounds of a generally rectangular light-transmitting section or exposure aperture **26**. Depending from three sides of forward wall **22**, respectively, are a pair of side walls **28** and a trailing end wall **30** which serves to space a substantially planar rear wall **32** from forward wall **22**.

The leading ends of forward wall **22**, side walls **28**, and rear wall **32** cooperate to define an elongated rectangular opening at the leading end of container **12** through which the contents may be inserted. After insertion, a leading end wall **36** which is preferably coupled to the leading end or rear wall **32** by integrally formed flexible hinges **38**, may be rotated 90° and joined to the leading ends of side walls **28** and rear wall **32** by any suitable method such as ultrasonic welding.

It will be noted that when leading end wall **36** is located in its closed position, its top edge **39** is spaced from the leading end of forward wall **22** such that an elongated withdrawal slot **40** (not shown) is formed therebetween through which dark slide **14** and film units **16** may be sequentially removed from container **12**.

The contents of film container 12 are preferably arranged as in the stacked relation shown in Figure 6.8d. Hermetically sealed battery 20 is positioned over the interior surface of rear wall 32 such that two electrodes on the underside of the battery (not shown) are aligned with a pair of teardrop-shaped openings 42 in the rear wall 32. When container 12 is operatively positioned in a suitable camera, a pair of electrical contacts mounted therein are adapted to extend through openings 42 for coupling battery 20 to the camera's electrical system.

Positioned over battery 20 is the film support member or spring-biased platen 18. Member 18 preferably includes a generally rectangular platen or open support frame 44 dimensioned to support the peripheral margins of the rearward-most film unit 16 in the stack. Integrally formed with frame 44 is a generally H-shaped center section 46 which includes resilient spring legs 48 that bear against container rear wall 32 and/or battery 20 to urge platen 44 upwardly.

Resilient pad member 17 is located between frame 44 and the most rearward film unit 16. The film units are integral negative positive film units which are substantially flat and preferably rectangular. They preferably include a rectangular or square photosensitive image-forming area 50, which is surrounded by opaque margins and a rupturable pod or container 52, containing a fluid processing composition, located at the leading end of the film unit outside of the bounds of the image-forming area. Details relating to suitable film units can be found in U.S. Patents 3,415,644; 3,415,645; 3,415,646; and 3,647,437.

Dark slide 14 is formed of any suitable opaque material such as cardboard, paper, or plastic and is initially positioned between the forwardmost film unit 16 and the interior surface of forward wall 22 to light seal exposure aperture 26 and opening 53. Once film container 12 is located at its operative position within a camera slide 14 may be removed through withdrawal slot 40 in the same manner as the forwardmost film unit 16. The film units are arranged in stacked relation on top of resilient pad member 17 with their image-forming areas 50 facing towards the exposure aperture 26 in container forward wall 22 and their containers of fluid 52 adjacent end wall 36. Subsequent to the removal of dark slide 14, the forwardmost film unit 16 (closest to wall 22) bears against the interior surface of forward wall 22 and is in position for exposure to actinic radiation transmitted through exposure aperture 26.

It will be noted that after dark slide 14 is removed from container 12 through slot 40, the leading end of the forwardmost film unit 16, containing pod 52, is also aligned with withdrawal slot 40 at the leading end of container 12. Subsequent to exposure, the forwardmost film unit is adapted to be engaged by a film advancing mechanism in the camera and moved out of container 12 through slot 40 in a direction substantially parallel to forward wall 22, for processing. Access for engaging the trailing end of the forwardmost film unit to move it forwardly through slot 40 is provided by an opening 53 located in forward wall 22 and the trailing end wall 30 of container 12.

In order to light-seal withdrawal slot 40, container 12 is preferably provided with an opaque flexible sheet 55 which is secured at one end to an exterior surface of leading end wall 36 and is disposed in closing relation to slot 40. This sheet forms a primary light seal for blocking light when container 12 is located within a camera. A secondary light shield for blocking slot 40 prior to inserting container 12 into the camera may also be provided in the form of an end cap member **58**. Member **58** is coupled to leading end wall **36** and includes an end cap **60** which is initially positioned in closing relation to the primary light seal **55** and withdrawal slot **40** and may be pivoted to an open position in response to inserting container **12** into the camera.

From the foregoing description, it will be apparent that the process presents to the art distinctive film packs for advanced photographic systems providing improved performance by reason of the integration of encapsulation means with the battery contained in such packs. In addition to hermetically sealing the battery, the laminar assemblies of the encapsulation means are particularly adaptable to high-speed, continuous heat lamination techniques. This adaptability is due in part to the integration of the plastomeric layer and the heat activable adhesive layer with each of the assemblies. This distinctive function of the plastomeric layer and adhesive and advantages obtained therefrom will be better appreciated by reference to Figure 6.8a.

As shown in Figure 6.8a, flat planar batteries are hermetically sealed between first and second laminar assemblies **100** and **200** by the application of heat and pressure to peripheral portions of the assemblies (**150** and **150A**) and to area **210**. Obviously, in such a heat-sealing operation maximum bonding efficiency is desired between the assemblies but dimensional constraints are also involved especially in the width of the seals about the peripheral portions of the assembly. In terms of optimum performance characteristics, the heat-sealing operation should provide maximum bonding between the assemblies but the width of the peripheral seals should be kept at a minimum especially along portion **150**. In turn, one way of minimizing the overall width of the peripheral seal is to seal the assemblies as close to the peripheral portions of the encapsulated battery as possible.

In hermetically sealed products of the process, however, the battery to be encapsulated is a flat planar battery having metal current collectors sealed about the other elements of the battery (Figure 6.8d). Under such conditions, encapsulation of batteries of the process requires sealing of the laminar assemblies about hard metallic edges defining the periphery of the rectangular battery and the corners of the battery are relatively sharp. Because of these hard edges and especially the relatively sharp corners, sealing of the first and second laminar assemblies close to the edges or corners of the battery can cause cutting or puncturing of one or both of the assemblies and shorting can result.

Flat Battery Production

A process described by *L.O. Bruneau; U.S. Patent 3,988,168; October 26, 1976; assigned to Polaroid Corporation* relates to electrical energy power supplies and more precisely, to specialized batteries and processes for producing such batteries which have special utility in packs or assemblies comprising a container holding self-developing film unit(s) integrated with an electrical power supply system.

The distinctive feature of the products and process involves a unique selection of the overall dimensions of separators which are involved in the fabrication of the batteries to which the process pertains. Essentially, the dimensions of the separator are selected so that the overall peripheral dimensions of the separator exceed the overall peripheral dimensions of sheet elements or electrochemical components integrated therewith. By employing separators of preselected dimensions, the overall efficiency of the assembly process is improved especially with respect to heat-sealing operations involved in the assembly operation and with respect to obtaining maximum efficiency for the function assigned to the separator both during and after the assembly process. For example, the improved assembly process provides maximum production of batteries exhibiting the desired performance characteristics and batteries so produced exhibit the desired balance of performance characteristics over extended times.

A multicell flat battery structure is presented generally at **10** in Figure 6.9a as it would appear in an electrochemically active state following component buildup in accordance with the process. Note, for instance, the presence of an electrically insulative sheet **12** located at the bottom of the pile structure. The length of insulative sheet **12** as well as its widthwise dimension are selected such that it extends slightly beyond the peripheries of the electrically active laminar components of battery **10**. The upwardly facing surface of battery **10** is present as the outer metal surface of an anode electrode current collector assembly **14**. The electrode current collector assembly is folded about one side of battery structure **10** to present a downwardly facing metallic surface portion as at **16** which is utilized to provide a terminal-defining surface for the battery.

An opening shown in dashed fashion at **18** and formed in insulative sheet **12** provides access to the downwardly facing metal surface of a cathode electrode collector assembly (Figure 6.9d) which is attached to the upward facing surface of sheet **12**. With the arrangement, cathode and anode terminals may be provided on one flat surface of the battery structure **10**. Also revealed in Figure 6.9a are the peripheral edges of electrically insulative separator elements **20**. The slightly depressed peripheral portion **22** of the assembly is occasioned from peripheral sealing procedures provided in the course of assembly of the structure **10**. This depression necessarily becomes more exaggerated in the sectional views of the battery.

Looking now to Figures 6.9b, 6.9c and 6.9d, the structure of battery **10** is revealed in more detail and the method of the process for fabricating it utilizing appropriately high-volume production techniques is shown. The preferred production technique is typified in the utilization of a web-type carrier **12** preferably an electrically insulative sheet. Sheet **12** may be drawn from a suitable roll-type supply **24** and introduced to a fabricating industrial line through appropriate web drives including tension adjusting rolls as at **26** and **28**. Web **12** preferably is introduced having a width selected to achieve the geometry described in connection with battery **10** of Figure 6.9a.

Accordingly, its width is selected as being slightly greater than the electrically active components of the battery structure or at least as wide as separator components **20**. Materials selected for the web, in addition to being electrically insulative, should be chemically inert. Suitable materials are Mylar or Estar, which is a film of polyethylene terephthalate or an unfilled polyvinyl chloride or the like. For photographic applications, sheet or web **12** may be coated with a substance opaque to actinic radiation to aid in maintaining the light-tight integrity of any film container within which the batteries are incorporated. A black "Mexican lacquer" coating is found to be suitable for this purpose.

FIGURE 6.9: MANUFACTURE OF FLAT BATTERY

a.

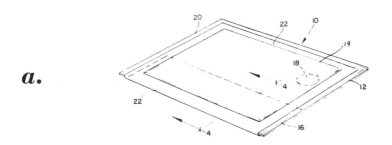

b.

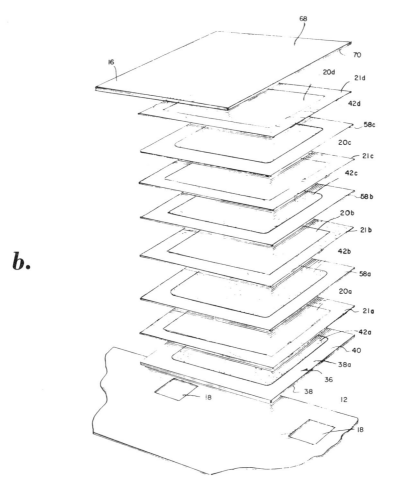

(continued)

FIGURE 6.9: (continued)

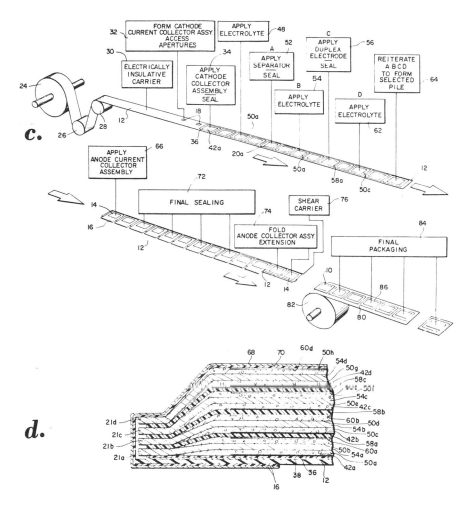

(a) Perspective view of a flat primary battery structure as it exists prior to packaging

(b) Exploded perspective representation of the components of a battery structure revealing the relative orientation of the sheet-type components

(c) Schematic diagram showing, in block fashion, the functions of fabricating stations along an assembly line for producing the battery structure

(d) Sectional view of a battery structure taken through the plane 4-4 of Figure 6.9a

Source: U.S. Patent 3,988,168

Web or carrier **12** is maneuvered through the production or assembly area with an intermittent motion in accordance with the spacing of individual multicell battery or pile assemblies which are made thereon. The provision of the insulative carrier **12** is revealed by function or station block **30** in Figure 6.9c, while the initial pitch or spacing between the battery units on the carrier web is established by the formation of access apertures **18** as revealed at station or function block **32**. By photodetecting an edge of these apertures **18**, suitable spacing and registration controls may be provided throughout all of the assembly stations of the production line. While openings **18** may take a variety of shapes, a convenient arrangement is an opening having a rectangular periphery formed forward one side of web **12** in a position such that a terminal-defining surface is in position for appropriate contact with the instrumentalities within which the battery is utilized as a power source.

As revealed at station function block **34** in Figure 6.9c, the component buildup upon web **12** is commenced by positioning a discrete cathode current collector assembly **36** in appropriate registry over access opening **18**. Assembly **36** is a discrete laminar structure including a metallic sheet or foil current collector **38** (Figure 6.9d), preferably an annealed tin-coated steel or an aluminum or lead sheet material on the order of less than 10 mils in thickness, which is laminated to a polymeric current collector **40**. Collector **40**, in turn, preferably is a sheet of electrically conductive carbon-impregnated vinyl film Condulon, having a thickness in the order of about 2 mils and configured having the same relative external or peripheral dimensions as metal surface **38**.

Predeposited over film surface **40** is now dry active positive material **42**, i.e., a cathodic paste deposition. Cathode **42** is positioned inwardly from the periphery of collector **40** to provide clear surfaces for subsequent sealing procedures and, for providing a Leclanche electrochemical system, may be present as a mixture of manganese dioxide (depolarizer) and electrically conductive carbon dispersed in a polymeric binder.

Of course, the battery structure may be designed to utilize any of a variety of known positive electrode materials such as other inorganic metal oxides, for instance, lead oxide, nickel oxyhydroxide, mercuric oxide, and silver oxide, inorganic metal halides such as silver chloride and lead chloride, and organic materials capable of being reduced such as dinitrobenzene and azodicarbonamide compounds.

When the preformed laminar collector assembly **36** is positioned as by "pick and place" technique common in the art, portions of the periphery of web **12** will extend beyond the widthwise periphery of the assembly to provide a geometry facilitating the insulation thereof. The peripheral area of assembly **36** is heat sealed to web **12** following its placement. As revealed at station function block **48** in Figure 6.9c, web **12** is driven to carry the subassembly including current collector assembly **36** to a position where electrolyte is applied coextensive the facing surface of cathode material **42**.

Represented in Figure 6.9d at **50a**, electrolyte **58** ordinarily will comprise a conventional aqueous solution of ammonium chloride/zinc chloride about 4:1 dispersed in a polymeric thickener such as hydroxyethyl cellulose, etc., e.g., on the order of about 5% or more. In addition, a small quantity of mercuric chloride may be present in the electrolyte as a corrosion inhibitor. Preferably, the electrolyte is deposited over cathode **42** utilizing positive displacement techniques as opposed to doctoring, silk screening and the like.

As revealed at function block **52**, the next station in the assembly procedure applies a discrete electrically insulative separator element **20a**. As is more clearly illustrated in Figure 6.9d and as has been described in connection with the separator elements described generally at **20** in Figure 6.9a, separator **20a** is formed having a rectangularly shaped periphery which extends slightly but importantly beyond all electrochemically active components of the battery. Note in this regard, that the separator element extends slightly beyond the periphery of cathode current collector assembly **36**.

Element **20a** as well as all separators within a pile structure may be constructed of and comprise a conventional battery separator material such as aqueous electrolyte-permeable fibrous sheet materials, e.g., fibrous and cellulosic materials, woven or nonwoven fibrous materials such as polyester, nylon, polypropylene, polyethylene or glass. A peripheral, frame-shaped portion of each separator, as at **21a**, is impregnated with a thermal-sealing electrically insulative adhesive such as a polymeric hot melt adhesive, e.g., a conventional hot melt vinyl adhesive suited to secure each separator with contiguous anode and cathode carrier sheets, such as polymeric current collector sheet **40**.

Note that sealing periphery **21a** is positioned well outwardly from the area of influence of electrolyte gel **50a**. Following appropriate placement of discrete separator element **20a**, its peripheral portion **21a** is thermally sealed to the subassembly including carrier sheet **12** and collector assembly **36**, a thermally induced bond being available between sealant **21a** and the noted polymeric current collector sheet **40**.

The dimensional configuration of separator **20** as well as the adherent capability provided by frame-shaped area **21** constitutes a distinctive feature of the process and provides special advantages in the assembly of batteries comprising such separators. For example, peripheral frame-shaped area **21** provides an effective adherent capability providing an especially effective bonding area presenting maximum available bonding surface thereby providing bonding surface tolerances which are especially advantageous in low-cost, high-speed assembly operations involving heat lamination or sealing steps and devices involved in effectively performing such steps.

Additionally, because frame-shaped area **21** extends beyond all electrochemically active sheet elements adhered thereto, contact between such elements that could occur because of the application of heat and/or pressure during the lamination or edge-sealing operations is effectively minimized, if not completely avoided. Accordingly, the integration of this feature presents to the art a low-cost high-speed battery assembly process providing an especially high degree of control for maximum production of batteries of the desired performance characteristics with minimal rejects.

As is represented at function block **54** in Figure 6.9c, the pile subassembly now moves to a second station for the application of electrolyte. Identified in Figure 6.9d at **50b**, this second electrolyte application is made over the central portion of separator **20a** within the porous area thereof not incorporating sealant **21a**. With this electrolyte application, a continuous electrolytic association between the cathode **42a** and a next adjacent anode may be effected.

The associated anode for the initial cell is provided, as shown at function block 56 in Figure 6.9c, with the positioning over the subassembly of a discrete duplex electrode 58a. Serving as an intercell connector, duplex electrode 58a may be prefabricated of a sheet or film of electrically conducting material, preferably a sheet of electrically conductive carbon-impregnated vinyl which, as described earlier, in connection with layer 40 may be Condulon or the like. Materials for the intercell connectors should be impervious to the electrolyte utilized within the cell structure, must provide a function for conducting electrical current between the positive electrode in one cell and the negative electrode in the next cell, should not create undesired electrochemical reactions with the electrodes or other components of the battery and should be heat-sealable.

Preformed centrally upon the lowermost side of the conductive sheet is a distribution of active zinc negative or anode material 60a which, in conventional fashion, is amalgamated with, for instance, mercury by contact with mercuric chloride within the electrolyte of the cell. Oppositely disposed upon the conductive sheet of the duplex intercell connector 58a is another dry deposition of active positive material 42b which is present, for instance, as a manganese dioxide/electrically conductive carbon mixture dispersed in a polymeric binder as described earlier.

The electrically conductive sheet forming the duplex electrode 58a is configured in discrete fashion having a peripheral dimension corresponding with that of cathode current collector assembly 36. As such, it may be observed in Figure 6.9d that separator 21a will extend beyond the periphery of electrode 58a. Following its positioning, the duplex electrode sheet is heat-sealed about its outer periphery to the sealant 21a of earlier positioned separator 20a. As a consequence, the first electrochemically active cell of the pile structure will have been assembled. Note at this juncture, that the active cell is carried by an electrically insulative and chemically inert carrier web 12 in spaced relationship and separated from other multicell components such that no electrical association of the discrete cells is derived along the production or assembly line.

As is revealed at the function block 62 in Figure 6.9c, the unit cell subassembly is moved by carrier 12 to a station applying electrolyte as identified in Figure 6.9d at 50c. This deposition in combination with cathode material 42b commences the buildup of the next serially coupled contiguous cell. Accordingly, the above-described sequence of construction may be reiterated as shown at block 64 of Figure 6.9c by select repetition of the series of station procedures employed in connection with blocks 52, 54, 56, and 62.

Note in this regard that these blocks, respectively, have been labeled A, B, C, and D, and identified in that sequence in block 64. In each reiteration the dimensional configurations of the discrete elements remain identical as well as their positioning in registration to achieve a requisite laminar structure. Turning to Figure 6.9d, note that a four-cell pile structure is illustrated as including separators 21a–21d, electrolyte depositions 50a–50h, duplex intercell connector elements 58a–58c, cathode electrode depositions 42a–42d, and anode depositions 60a–60d.

Following the final electrolyte deposition 50h (Figure 6.9d), the multicell subassembly is moved by web 12 to a station applying a discrete anode current collector assembly as shown in Figure 6.9c at block 66. Identified earlier at 14 in Figure 6.9a, and illustrated in more detail in Figures 6.9b and 6.9d, assembly 14 is configured in similar laminar fashion as earlier described assembly 36. In this regard, assembly 14 is constructed having a metallic sheet terminal surface portion preferably formed of annealed tin-coated steel sheet material on the order of less than 10 mils in thickness, as shown at 68, in laminar electrical and physical bond with a polymeric current collector sheet of electrically conductive carbon-impregnated vinyl film 70. Sheets 68 and 70 are coextensive in dimension; however, such dimension is extended as previously described at 16, to a widthwise dimension protruding beyond the adjacent edge of carrier web 12.

Centered within that portion of assembly 14 excluding extension 16 is active zinc negative material 60d predeposited thereupon in similar fashion as provided at anode deposits 60a–60c. Upon being positioned as shown in Figure 6.9b, the multicell pile subassembly is moved into a final sealing area defined by function block 72. Within this area, the peripheries of the discrete components of the multicell assembly are subjected to a series of heat-pressure sealing operations to assure the integrity of all peripheral seals within the assembly. Through the use of multiple-stage sealing, excessive heat buildup is avoided which otherwise may adversely affect the operative quality of the electrolyte depositions within a battery.

As revealed at block 74 in Figure 6.9c, following final sealing, the extension 16 is folded around the edge of the multicell assembly to a position wherein surface 70 thereof abuts against the underside of web 12. As described in connection with Figure 6.9a, this exposes the metallic surface 68 of assembly 14 to the underside of the battery in juxtaposition to the metallic surface 38 of collector assembly 36. A select portion of the surface of metallic layer 38 is accessed through rectangular access opening 18 initially formed in web 12. The terminals of the multicell battery 10, therefore, are on one side of the battery and in conveniently spaced juxtaposition.

Looking to Figure 6.9d, it may be observed that no additional insulative materials are required to accommodate for the noted folding of extension 16 inasmuch as separator peripheries 21a–21d extend beyond intercell connectors 58a–58c, while the complementing peripheral edge and exposed surface of electrically insulating carrier web 12 provides insulative protection for collector assembly 36. Accordingly, the structure is simply formed, retaining a high reliability through the geometry of its component discrete elements. Web 12 then carries the battery assemblies 10 to a shearing station depicted in Figure 6.9c at 76.

At this point, electrically insulative web 12 is sheared to provide discrete battery units such as that shown in Figure 6.9a. No electrical interconnection is formed between the web-connected multiplicity of subassemblies; therefore, no voltage buildup phenomenon is witnessed. Further, no electrically conductive material, e.g., the polymeric electrically conductive materials and metal foils, is cut or sheared within an electrochemical environment on the production line. Only the electrically insulative carrier 12 is sheared. As a consequence, edge shorting occasioned during the formation of discrete elements is substantially eliminated.

Following shearing of web 12, the discrete battery units 10 are positioned upon a continuous carrier roll of cardboard stock 80 shown extending from a supply roll 82 in Figure 6.9c. Stock 80 is provided having a widthwise dimension greater than that of the completed battery assembly 10 and an upward facing surface having formed thereon a low-temperature heat-sealing material upon which battery assemblies **10** are positioned. As depicted generally at function block **84**, final packaging of the battery units **10** may then take place upon this insulated carrier **80**. In one such packaging arrangement, the forward edge of battery elements **10** is heat-sealed to the adhesive coating of card stock **80** at the forward edge thereof through the use of a simple hot bar technique. The thus-attached battery units are then moved by the continuous web card stock through a station wherein a thin electrically insulative film having a heat-sealable coating, also electrically insulating, is positioned over the card stock battery combination in continuous fashion.

I. Erlichman; U.S. Patent 3,993,508; November 23, 1976; assigned to Polaroid Corporation describes a method for manufacturing flat or planar batteries wherein a carbon-impregnated polymeric carrier web which is selectively zonally prelaminated with continuous strips of metal ultimately serving current collector purposes is machined to form discrete current collector subassemblies as an initial step of production. Through the use of select laser cutting, geometries of individual battery elements are maintained. Chemical milling techniques may be utilized for the noted metal removal.

T.P. McCole; U.S. Patent 4,019,251; April 26, 1977; assigned to Polaroid Corporation describes a method for constructing flat batteries wherein the peripheral seal provided each battery exhibits an enhanced intercomponent lamination or seal integrity and wherein the extent of the peripheral seal advantageously is expanded. This improved method is characterized in the utilization of a cover sheet of material and polymeric connector materials are dimensionally unstable. This cover sheet is interposed between a fully built-up battery pile structure and thermal and pressure sealing device prior to the sealing operation of the manufacturing method.

This cover sheet is characterized by conducting transversely but not laterally, thereby permitting greater control of the actual seal width and the use of a wider seal, as discussed in more detail below. By dimensioning the cover sheet so as to extend over all seal configurations as well as the metallic surface of a collector assembly otherwise contacted by the thermal sealing elements, the rebound or memory effects of the metal components are somewhat restrained to avoid delamination effects occasioned with such sealing procedures.

In the process, the multicell batteries are formed by assembling a predetermined sequence of discrete-type components, electrolyte and electrode active materials to define a multicell pile. These components include current collectors which are externally disposed within each battery assemblage as well as electrically conductive polymeric intercell connectors and dye-impermeable but insulative separator sheets. These components are associated with a thermally activable material seal which is peripherally disposed about the corresponding peripheries of each cell unit within the multicell assembly.

Prior to introducing the assembled or compiled multicell structures to a sealing station, the noted cover sheet material is applied over the surface upon which seal contact is made and heat and pressure then are applied against the cover sheet to effect final sealing. In a preferred case, the method includes a final step of cold striking or the like wherein a cool platen acting as a heat sink is applied against the seal to effect a permanent set in the sealing material. Additionally, an adhesive may be provided intermediate the noted cover sheet and that component of the batteries with which it comes in contact prior to the heat-sealing procedure. In the preferred example the cover sheet material is glassine, a thin, hard and almost transparent paper made from well-beaten chemical wood pulp.

R.D. Fanciullo and L.G. Fasolino; U.S. Patents 4,028,479; June 7, 1977; and 3,907,599; September 23, 1975; both assigned to the Polaroid Corporation describe a thin battery of the Leclanche type which incorporates a multiplicity of electrically conductive layers and sheet separators arranged to cooperate in conjunction with electrochemical material to form a multicell battery. Continuous frame-type sheet seals are thermally bonded to the periphery of each of the separator and conductive layers to provide an insulative sealing arrangement. The outer peripheries of these frames also are heat-sealed to provide a rigid, secure border seal for the battery assembly. The sheet separators are positioned intermediate superpositioned ones of the electrically conductive layers.

Thermally Stable Insulating Layer

R.M. Delahunt; U.S. Patent 3,912,543; October 14, 1975; assigned to Polaroid Corporation describes an energy cell or battery which comprises one or more individual cells each of which includes a planar anode superposed substantially coextensive a planar cathode, at least one of the anode and the cathode comprising thermoplastic material, and possessing a planar separator including an electrolyte permeable central portion surrounded by substantially electrolyte-free marginal portions positioned between the anode and cathode.

Aqueous electrolyte is disposed in the central portion of the separator and in contact with opposed facing surfaces of both the anode and cathode and the battery specifically includes an electrically nonconducting adhesive extending coextensive and intermediate marginal portions of the separator and next adjacent facing surfaces providing adhesion between such surfaces and an electrically nonconducting layer extending intermediate the marginal portions of the anode and cathode which layer is thermostable at the temperature at which the thermoplastic electrode softens.

As seen by reference to perspective Figure 6.10a and cross-sectional Figure 6.10b, the cassette may comprise a generally parallelepiped container or box **10** for holding and enclosing a plurality of film units **11**, gas collector means **12** and a planar battery assemblage **13**. Container **10** is shown as comprising a forward wall **14**, side walls **15**, a trailing end wall **16**, a leading end wall **17**, and a rear wall **18** and may be formed of a resilient plastic material. Forward wall **14** is provided with a generally rectangular exposure aperture **19** for transmitting light for exposing film units **11** carried within container **10**.

Leading end wall **17** is provided with a generally rectangular slot or exit orifice to provide a passage **20** at the leading end of the container through which film units **11** carried by the container are adapted to be individually withdrawn. Container **10** may additionally be provided with a dark slide or cover sheet of any suitable opaque material such as paper or plastic sheet material positioned between the forwardmost film unit **11** and aperture **19** to serve as a light seal and which may be removed through withdrawal slot **20** once container **10** is located in its operative position within a camera apparatus.

FIGURE 6.10: PLANAR BATTERY

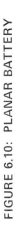

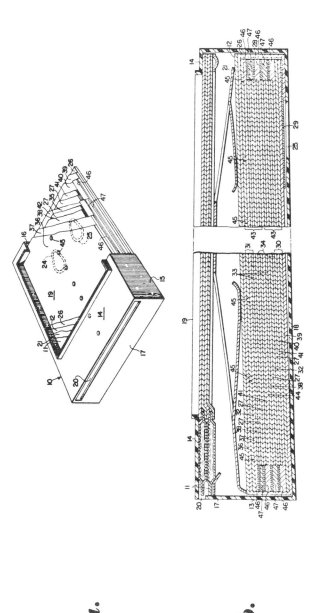

a.

b.

(a) Fragmentary perspective view of one form of a photographic film assemblage including a cassette, photographic film units, spring biasing member, gas collector, and planar battery

(b) Diagrammatic enlarged cross-sectional view of a film assemblage of the type set forth in Figure 6.10a, illustrating the association of elements comprising the photographic film assemblage

Source: U.S. Patent 3,912,543

The stack arrangement within container 10 of a plurality of film units 11 (one is shown in Figure 6.10a and two are shown in Figure 6.10b), gas collector 12 and planar battery assemblage 13 is specifically illustrated in Figure 6.10b. Each film unit 11 is arranged in overlying relationship with its exposure surface facing in the direction of exposure aperture 19.

As described in U.S. Patent 3,543,662, the cassette additionally includes a spring-loaded platform 21 positioned between battery assemblage 13 and next adjacent film unit 11 for compressively retaining the battery terminals next adjacent cassette terminal ports 24 and 25 for interengagement with camera electrical leads, and for biasing film units 11 in the direction of exposure aperture 19. The completed film cassette assemblage including film units 11, spring member 21, collector 12 and battery 13 shown in Figures 6.10a and 6.10b is adapted for direct employment in camera devices of the type mentioned in U.S. Patents 3,643,565 and 3,714,879.

As shown in Figures 6.10a and 6.10b, battery 13 may include gas permeable envelope or container 26 retaining the superposed electrical energy generating components of the battery disposed within, e.g., an electrically nonconducting water-vapor impervious thermoplastic envelope extending around and coextensive the external or exterior surfaces of the generating components. Envelope 26 acts to encapsulate the energy components to further prevent escape of aqueous electrolyte 27 and/or electrolyte solvent or vapor from its predetermined spacial location within battery 13's structure. Electrical leads 28 and 29, respectively, extend from the interior of the battery and specifically, individually from cathode or positive electrode 30 and from anode or negative electrode 31 of battery 13 for electrical interconnection with the intended device to be powered by the cell.

As seen by reference to Figures 6.10a and 6.10b, the electrical energy generating components of battery 13 comprise, in essence, planar anode 31 in superposed relationship with planar cathode 30 having separator 32 disposed intermediate facing surfaces 33 and 34, respectively, of anode 31 and cathode 30, within the confines of envelope 26. Aqueous electrolyte 27 is disposed in the central portion 35 of separator 32 and in contact with each of facing surfaces 33 and 34.

In the preferred example, the battery will ordinarily comprise a Leclanche electrochemical system including a zinc negative or anode system 31 and a manganese dioxide positive or cathode system 30. The aqueous electrolyte 27 will generally comprise an aqueous ammonium chloride-zinc chloride electrolyte and in addition, a small amount of mercuric chloride which will be disposed between and in contact with the facing surface of each of the anode and the cathode and in contact and impregnated into electrolyte permeable central portion 35 of separator 32.

As shown, the preferred anode 31 may itself advantageously comprise, in combination, a composite structure including metallic sheet current collector plate 36 preferably an aluminum, lead or steel, e.g., tin-plated steel, sheet material on the order of less than 10 mils in thickness possessing a plurality of gas ports or orifices 45 adapted to allow exit from the battery of gas generated or liberated within the energy cells; polymeric current collector 37 preferably a sheet of electrically conductive carbon-impregnated water-vapor-impermeable thermoplastic polyvinyl chloride film sold under the name Condulon, possessing the same relative external dimensions as the anode and in the order of about 7 mils in thickness, and a distribution of active zinc negative material **38** either as a zinc paste carried on the conductive polymeric collector **37** or active sheet zinc secured to collector **37**, in each instance amalgamated in the conventional manner as, e.g., with mercury by contact with mercuric chloride. The preferred cathode **30** may itself comprise a metallic sheet current collector plate **39** analogous in construction to collector **36**; polymeric current collector **40** also analogous in construction to collector **37**, and active positive material **41** as a cathodic paste deposition on collector **40**, e.g., the manganese dioxide/electrically conductive carbon mixture dispersed in a polymeric binder which is employed in the manner conventional in the battery art.

Electrolyte **27** will ordinarily comprise conventional aqueous ammonium chloride-zinc chloride about 4:1 dispersed in a polymeric binder such as hydroxyethylcellulose, polyacrylamide, etc., e.g., on the order of about 5% or more applied to and impregnated in central portion **35** of separator **32** and in contact with the facing surfaces of active positive material **30** and active negative material **31**. In addition, as previously mentioned, a small quantity of mercuric chloride will be present in the electrolyte as a corrosion inhibitor for the zinc anode system.

In general, marginal portions **42** of separator **32** will be maintained free of electrolyte **27** and may be coated on each surface with and impregnated by an electrically nonconducting, preferably water-vapor-impervious, adhesive adapted upon application of thermal energy to secure the separator to the anode and cathode, respectively, e.g., marginal sections **42** of separator **32** to the facing marginal sections of electrically nonconducting insulator layers **46** carried on the marginal portions of polymeric collectors **37** and **40**.

Planar Leclanche-type batteries generally measure about 2¾ x 2⅜ inches in terms of their major dimensions and the marginal areas of such batteries will generally comprise about ¼ inch. Accordingly, the adhesive may be located on each marginal facing surface at a coverage of about 3 mils and thermally sealed by the pressure application of a heated die possessing the contact surface configuration of the marginal areas to be secured.

In accordance with the process, electrically nonconducting insulator layer **46** is disposed intermediate the facing marginal sections of thermoplastic polymeric collectors **37** and **40** and, in the multicell batteries, connector **44** to specifically prevent electrical short-circuiting of the battery resulting from the thermal flow of collector **37** and/or **40** and/or connector **44**'s electrically conductive material, and thereby preventing direct electrical interconnection, between the anode and cathode during thermal sealing of the sealant forming adhesive layer **47** and/or incomplete electrically nonconductive adhesive sealant coverage intermediate the separator's marginal portions and the facing surfaces of the anode and cathode next adjacent.

The insulator layer is accordingly constructed to exhibit thermal stability at the temperatures employed to effect sealing of the battery's planar elements at their margins and may thus comprise any electrically nonconducting material which is thermostable at a temperature at which the thermoplastic adhesive is fluid or melts and may be employed to effect sealing of the battery elements. Although substantially any thermally stable electrically nonconductive insulating material may be chosen for employment in the process and may be positioned intermediate the facing surface of the polymeric collectors and/or connectors as a preformed sheet laminated or adhered to the separator or to the collector, or as a constituent of or as the marginal portion of the separator, and/or as a composition coated on a facing surface or surfaces of the separator and/or collector, preferred insulating materials comprise polymeric compositions which are adapted to be efficiently and economically coated, e.g., coated on the marginal surface of a collector sheet and which in the coated state comprise a continuous polymeric layer exhibiting the required electrical and thermal properties. A preferred class of such polymeric compositions have been found to comprise polyacrylate materials.

As examples of polyacrylate including methacrylate, etc., compositions comprising a significant proportion acrylate, preferably greater than about 50 wt % acrylate, particularly adapted for employment in the process to provide insulator layers **46**, mention may be made of, e.g, butyl acrylate/diacetone acrylamide/-styrene/methacrylic acid (60/30/4/6); butyl acrylate/styrene/methacrylic acid (60/34/6); butyl methacrylate/butyl acrylate (80/20); vinylidene chloride/methyl methacrylate (80/20); and the like, and which optionally may include crosslinking adjuvants such as phenol-blocked diisocyanates, etc.

In the preferred form, the polymeric composition adapted to provide insulator layer **46** may be applied to the marginal sections of the separator and/or corresponding opposed facing surfaces of the next adjacent anode and/or cathode member as, e.g., the pertinent opposed marginal surfaces of polymeric collector **37** and/or **40** and/or intercell connector **44**, by any conventional coating technique adapted to effect selective deposition of a polymeric composition on a sheet material and may be coated from an organic solvent system employing a dischargeable organic coating solvent specifically selected for dissolving the specific polymer selected or, in preferred embodiments, from an aqueous coating system as a latex composition preferably possessing a relatively high solids content, e.g., about 30 to 50 wt %, coalescing to provide a continuous film upon discharge of the water solvent.

In a particularly preferred example, a thermoplastic polyacrylate electrical insulator coating of the type described above is applied to the marginal surface or surfaces of the separator and/or anode and/or cathode member, e.g, as an approximately 0.2 to 0.5 mil thick coating. Such coating may be expeditiously applied by high-speed gravure coating techniques, and the respective facing surfaces thermally sealed employing a cooperative thermoplastic adhesive possessing a softening or melt temperature below that of the thermoplastic insulator layer generally of the hot melt type, e.g., those commercially distributed for electrically nonconducting, water-vapor-impervious sealing purposes such as the hot melt adhesive Versamide, which comprises thermoplastic polymers with molecular weights ranging from about 3,000 to 10,000 and softening points of from about 43° to 190°C prepared by the condensation of polymerized unsaturated fatty acids with aliphatic amines such as ethylene diamine.

In general with respect to batteries possessing the parameters set forth above, such cooperative adhesive may be coated at an independent coverage of about 3 mils for effective sealing in conjunction with the polyacrylate insulator coating.

Aqueous Slurry Electrodes

E.H. Land; U.S. Patent 4,042,760; August 16, 1977; assigned to Polaroid Corpo-

ration describes a planar primary battery particularly suited for a photographic application which is characterized in exhibiting high current drain capacities and improved manufacturability. The battery structure incorporates at least one electrode present in aqueous slurry form as a particulate dispersion of active material and an inorganic particulate additive in combination with electrolyte. This slurry combination, functioning in the absence of dispersing or binding agents present in the form of hydrophilic polymers and the like, provides for batteries exhibiting enhanced performance characteristics.

Incorporating at least one electrode having a slurry component, the structure is characterized by the absence of hydrophilic polymeric binders and the like within the slurry format as well as the absence of specific adhesive coatings intermediate the active component and an associated collector surface. The enhanced characteristics of the battery cells have been achieved as a result of the discovery that a substantially nonconductive inorganic particulate additive, when dispersed within the active material-electrolyte slurry of the structures, serves to promote the adhesion of the slurry format with associated current collector surfaces while imposing no hindrance to ionic conduction within the electrochemical system. An inorganic additive found highly successful in carrying out the process is titanium dioxide.

Another feature and object of the process is to provide a planar primary battery of a variety having at least one cell including positive and negative electrodes and a planar electrolyte ion permeable separator positioned between. In one example, the negative electrode of the cell structure includes a planar anode current collector carrying, on one surface, an aqueous slurry of active material particles and an inorganic, substantially nonconductive particulate additive present in uniform distribution with aqueous electrolyte, no polymeric dispersants or binding material being present in the structure.

Alternately, the negative electrode of the cell may be formed including a planar anode current collector carrying on one surface, in order, a first anode particulate dispersion of metallic anode particles in a binder matrix, and a second anode particulate dispersion of metal anode particles as well as a particulate, substantially nonconducting inorganic additive, the dispersion being disposed within aqueous electrolyte. The following formulation is exemplary of one positive slurry electrode:

	Grams
Titanium dioxide	100
Magnesium dioxide	200
Ammonium chloride	33
Zinc chloride	15
Water	100
Shawinegan Acetylene Black	25

Negative slurry electrodes may, e.g., be formulated by blending 4 g of mercuric chloride with 200 g of an aqueous solution of about 33 wt % ammonium chloride and about 15% zinc chloride. To this mixture is blended 125 g of titanium dioxide and 300 g of powdered zinc. Optionally, about 8 g of carbon black may be added to the mix.

Typical of dry patch electrode formulations are those on the following page.

Negative Dry Patch Electrode	Grams
Powdered zinc	1,000
Carbon black	5
Polytex 6510 (an acrylic emulsion resin)	39.1
Bentone LT (an organic derivative of hydrous magnesium aluminum silicate)	0.62
Tetrasodium pyrophosphate	0.25
Water	150

Positive Dry Patch Electrode	
Particulate magnesium dioxide	1,000
Shawinegan Black (a carbon black)	40
BP-100 (a latex, Exxon Chemical Co.)	67.68
Ethylenediaminetetraacetic acid (Versene)	4.12
Tetrasodium pyrophosphate	10
Lomar D (a dispersing agent)	3.85
Water*	

*In sufficient quantity for deposition

Subsequent to disposition of the initial material upon a collector surface the aqueous phase of the mixture is removed preferably by heat-induced vaporization. The gel electrolyte may be provided from formulations as follows:

	Percent by Weight
Water	58
Ammonium chloride	20.92
Mercuric chloride	2.09
Zinc chloride	15.78
Hydroxyethyl cellulose	3.20

It has been determined that somewhat optimum dry patch type cathode structures are provided having manganese dioxide and carbon present, respectively, in a weight-to-weight ratio of 25:1. Where slurry cathode structures are provided, this ratio may vary from about 6:1 to 12:1, a preferred ratio being 8:1. As is apparent from the above formulations, it is desirable to provide the titanium dioxide additive in a 4:1 weight-to-weight ratio, respectively, with the carbon constituent.

Tetraalkyl Ammonium Chloride Electrolyte

A process described by *A. Hoffman; U.S. Patents 3,953,242; April 27, 1976; and 3,945,849; March 23, 1976; both assigned to the Polaroid Corporation* is directed to a galvanic cell employing as essentially the sole electrolyte a compound of the formula NR_4X where R is an aliphatic or aromatic group, preferably an alkyl group, more preferably a 1 to 10 carbon alkyl group and X is a halide, e.g., bromide, chloride. The particularly preferred compounds are those containing 1 to 3 carbon atoms in the alkyl groups. Thus, it has been found that the greatest efficiency is achieved employing quaternary ammonium salts of relatively high solubility. The more soluble the compound, the greater the capacity of the cell.

However, the quaternary ammonium halide, which is also a corrosion inhibitor, also adsorbs onto the electrode, thereby increasing internal resistance. Quaternary ammonium compounds which are most effective as corrosion inhibitors are less soluble and therefore, would be lower in ionic conductivity. Thus, the selection of the particular quaternary ammonium compound should be governed by a consideration of the battery performance and corrosion resistance desired. A particularly preferred case contemplates the employment of a relatively soluble and a relatively less soluble compound to achieve a balance of properties. For example, tetramethyl ammonium chloride and tetraheptyl ammonium chloride constitute a particularly preferred combination.

The preferred compounds comprise the tetraalkyl ammonium chlorides. Such compounds not only provide electrical properties in cells equivalent to prior art electrolytes, but also are particularly suitable for use as corrosion inhibitors in cells employing metal anodes, such as zinc, lead and the like where corrosion of the anode is a problem, by providing in effect an infinite source of corrosion inhibitor, avoiding the problem of exhaustion of inhibitor which may occur in conventional inhibitors. This is especially so if a flat anode is employed composed of fine powder zinc, rich in zinc oxide. Electrolyte placement in the zinc anode and slow dissolution of zinc oxide relative to amalgamation require the above-described infinite source of corrosion inhibitor.

In still another advantage, the electrolyte of the process permits the use of thin foil, e.g., 2 to 10 mils, as the anode in cells because no embrittlement takes place. Thus, such foil can be employed in flat batteries of the type described in the aforementioned patents which previously required the use of zinc powder as the anode to avoid the embrittlement problem.

Example: Cells were assembled employing a powdered zinc anode, a manganese oxide-carbon slurry cathode and an electrolyte designated below. The following table shows the electrolyte employed and the open circuit voltage (OCV) and the closed circuit voltage (CCV). The load resistance in all cases was 0.83 ohm and the CCV was measured 1 second after load was imposed.

	Electrolyte	OCV	CCV
Control	22 g ammonium chloride	1.85	1.55
	10 g zinc chloride		
	2 g hydroxyethylcellulose		
	2 g mercuric chloride		
	66 g water		
Example 1	22 g tetramethylammonium	1.85	1.40
	chloride		
	2 g hydroxyethylcellulose		
	26 g water		
Example 2	11 g tetramethylammonium	1.65	1.0
	chloride		
	1 g tetraethylammonium		
	chloride		
	36 g water		
	1 g hydroxyethylcellulose		

Planar batteries of this type are particularly suitable for employment in a film pack.

COMPANY INDEX

INVENTOR INDEX

357

U.S. PATENT NUMBER INDEX

3,963,518 - 266	3,994,747 - 210	4,021,597 - 227
3,963,520 - 19	3,996,066 - 316	4,021,598 - 112
3,963,522 - 216	3,996,068 - 5	4,022,949 - 113
3,964,931 - 123	3,996,069 - 146	4,022,952 - 113
3,964,932 - 60	3,997,362 - 136	4,024,322 - 122
3,966,496 - 120	3,998,658 - 131	4,024,420 - 299
3,966,497 - 218	4,001,043 - 235	4,024,422 - 241
3,967,292 - 332	4,001,044 - 25	4,024,953 - 69
3,967,977 - 29	4,002,497 - 285	4,025,696 - 48
3,969,142 - 317	4,002,498 - 275	4,025,700 - 297
3,969,147 - 16	4,002,808 - 72	4,025,702 - 96
3,970,473 - 202	4,003,757 - 84	4,026,725 - 269
3,970,476 - 19	4,004,946 - 201	4,026,726 - 319
3,970,477 - 73	4,005,246 - 248	4,027,076 - 194
3,970,479 - 310	4,007,054 - 5	4,027,077 - 272
3,970,480 - 120	4,007,057 - 234	4,027,078 - 32
3,971,671 - 295	4,007,316 - 237	4,028,138 - 173
3,971,673 - 183	4,008,357 - 124	4,028,478 - 48
3,972,730 - 271	4,009,053 - 67	4,028,479 - 347
3,972,734 - 279	4,009,055 - 292	4,031,295 - 72
3,973,995 - 33	4,009,056 - 90	4,031,296 - 254
3,977,899 - 208	4,010,043 - 317	4,032,624 - 162
3,977,900 - 264	4,010,534 - 299	4,032,696 - 138
3,980,498 - 236	4,011,075 - 73	4,034,598 - 63
3,980,499 - 213	4,011,103 - 68	4,035,552 - 146
3,981,744 - 317	4,011,371 - 144	4,035,909 - 173
3,981,748 - 165	4,012,563 - 194	4,038,466 - 14
3,982,958 - 149	4,012,564 - 152	4,038,467 - 80
3,985,573 - 133	4,015,055 - 74	4,041,211 - 119
3,985,574 - 167	4,015,056 - 91	4,041,213 - 100
3,986,514 - 310	4,016,338 - 160	4,041,214 - 104
3,986,894 - 32	4,016,339 - 224	4,041,217 - 269
3,986,895 - 245	4,019,251 - 346	4,041,219 - 91
3,988,164 - 205	4,020,240 - 156	4,042,756 - 136
3,988,168 - 338	4,020,241 - 42	4,042,760 - 351
3,992,225 - 288	4,020,242 - 141	4,042,761 - 113
3,992,228 - 72	4,020,246 - 206	4,043,113 - 101
3,993,501 - 163	4,020,247 - 230	4,044,192 - 275
3,993,508 - 346	4,020,248 - 135	4,049,980 - 317
3,994,746 - 93		

NOTICE

Nothing contained in this Review shall be construed to constitute a permission or recommendation to practice any invention covered by any patent without a license from the patent owners. Further, neither the author nor the publisher assumes any liability with respect to the use of, or for damages resulting from the use of, any information, apparatus, method or process described in this Review.

FUEL CELLS
FOR PUBLIC UTILITY
AND INDUSTRIAL POWER
1977

Edited by Robert Noyes

Energy Technology Review No. 18

Fuel cells are generators of electricity containing no moving parts except small extraneous pumps for the movement of fuel and oxidant into the cell and the products of oxidation out of the cell.

Public utilities and industrial consumers require high voltage, three-phase alternating current. In this application fuel cells must compete with turbine-driven generators which provide such current. While the output of a fuel cell is low voltage DC power, cells may be connected in various series and parallel arrangements to give whatever voltage is desired, but mechanical rotary converters or delicate electronic inverters must then provide conversion to AC (60 cycles for each phase in the USA).

The advantages of a fuel cell system over turbine-driven generators lie in greater efficiency at full load which even increases as the load diminishes, so that inefficient peaking generators are not needed. There are considerable pollution control advantages to be gained as well. Because the various suitable fuels react electrochemically rather than by burning in air, no nitrogen oxides are formed. For the same reason, emissions of unburned or partly burned gaseous and particulate products are practically nil.

Fuel cell installations using inverters have a long life with relatively little maintenance. It is now entirely feasible to have small, completely unattended fuel cell power plants using waste fuels on location (such as hydrogen from chlorine production) to produce convenient power.

This book, based on information derived from U.S. government-contracted studies, contains considerable practical down-to-earth technical information relating to fuel cells for power plants. A partial and condensed table of contents follows.

ISBN 0-8155-0676-7

325 pages

THERMAL INSULATION 1978
Recent Developments

by Joseph B. Dillon

Chemical Technology Review No. 99

Energy Technology Review No. 23

This very timely book provides a detailed, practical view of the research and product developments, as well as the actual applications technology of insulation systems which are now an integral part of construction planning in the industrial and private sectors.

Individual homeowners, faced with staggering increases in monthly fuel and energy expenses, are installing thermal insulation at a rapid rate. In many countries, particularly those hard hit by oil shortages, government policies have been implemented to absorb some of the cost of improving the insulation systems in private and industrial buildings. In a very significant way, the saving of energy can now influence the standard of living in most industrial countries and grossly affect the balance of payments position of a nation.

Common materials which provide varying degrees of insulation because of their low thermal conductivity, are cork, mineral wool, fiber glass, perlite, vermiculite, clay, other ceramics, wood fibers, boards, felt, many foamed plastics, macerated paper treated with fire-retardant chemicals, reflective metal foils, silicates and cellular glass. Many of these materials are used in combination, and in recent years, foamed low-density inorganic and ceramic systems have been utilized more and more.

This book is a survey of the patent literature. A partial table of contents follows here. Numbers in parentheses indicate a plurality of processes per topic.

ISBN 0-8155-0687-2

339 pages

MAGNETOHYDRODYNAMIC ENERGY FOR ELECTRIC POWER GENERATION 1978

Edited by R. F. Grundy

Energy Technology Review No. 20

The production of electricity by magneto-hydrodynamic (MHD) processes is classed as direct energy conversion.

In an MHD generator the expanding working fluid (hot ionized gas or plasma) interacts with a magnetic field to produce electricity. The amount of current produced is set by the velocity and temperature of the ionized electrically conductive gas.

Motional electromagnetic induction (without mechanical contrivances) gives the MHD generator the unique capability among heat engines of delivering its output directly in electrical form.

The electrical power level at which MHD generators become more economical than mechanical turbines is generally considered to be over one megawatt. Success depends on the intended application and the availability of high-temperature heat sources required by MHD technology.

At this time the addition of an MHD stage to coal-fired power plants seems attractive, since the MHD stage may be able to operate directly on the high temperature products of coal combustion. Other sources of high temperatures are, of course, fusion and fission type atomic reactors.

This book is limited to the potential of MHD in the production of the much needed 60 Hz alternating current at less cost and less fuel consumption than by conventional power-generating systems. As such it presents an accurate status report based largely on federally-funded studies. A partial and condensed table of contents follows here.

ISBN 0-8155-0689-9

230 pages

HYDROGEN MANUFACTURE
BY ELECTROLYSIS, THERMAL DECOMPOSITION AND UNUSUAL TECHNIQUES 1978

Edited by M. S. Casper

Chemical Technology Review No. 102
Energy Technology Review No. 21

This book deals with alternate sources and processes for the production of hydrogen. Today, hydrogen is made mainly from natural gas and by petroleum refining. These sources are dwindling and it is an appropriate time to develop other means for obtaining this essential energy carrier.

Coal is the cheapest, nearest-term, large scale source of hydrogen. Coal and water are the basic feedstocks needed to produce coal-derived hydrogen by well-understood processes.

Nuclear power is the leading intermediate-term source of energy for hydrogen production, either by proven electrolytic means or by proposed thermochemical means. At present substantial technological improvements are required to compete economically with hydrogen derived from coal, but nuclear energy is a likely long-term source of energy for hydrogen production.

Solar energy is also a long-term candidate for the production of hydrogen, either through electrolytic or thermochemical means, and ultimately may become competitive with other methods.

As we seem to be headed toward a possible hydrogen economy, many other methods and an overview of the foreseeable economics are presented also.

Various technological studies were the basis for this review. The available material was organized and excerpted. A complete bibliography of the source material is appended. A partial, condensed table of contents follows here:

ISBN 0-8155-0691-0

360 pages

HOW TO SAVE ENERGY AND CUT COSTS IN EXISTING INDUSTRIAL AND COMMERCIAL BUILDINGS 1976

An Energy Conservation Manual

by Fred S. Dubin, Harold L. Mindell and Selwyn Bloome

Energy Technology Review No. 10

This manual offers guidelines for an organized approach toward conserving energy through more efficient utilization and the concomitant reduction of losses and waste.

The current tight supply of fuels and energy is unprecedented in the U.S.A. and other countries, and this situation is expected to continue for many years. Never before has there been as pressing a need for the efficient use of fuels and energy in all forms.

Most of the energy savings will result from planned systematic identification of, and action on, conservation opportunities.

Part I of this manual is directed primarily to owners, occupants, and operators of buildings. It identifies a wide range of opportunities and options to save energy and operating costs through proper operation and maintenance. It also includes minor modifications to the building and mechanical and electrical systems which can be carried out promptly with little, if any, investment costs.

Part II is intended for engineers, architects, and skilled building operators who are responsible for analyzing, devising, and implementing comprehensive energy conservation programs. Such programs involve additional and more complex measures than those in **Part I**. The investment is usually recovered through demonstrably lower operating expenses and much greater energy savings.

A partial and much condensed table of contents follows here:

Much of the technology required to achieve energy savings is already available. Current research is providing refinements and evaluating new techniques that can help to curb the waste inherent in yesteryear's designs. The principal need is to get the available technology, described here, into widespread use.

ISBN 0-8155-0638-4   **725 pages**

ENERGY FROM BIOCONVERSION
OF WASTE MATERIALS 1977

by Dorothy J. De Renzo

Energy Technology Review No. 11
Pollution Technology Review No. 33

One of the chief gaseous products of the anaerobic decomposition of organic matter is methane, CH_4. This is how natural gas was formed in prehistoric times along with other fossil fuels.

By applying this principle today in environmentally acceptable fashion it is possible to bioconvert municipal solid sewage, animal manure, agricultural and other organic wastes into substitute natural gas (95% CH_4). In its simplest essentials the process consists of loading the material into a digester (a closed tank with a gas outlet). Given favorable thermal and chemical conditions, the appropriate biological processes will then take their course.

The bioconversion of waste materials to methane provides at least partial solutions not only to the energy problem, but also to the solid waste disposal problem. The harvesting of heretofore undesirable vegetations, such as algae, water hyacinths, and kelp as "energy crops" offers unconventional opportunities for supplementary utilization of natural resources.

This book describes practical methods for the bioconversion of waste matter. It is based on reports of academic and industrial research teams working under government contracts. A partial and condensed table of contents follows here. Chapter headings and important subtitles are given.

1. SOURCES OF WASTE MATERIALS
Suitability & Characteristics
Quantities & Availability
Agricultural Crop Residues
Forests
Urban Wastes
Rural Wastes
Strictly Animal Wastes
Industrial Wastes
Cost Considerations

2. MECHANISMS & PATHWAYS
Anaerobic Decomposition Processes
Terminal Dissimilation of Matter
Degradation of Cellulose
Bacteria and Protozoa
Cellulases & Other Enzymes

Controlling Factors in
  Methane Fermentations
Acetate Utilization
Trace Organics
Effect of Temperature Changes

3. SOLID WASTE & SEWAGE SLUDGE
U. of Illinois Studies
Pfeffer-Dynatech Anaerobic
  Digestion System
Supporting Studies
Addition of Coal to Sludge
Synergistic Methane Production

4. METHANATION OF URBAN TRASH
Digester Feed Preparation
Digester Design
Gas Production
Gas Scrubbing Technology
Desirable Gas Characteristics

5. ANIMAL WASTE DIGESTION
Oregon State U. System
Animal Waste Management
Berkeley Conversion Studies
Digestion for Disposal
Digestion plus Photosynthesis
Dept. of Agriculture Study
Other Studies on Animal Waste Digesters
  Useable for Energy Purposes

6. INDUSTRIAL WASTES UTILIZATION
Petrochemical Wastewaters
Distillery Slops Digestion
Rum Distillery Slops
Process Flow Sheets
Design Criteria
Economic Analysis
Use of Biogas in the Sugar Industry
Winery Waste Treatment

7. METHANE FROM ENERGY CROPS
Integrated Conversion Systems
Digestion of Algae
University Studies
Single Stage vs. Two Stages
Mariculture Investigations
Conclusions and Drawbacks
Storage Difficulties

Note: Each chapter is followed by bibliographic reference lists in order to provide the reader with easy access to further information on these timely topics.

ISBN 0-8155-0656-2

223 pages

SOLAR HEATING AND COOLING 1977
Recent Advances

by J. K. Paul

Energy Technology Review No. 16

The technology for solar energy utilization is becoming increasingly available, as indicated by the large number of patents issued in the past several years. Recent developments encompass a number of areas. The emphasis in this book is on low temperature (to $+90°C$) solar collector construction and heating and cooling systems which use these low temperature collectors. The material discussed here is based on 175 U.S. patents, issued since 1970, which illustrate 157 processes.

In its simplest form a collector consists of a sheet of glass or other transparent material situated above a flat plate so constructed that it acts as a black body to absorb heat. The sun's rays pass through the glass and are trapped in the space between cover and plate. The heat may then be utilized by passing a fluid through a conduit system located between the cover and absorber plate; the heated fluid subsequently being used to heat a home, water supply, or swimming pool, or even run a heat pump for cooling.

Focusing collectors use curved or combinations of flat devices to reflect solar rays onto an absorber surface to achieve greater concentration of energy (higher temperatures).

Information has been included describing suitable coatings used to improve absorption properties and detailing a number of devices which employ liquids, crushed rock or other media for the storage of absorbed energy when the weather is cloudy or hazy, or at night. A partial table of contents follows here. Numbers of processes are in parentheses.

ISBN 0-8155-0674-0

485 pages

OFFSHORE AND UNDERGROUND POWER PLANTS 1977

Edited by Robert Noyes

Energy Technology Review No. 19
Ocean Technology Review No. 6

Uncontroversial plant sites for generating electric power by any means are becoming increasingly scarce. Both nuclear and fossil fuel power plants require nearness to large quantities of cooling water and to load centers, but not on land prone to earthquakes. Consequently, utility companies are looking out to sea or beneath the surface of the earth for suitable power plant locations.

This is a most exhaustive and detailed treatise: the first six chapters being devoted to nuclear and fossil-fuel fired power plants, while the remaining six chapters deal with power generation of a more esoteric variety.

The relative costs of fixed and floating offshore plants depend on the depth of water and on the degree of earthquake resistance required. For shallow water, the fixed breakwater offshore floating station is of the most interest.

Most of the information presented in this book is based on federally-funded studies. Each chapter has its own bibliographic list, while the references at the end of the book give the full titles of the government reports on which this book is based, together with a source of their purchase. A partial and condensed list of contents follows here.

ISBN 0-8155-0680-5   **309 pages**